From Cosmos to Creature:
The Origins of Human Biology

Embryogenesis

RICHARD GROSSINGER

illustrated by
JEFFREY PARKIN

NORTH ATLANTIC BOOKS, BERKELEY, CALIFORNIA

Embryogenesis
From Cosmos to Creature: The Origins of Human Biology

Copyright © 1986 by Richard Grossinger

ISBN 0-938190-81-4 (paperback)
ISBN 0-938190-82-2 (cloth)

Published by North Atlantic Books
2320 Blake Street
Berkeley, California 94704

Cover art by Jeffrey Parkin
Cover design by Paula Morrison

Embryogenesis is sponsored by the Society for the Study of Native Arts and Sciences, a nonprofit educational corporation whose goals are to develop an ecological and crosscultural perspective linking various scientific, social, and artistic fields; to nurture a holistic view of arts, sciences, humanities, and healing; and to publish and distribute literature on the relationship of mind, body, and nature.

Library of Congress Cataloging-in-Publication Data

Grossinger, Richard, 1944–
Embryogenesis, from cosmos to creature.

Bibliography: p.
Includes index.
1. Human evolution.   I. Title.
GN281.G76   1986   573.2   86–8571
ISBN 0–938190–82–2
ISBN 0–938190–81–4 (pbk.)

Embryogenesis,
Being a description not of *who* we are but of *what* we are.

Table of Contents

For my unknown ancestors and descendants

Preface

This book is built around a detailed narrative account of the formation of an embryo—the human embryo itself and the sequence of nonhuman embryos on which it is based. The account brings together ontogeny and phylogeny, the two histories which converge in our formation: the former describing the development of an individual man or woman, and the latter describing the evolution of our species.

But *Embryogenesis* is also a book about life and mortality. It is a long essay exploring the means of our coming to be, not only physically but psychologically, not only psychosomatically but spiritually and epistemologically.

I have assembled this book from dozens of texts, lectures, and discussions with embryologists and doctors. I have attempted to put together the information in a way that is experientially (rather than scientistically) meaningful and that speaks to the important issues of psychology, anthropology, politics, philosophy, and religion. In the simplest sense, I have reconstructed the twentieth-century biological consensus of what we are and how we are made, while at the same time asking: What is the meaning of an existence arising from such a process? I have searched not for the more apparent revelation of modern science but for its shadow.

Throughout the writing of the text, I have gone by the premise that opposing viewpoints, values, and cosmologies shed more light on the true nature of things by their contradictions of each other than by their assertions and lucidities. For anything we might be tempted to accept at face value, there is also an opposite with its own reality to express. The truth supposedly revealed by the material world masks a truth forever concealed by the appearances of that same world. The so-acclaimed spiritual world is also a mask—a temporal one obscuring another spiritual realm.

xi

At this stage of human history, the physical/spiritual split is merely a symptom of our social and ideological failure. Rigid adherence to one position or the other may sustain a career, but not a whole life. If, on the one hand, we accept uncritically the physical laws and statistical facts culled from nature, we will lose the actual thread of objective inquiry and become hollow scientists. If, on the other hand, we adopt theosophical systems and their landscapes of eternal life without experiencing the actual genesis of those systems in our personalities, we will find ourselves nihilists again at the end despite the years of belief. In *Embryogenesis* I have taken a different path in place of a choice between these two (or a modernistic synthesis *of* them). I have tried to recognize an actual experience that occurs outside the ordering of science and religion yet with reference to their roles in forming our collective phenomenology.

The narrative account in this book is compiled from many sources (as spelled out in the *Notes* at the back). The technical aspects have been corrected by a number of scientific readers; Dr. Stephen Black of the Department of Embryology, University of California at Berkeley, has been a consultant and advisor from beginning to end. I would also like to acknowledge and thank Dr. Barry Coller of the Department of Hematology, State University of New York at Stony Brook, for his help in the area of blood-cell formation.

I encourage readers to use their own judgment in going through the pure embryology in this book. The technical sections can be dense and difficult. Skim where you get bogged down. I wrote this book not to teach scientific terminology but to give a sense of our situation in the physical universe and to represent the twentieth-century version of human reality in all its precision and determinism. When I considered various alternative degrees of detail, I finally did not feel it was sufficient to say such things as "cells move. . . ." and "cells follow paths induced by other cells. . . .," or to settle for a summary description of biological fields. These are all abstractions. I wanted to show how the actual heart and lungs and genitals are formed, even if the words are merely a different level of abstraction. There is no need to remember the names and details of every stage and organ, but they give the reader a sense of the physical reality underlying existence.

I have also had various "nonscientific" readers give their thoughts about the book at various stages and drafts, and their additions and criticisms have been invaluable. Charles Poncé, Danny Moses, Jeffrey Auen, Laura Lederer, Lindy Hough, Herbert Guenther, Richard Heckler, and Randy Cherner have all contributed insights to this work. Additionally, I would like to thank Stanley Keleman of the

Institute for Energetic Studies in Berkeley, California, for his introduction to the role of embryo development in bioenergetic etiology
and diagnosis.

I would also like to acknowledge Jeannine Parvati, a spiritual midwife, herbalist, and counsellor. Her notes on the manuscript led me to
reevaluate several sections. I have tended to trust her direct experience of pregnancy, childbirth, and women's mysteries. Embryology
may be one of Apollo's sciences, but the phenomena it entails arise
first in Artemis' queendom, and to underestimate them is to pretend
to be objective overseers of the universe rather than mortal life forms
embodying profound transformations. *We* are embryos, so we are initiates in a mystery.

Parvati, though a strong supporter of this text, believes I have been
too undiscriminating in my use of scientific observation at the expense of intuition and worship. She challenges my "literal embryological narrative," on the basis that it is a dangerous illusion, drawn
from the work of morally corrupt scientific researchers who, without
compassion, tortured embryos for their raw information. She feels
that the textbooks deriving from such research are only the symptoms
of a sick society.

During the writing of *Embryogenesis* I thought repeatedly about
the life-styles and ethics of the scientists whose work made possible
the "facts," and although I experienced an ongoing discomfort (usually unacknowledged and on an almost subliminal level), I cannot re-
ject all of experimental science out of hand. If we could end our
exploitation of sentient life forms as part of an overall change of consciousness and planetary politics, I would be glad to support and participate in a compassionate and visionary path to knowledge.
However, to ignore the science of biology because it is based on mutilation of creatures would be, for me, an ideological position which
would prevent the writing of this book.

I agree that knowledge gained from the severing of brain lobes of
octopi, squirrel monkeys, etc., and the induction of tumors in helpless rabbits, chickens, and the like must, in some way, be sullied and
distorted by the experiments themselves. Even the "torturing" of
worms and insects is an ongoing crime against nature and against the
sacred power of the universe. But to boycott such knowledge is to
leave the twentieth century. That might not be a bad idea (as Parvati
has shown by her life as healer and yogini), but my path at the moment is a different one. As far as I can see, the damage has already
been done; we might as well examine the golden eggs for which we
cut open the goose. At the same time, we must not pretend our
knowledge is innocent or bloodless. Ambivalence and treachery lie at

the heart of this book. I have written in order to confront the changeless addictions of our society and to challenge the tyranny of our self-images and self-definitions.

Embryogenesis completes an informal trilogy begun with *Planet Medicine: From Stone-Age Shamanism to Post-Industrial Healing* (Doubleday/Anchor Books, 1979; revised edition, Shambhala Publications/Random House, 1983) and continued with *The Night Sky: The Science and Anthropology of the Stars and Planets* (Sierra Club Books/Random House, 1981). The books are clearly individual works and may be read in any order. The trilogy is the result of an eight-year inquiry into the meaning of origins: how we define ourselves and our society in the universe at large. The particular order in which they occur traces my own path of insight. I began with medicine as our unintentional self-diagnosis of a collective disease and a statement of our need to cure ourselves and our society. From there I went on to explore how we form images of the boundaries of our existence—matter, space-time, and the creation itself; what we think the stars are often determines what kind of society we generate. *Embryogenesis* contains a more explicit scientific narrative than either of the other books, partly because scientifically derived images of cell morphogenesis and embryo formation are not as available in our popular culture as equivalent images of systems of medicine and of stars and planets. Such a painstaking biological account also grounds the trilogy in protoplasm and living fields.

Recently, after finishing this book and taking a two-month break, I went on a long walk with my fourteen-year-old son around the Echo Park area of Los Angeles. Reaching the top of a steep hill we found ourselves both staring at pigeons seated on three vertically separated telephone wires above our heads. In the bright sun every feather and ruffle and coloration was etched in each bird, and each bird was different. In the rush of the moment I said: "That's what my embryology book is about—how did those birds get there?" We looked at the birds and he thought about it. Then, a few minutes later, after we had passed them, he said that his ten-year-old sister had asked him if the universe went on forever, and he didn't know what to tell her.

The difference between *The Night Sky* and *Embryogenesis* is spelled out in the tension between those two questions. *The Night Sky* addresses: Does the universe go on forever?; and *Embryogenesis* asks: How did those birds get there, in every aspect of mind, and ruffle of feather?

I would choose at this point, after eight years of writing such books, to pull out of the tar-baby. The questions are unanswerable and will always be unanswerable, and it is a danger to give one's life to them. They will devour everyone and everything. When addressed by a Western mode of analysis, they merely double back on the author with mirages instead of new insights. I fear this would happen if I went on, so *Embryogenesis* is my last work in this mode. I will attempt other works of a different nature now.

What I finally seek is not a lifelong accumulating opus on the mysteries of physical and spiritual science, but a mode of transformation through the work itself. *Embryogenesis* is the best and clearest acknowledgment of the mystery that I can make.

(One other note to the reader: For me this book begins with Gene McDaniels singing ''A Hundred Pounds of Clay.'' Each time I got stuck in ideas I went back to my old 45 record of that song. What this tells me is that it is not finally a book of facts and conclusions; it is a pop theme, a melody taken right off the surface of America. The obscurity of the simplest feelings finally outweighs the most complex metaphysics.)

—Richard Grossinger
Berkeley, California
March, 1984

The Ceremony of the Animals

In this world, we are the animals. For sure and for certain. We may kill them, eat their remains, ignore them, or judge ourselves by consciousness above them—but we *are* them.

If the planet is a temple, blue skies keeping the ceremony within, we are priests, Aztec in our famous cruelty, Aztec in our clarity. We carry out finite law.

They do not have personalities as we do, but they do not have the scourge that we do. They are not diseased. They bear no grievance. They are there until the absolute last moment, then they are not.

It is wrong to think of us as the bane of the animals of this world. We are their completion, their ritual. They did not intend us.

We suffer consciousness that they may be fleet and light.

We consider and judge that their ferociousness and hunger are unabated.

We dream, and they are dreamless night.

We make a text, but their bodies and footprints lie in margins we can never clear.

We make language, they are outside language.

We think. They pray. We are their unspoken intention to speech. The truth we speak they are.

We suffer disease and madness. They suffer.

Everything we do, our cities, billboards, poems, wars, machines, houses in which they build nests, they allow. Their pure reception makes our doing it possible.

Even Roger Miller singing "King of the Road" on the radio is the ceremony of the animals. Raccoons, and starlings, and fish in the river the melting snows fill, fly buzzing in the room, tapping the windows. He sings: "Trailers for sale or rent/Rooms to let fifty cents."

Foreshadowing of this book in unpublished journal notes, 1977.

1. Introduction to Embryogenesis

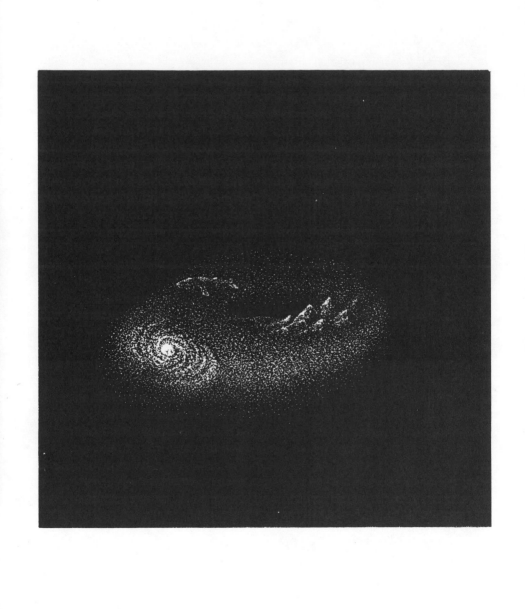

What is the life that brings us here? How is the loose energy of the cosmos snared in our tissues and our personalities? By what agency do single entities, awake and aware, escape the vast homogeneous current?

These questions generally go unasked because there are no answers. Consequently, we distance ourselves from the intuition of our own life span and its mortal consequences. We project the death within us to an abstract spatiality without, and we placate it with metaphors and relativities, as if it did not swallow our destiny into its otherness. Not only our minds but our nerve cells, guts, lung tissue, and hearts prefer "business as usual," so it *is* business as usual, right up to the end.

We pass through this world as a shadow swims through light. It is around us, in us, inside our inside; yet we do not contain it and cannot wholly grasp it. We become alive, bringing a distant intuition of our own presence into being with us, and we sustain its fragile range of personalities all our days.

There is a spirit within us that approaches life as a limitless possibility; that expects to be surprised, forever; and that labors to make us somehow real to ourselves. Immortality seems out of the question at this stage of things—that is, the conventional linear immortality of Western cosmology—but everything we are, including the part of us that was immortal before the ascent of science, arises in an embryonic process whose origin and principle lie outside the present economy of nature. It is to the mystery we should look both for meaning and the threat of no meaning at all.

For most of our sentient history, human beings have been considered finished and complete creatures—final causes of the deific or natural agency of creation. But nothing in the world described by modern science is finished or final; all are fleeting realities assembled in the collision of time and space by events rushing to other resolutions. The *physical* basis of life is a field of constantly changing atoms, pouring through the body like light through a crystal. It takes but five years to replace every atom in us with another atom. When we meet again after a long interim, we appear as new assemblages of

5

particles. One so closely resembles the other (and bears its memories) because prior atoms make room for new ones only in positions equivalent to the ones they replace. Each of us contains atoms that were used by Homer and Buddha, as well as by billions of worms, fish, birds, extinct crustaceans and corals, and Stone Age men and women. The *biological* basis of life is a sequence of cellular fields, each one nested upon a previous one, so that the shapes of plants and animals emerge from the configurations of prior species, from a beginning in simple inanimate crystals which themselves originated in unidimensional chemical clusters.

Viewed physically, human existence is a temporary lattice of star currents and cosmic dust. What feels whole to us is what we make whole by being; otherwise it is a heap of atomic and cellular contradictions. The objective facts, if facts they are, lead (one way or another) to a view of life as an abnormally organized zone of molecular debris. Scientists have their own elegant aphorisms for this predicament. Frederick Hopkins pronounced, "Life is a dynamic equilibrium in a polyphasic system."[1] Other scientists have ordained that life is a partial realization of the informational potential in atoms and molecules.

In our modern revised world view, life is neither inherent nor inevitable, and, if circumstances had gone differently, there would be none on Earth (or perhaps in the universe at large) now or forever. The same elements have the potential for lifelessness, and there is nothing we know that predisposes them to making living systems. We must concede that, once these systems have been made, atoms and molecules sustain them millisecond to millisecond.

What we have said about life, in general, is even more true (if possible) for human life. Consciousness is a unique realization of the informational potential of atoms and molecules organized in cells (at least in our version). Most scientists find life so unlikely that, to them, intelligence is a mere elaboration upon the initial marvel. Biochemists see no order in cells or simple animals that guarantees the later emergence of symbols and language. It is fortunate that our own assessment of the odds against our coming into being has no effect upon the present fact of our existence.

We are apparently the conscious offspring of the unconscious struggle of matter, which probably felt about the same as water does running along a rock, until the dying and devoured trillions ascended to the fierceness of the scars of their ancestors. We are now stuck at a curious place in our own history: our search for origins has left us more and more alone in an alien place. Although we can continue with the so-called utopian program of our civilization, that civilization is coming to exist in the middle of nowhere. We are free of the promises and threats the gods have made throughout history, but it is only the same freedom that the raw stellar elements had in the beginning, to make creatures or not. Since we are mere elemental accidents, who would bother to speak for or against us, and who could possibly protect us?

There is no conventional way out of this dilemma, so we protect ourselves within hierarchies and false historic domains. We are now the victims of history posing as statistical laws and scientific governments even as we were once the victims of the divine right of kings and the proletarian revolution. Finally our experience is more authentic than our institutions, and we are attracted to another unknown thing.

We do not experience our physical makeup directly; we know it only as a component of some other awareness—the desire for food arising in our stomachs or the brightened contour of the world igniting in our eyes. We lose the individual foci in a cloud of continuous being that seems to originate in our heads as a collection of thoughts and memories, feelings and beliefs. We breathe this cloud into being; our heartbeat steadies it; and our senses knit it together. We may experience pain, but no needle ultimately penetrates the haze. The wholeness of our being engulfs every torture and trauma. The pierce of a dagger wound, the drilling of a frozen or crushed body part may momentarily deform the cloud and push it inward to its physical basis, but before the roots of bone and nerve are reached, the abstraction of existence always absorbs it.

"Today while the blossoms still cling to the vine," the pop song begins. We are the blossoms and we are the vine too, and this is what brings the song into being. We see the pinkness of the petals, the fragility of their connection to life, and we hear the madrigal and know it as ourselves. The vine grows up the brick wall behind the Japanese restaurant, its blooms opening umbel by umbel. We project ourselves into an image that expresses the sweetness and luxuriance we feel within—also, always in the context of peril. There is no escape for those who are chosen in the light, either from being born or dying. Our bodies open by the same hidden geometry (our minds), bud by bud, flower and then fade. "I'll taste your strawberries, I'll drink your sweet wine." The song will become a conventional love ballad, the summer will pass, but the connection to something eternal and mysterious remains, something which the lover momentarily embodies. "A million tomorrows may all pass away, ere I forget the joy that is mine today." The protozoans in the pre-Cambrian mud may have felt that too, but they couldn't have known it, and it was certainly left for us to express it for an entire planet, and to leave our works to go on beyond us, even before we are dead.

"Who are these coming to the sacrifice?" asked John Keats. "To what green altar, O mysterious priest,/Lead'st thou that heifer lowing at the skies?"[2]

Our own shadows haunt us, be they though the play of light and darkness through the lattice of molecules that has swallowed us now up to our tongues.

After the fact, we can see that we have used scientific inquiry to abstract ourselves from the cloud of self-complete identity that surrounds any living creature, and, in so doing, we have discovered the composition and some of the dynamics of that cloud. We have come to understand, in part, how it is that we exist and what our existence obscures about itself. We have reasonably assessed the thermodynamic aspects of suns, and atoms, and of the merging life protein in their fields. We have objectified our own position in the universe, removing ourselves imaginarily from our actual place and reconsidering our existence in the context of raw matter, the physiology of our senses, and the mathematical properties of our minds.

Science describes us, yet it stands against us. Our fantasies and hungers do not match up with the physical reality of our being, except as we have chosen to call them "instincts"—in which case we seem to find their forerunners in the animals and their volatile aspects in the chemistry of carbon. But we have not only missed the core of creation, we have missed ourselves missing it. *We* must be the basis of our own existence—not as collections of elements in fragile harmony but as a direct experience of creation itself. Scientific discourse cannot recover this primal clarity without first breaking with its own objectivity—which would be a dangerous detour inasmuch as it could lead only to the reinvention of science again and again until the end of our society. The compulsion to save ourselves through externalization seems to be inherited with mind itself.

By now, we have science and its collection of experiments and facts in one place, ourselves and our desire to know who or what we are in another, and only tenuous, tantalizing bridges exist between the two. It is the dilemma that constitutes this book.

Embryogenesis is the active process of transformation that brings each of us into being. Even if we propose that a spirit or soul exists apart from matter, the embryo weaves the two entities together as organism and psyche. Embryology is the branch of biology and social science that locates living and symbolic systems, animals and institutions, as passages between conception and dying. Philosophers may transcend the rationale of mere physical organs, but their bodies and minds come into being ontogenetically, and their mortal lives express an epistemology in terms of that genesis. Psychologists may propose a mind epiphenomenal to the flesh, but that mind must be located somewhere in the configuration of cells. Physics and chemistry may insist that matter is prior to life, but embryology explains how molecules are arranged into animate cells.

We do not *have* children; rather, they pass through our tissue uncognized and inalterable. We are the receptacle for their germ, a capacity imbedded in us by a prior receptacle in which we were seeds. Both parents and children are by-products of a transmission through biological time. The question: "Which came first: the chicken or the egg?" is more a dilemma of nomenclature than of real priority, for the chicken never stops being a developing egg, and the egg is never anything other than a chicken developing.

Early in the embryonic life of most creatures, cells providing the blueprint for another generation migrate to a region of tissue which will become genitalized around them. The codes that reside in the nuclei of those cells are translated sequentially in the zygote, and new cells are spawned, giving rise to a sequence of living beings, each one constructed out of the prior one with the consignment of the genetic information. Cells continuously differentiate from one another and are reorganized in systems which lay the foundation for more complex systems. Only one continuum of living entities is provided by the nuclei of a single cell—a microlith combining elements of both parents, which will be identical to them in its general aspects since they are both members of the same species. The embryo can become only the thing its parents are, or perish. It is one of the mysteries of nature that this law, which cannot be broken from generation to generation, is somehow broken cumulatively over many generations; otherwise, how would new species arise?

The quantitative aspect of embryogenesis is simple: Each cell arises from a previous cell, by mitosis, or, if it is to become a germ cell, by meiosis, a similar process of fissioning. As these cells come into being, they interact with organic material, the environment, and all the other cells in their organism to mold a living creature. A full-grown human consists of uncountable trillions of such cells, all dependent upon each other for survival. A rotifer consists of several hundred cells, and an amoeba consists of one cell. All of these are individual complete life forms bearing germ nuclei for the continuation of their lineages.

The qualitative aspect of embryology is a conundrum. One living being after another swallows its predecessor in itself, and more and more complex functions are integrated into the overall developing organism. Not only are cells spawned and differentiated but they are organized in layers. No sooner are these layers formed than other layers are occurring within them. As new layers are absorbed, the embryo sequentially changes its character. Structure is hidden within structure, and disassembles existing configurations as it forces its way through the tissue that surrounds it. In a dramatic case that has been recognized from ancient times, the growing butterfly in the coccoon melts down the caterpillar until its integrity disappears within her own.

Living forms and their artifacts are always continua; there is no other way they could come into existence. This sequencing must recur from level to level within hierarchies of tissue and sentience. Just as the states of the embryo are absorbed in each other, species of plants and animals disappear into their descendants. As we look back into our chain of grandfathers and grandmothers we find only a few thousand wise old seers; the further back we go, the more they look like primates and then like moles. Our ancestors lose speech, lose consciousness, and ultimately lose all texture. Our ultimate great-grandparents, of course, are not even human.

Cultures, equally, must obey embryogenesis. Palaeolithic bands vanish into Mesolithic clans, tribes into villages, kingdoms into empires, and from Mediaeval cities arise Toltec glyphs and Flanders trade fairs. Just as one jellyfish originated in another in the ancient seas, so do pots, and harpoons, and myths have offspring. Musical themes snarl and simplify and entangle again like spiderwebs or songbird plumage. Languages no longer spoken are hidden within present tongues, but by now every phoneme and syllable may have changed to another, so we must speak unknowingly for the dead and about things we are not aware of, even as we must use the tissues of the dead in bodies that would be strangers to them.

Once, there was no such plan of things, and all of this had to be invented, to invent itself. There is only one way we can imagine complex organisms being formed from simple cells: The earliest one-celled animals, which no longer exist, were imbedded in life forms which also no longer exist. From the beginning of life, creatures have continued to be imbedded in creatures billions of times over in sequences of organisms ancestral to every creature alive today. The vast majority of them have disappeared from the Earth, and their traces in other life forms have been unrecognizably condensed and merged with later forms which have also been condensed and combined. The plans of ancient animals are built upon the plans of primeval animals; and the plans of modern animals incorporate all the blueprints which precede them in their lineage.

Phylogeny, functioning as embryogenesis, builds creatures out of prior creatures even as ontogeny does. But ontogeny does not recapitulate its whole comprehensive phylogeny. Since every act of procreation embodies and condenses the history that preceded it, the sequential bodies of an embryo recall not its ancestors but its ancestral ways of surviving at every crisis of development, many of which occurred larvally or foetally. The stages of the embryo must adapt even as the species of a phylum; they cannot be simply nonfunctioning stepping-stones to more complex organisms.

Only small changes are incorporated in one generation—infinitesimal changes. If these are sustained, new changes may be made from them. Leg rudiments form in a legless creature; gill slits swell into primitive lungs. Collectively, these changes bring whole new categories of experience into the world. Yet, paradoxically, an offspring can never be of a different species from its parents.

Creation cannot subtract from its blueprint; it can only add, using what already exists as the basis for new patterns. In order to arrive at a new human being, the embryo must follow the same procedure and design by which it became human. The same steps, however, are never precisely identical. Every stage and configuration is preserved historically, but whatever is new (in nature or in the stars) is clandestinely added to each unrepeatable being, to appear in its offspring.

What we inherit today, in our special moment as animals who experience the creation (and have taught themselves to reconstruct the origin of species), is a series of designs that go on happening by precedent through a continuity of the basic thermodynamic frame in which they arose and became joined. They go on happening not because anything requires them (not on this plane of existence anyway); they go on because, at this threshold of sequentiality, they are bound to replicate, they could not do otherwise. If life continues because of the desire of living creatures to be here, then that desire has found a way of being that does not require it mechanically for its own entelechy. For instance, we can claim that desire draws conjugal pairs together, but this seeming romance is merely a symptom of life; it is really the fissioning and joining of cells that sustain the compulsion to live.

The way in which embryogenesis has assembled us is the way in which we must be shaped by our lives. We take on images, experiences, knowledge; yet we change them or make them part of new instances. Even when they are totally forgotten, they continue to be contained within the fabric of being. The world of a two-year-old has been lost by a six-year-old, and her world is forgotten by the teenager. Shirley Temple came to experience her childhood roles as the performances of another little girl; William Faulkner had forgotten the characters of *Absalom, Absalom!* by the time he reincarnated them in *The Hamlet.*

The disappearance of the precise memories, like the tissues of embryonic layers, does not prevent their contribution to the whole. In fact, the whole being is made up, both in body and personality, of things which cannot be remembered but are wrapped together in the life fabric. That is where the blue and red boat of the two-year-old went; lost but not obliterated. If we remembered everything and lived millennia, time itself would freeze, so we need some other method of preserving essence. Our memories are swallowed by each other to preserve the unity of our existence.

How many discrete days can you remember? Is there even one moment left from the third week of August 1958? Yet, surely you lived it once, and believed it, and promised to stay loyal to it. Each day stood out then with brilliant clarity; each was experienced as the eternity of being.

When memories are lost, what takes their place is an aggregate meaning. Likewise, when the embryonic stages are absorbed, they are replaced by the aggregate being. Poems are forgotten by the poet in her old age, and then lost by the culture; ultimately the language in which they are written becomes extinct. It seems terribly sad that what was precious once is always eclipsed, both in psyche and in tissue. If it meant something, it should not be abandoned. If it lived, it should not merely have its reality incorporated in subsequent creatures. Love should stay true and remain, and fuel the unfolding of the universe.

In moments of extreme happiness, in the calm of the summer day, golden blossoms, wild birds, sound of a brook, it doesn't matter that it's not perfect because there is nothing else to replace it, no other way for things to be.

The original Freudian insight was an embryological one: We are contained in layers of tissue and symbols; most of them have become unconscious, but they have not been lost.

We see that only in its most limited usage is embryogenesis limited to the womb. The grasshoppers in the weeds, the schools of fish in the thermal, the children in the kindergarten, and even the aged eagles are still embryonic. After birth they continue to grow and change (usually more slowly) by the same processes that preceded their birth. Parturition is a transition from one phase to another, but it is not a termination.

Termination is impossible. An organism can exist in time only as an unravelling thread of cells. Each level of being generates the next level until the creature is killed by external intercession or withers from within. These single discontinuities, however, do not break the thread itself. So far as we know, no plant or animal lives forever, but all of them produce seeds with the same potential vitality as themselves, and these seeds carry the lineage from lifetime to lifetime.

For us as organisms, there is a crucial difference between the embryogenesis of the individual, which terminates with the death of the organism, and the embryogenesis of life, which continues, hypothetically, for unlimited generations despite the fate of individuals or even species. We tend not to think of our own deaths as the death of all consciousness, for we imagine future generations of intelligent animals continuing the making of the Earth. The terror of nuclear holocaust is, in part, the fear that this long embryogenesis will terminate. Although our cosmology now requires that the universe itself terminate violently someday (and our own Sun and world terminate well before that), we accept that as a natural cosmic rhythm; whereas planetary suicide appears to condemn our whole adventure in the light to an aberration or a pathology.

A thread of tissue touching one life always touches at its other end the dawn of life, and it takes all of creation working from then to spawn any one unique being. It is this delicate history we squeeze out of the fly when we crush its filaments and jellies. Individually, we experience the fate of single embryos. We die when the blueprint ends, whatever we may believe about transmigration or reincarnation. The blueprint may pass through us to other unique beings, but we can never exist again, for we require the whole path back to the first cells and the unknown genesis of protoplasm. So William Beebe wrote: ". . . when the last individual of a race of living things breathes no more, another heaven and another earth must pass before such a one can be again."

In knowing what it is like to exist, to be specific, to live out the desires of a lineage, all creatures, however short their lives, share an event. Human beings know and manifest eternal things because they are made of them, and if there are no archetypes, then they manifest whatever there is, as "archetypes." The animals touch and map the eternal thread too, though they realize it as something else in their living and dying, and in the hollow lucidity of the land.

* * *

However we come to understand the world of nature, we ourselves are a mute philosophy. We embody the philosophical system all other systems seek to explain. We look back through the specks of our body, the fragments of our thoughts and dreams, through the gauze of material making us up, and we are travelling back through the history of suns to the beginning of the universe. The cells are embryonic stars in galaxies. A single charged ion is the skin of creation; but this is also the myth of Uranus and Zeus, and is true only in some other way than the mist over the waters in morning. All beginnings are folded into other beginnings, and they gyrate backward and forward into streams that precede them and arise from their having been.

"The voice I hear this passing night was heard," John Keats, "In ancient days by emperor and clown."[3]

So for all the frenetic talk to establish even the vaguest chatter that sounds right, silence would have the same effect. Our mere being is the origin and the conclusion of this noisy book. It is our goal to arrive back there after confronting the static we have given rise to as we thrash about in the pond of creation asking frantically who we are.

Leo Tolstoy's final words were: "I don't understand what I'm supposed to do." But as Gertrude Stein told us in another way, that is not the question, that is the answer.

2. The Original Earth

A long silence preceded the Earth. It is a silence we have not forgotten. It would seem to be behind us forever. But we have distorted time and space by our presence. The silence of the beginning now lies within us, and much of our being still addresses it.

At one point in another time, there was no Sun; there was no Earth. The starry clusters had motion and form, but this was a kind of cosmic clatter having nothing to do with us. If it existed prior to us (a thing we presume), it did so in a different dimension: the preterrestrial heavens are a posthuman image. So too is the intuition that they existed in a different dimension.

Even when the primal Earth had formed and the stormy atmosphere triturated water and hydrocarbons out of the spoor of cosmic gas, there was no one to listen to the rain. It rained for a thousand years, a hundred thousand, a billion years, but it rained only for an instant before something interrupted its silence.

In *Genesis,* the "Book of Moses" in the *Bible* of the West, we are told, "In the beginning . . . the Earth was waste and void; and darkness was upon the face of the deep." This is before the Spirit of God moved upon the face of the waters and said, " 'Let there be light,' and there was light."

"Let there be light" preceded the Sun, the Moon, and the stars. The animals followed. "Let the waters swarm." We know they did, even down to the smallest water droplet. The prehuman Angel filled the sky with birds, and we still see the dense flocks of them above land and sea.

This description is itself a mystery masking another mystery masking still another and another. It is not within our possibility to get to the bottom of these. We understand that silence ended (though the silence still haunts us). We embody a cosmic silence that is also light trapped within darkness, but we understand it as silence because sound is slower than light, and in the beginning also was the Word.

There was nothing here at the beginning—not Christianity, not Buddhism—only primal life force. It was Apache wind and lightning in a field of Aranda dream-ghosts. It was a maelstrom of *yin* and *yang* igniting from the roots of unborn sea, dark cloud of Afro-Caribbean

17

voodoo splashing with volcanic Philosophers' Stone. The Dogon water *Nummo* bathed in a gold liquor of Zoroastrian *xvarenah*. We can't name it then because no words, no incantations have yet been uttered. We stand also before the first totems and the first temple mounds, and all the gods and spirits commingle in a pan-Gaian pagan rite.

The Biblical and scientific versions of the Creation differ in every conceivable way, but they describe the same mystery. The Biblical version, while attempting to depict vast externalizing acts, actually recounts an interior awakening, the birth of a sacred planet. The scientific version, although seeming to, cannot actually go back beyond the Word, for it uses language as a logical, objectifying tool. But it grasps a Gnostic element the literal *Bible* of the West omits—that the moment of creation continues to spread and the way in which the Void was originally breached occurs again each instant as new creatures burst into the world, from eternal darkness into the rain forest of sound and light. The Spirit itself is no longer visible; it is a seed which carries the creationary message, ostensibly from the beginning of time until the terminus of the life chain, whether that be in the near future or at the extinction of the universe itself.

(The conflict between creationism and evolutionism is a mirage. It is but a clash of the extremist preachers of opposing modern religions. God and Nature could not be at war.)

The Earth's silence came to an end when the rain was heard, but it was not first heard by us; it was recorded in the ocean by the primordial cells. These sensings were without knowledge, without memory, without desire. Even the newt knows little by our standards; yet its way of knowing must lie at the basis of ours; we must also inherit its "dumbness," its relation to the original silence.

The Earth that condensed from the starry hydrogen and helium was uninhabitable. Having just been part of a sidereal field, it was as much a sun becoming a planet as it was a place with its own topography and climate. Stars may be colossal dense objects, but they are extremely simple, elementally. Their conditions do not allow an array of different substances. Matter generally remains in its single-proton state—hydrogen. But on the far cooler and calmer satellite worlds attached to suns, the proton nets increase, with a fair abundance of many of the lighter elements and even a certain portion of heavier matter like gold, lead, and uranium. The smaller planets were unable to hold the bulk of their light gases from the original sidereal cloud; these simple elements have been sifted up through the Earth's atmosphere, leaving our world with an even greater percentage of rare medium-weight elements like oxygen, silicon, carbon, and phosphorus. But many potentially volatile gases have also been retained by becoming locked into minerals in the emerging planetary crust. The Earth's hot core is nickel-iron, for the most part, and its crust has the general composition of the mineral olivine, which includes iron, magnesium, silicon, and oxygen.

The development of the Earth as a world proceeds directly from the novel associations of its elements outside the severe stellar furnace. Although the elements' properties are determined by the protons and electrons in their shells, they are not predictable on a simple mathematical basis, either individually or in interaction with each other to form molecules. So the phenomena of the Earth arise on the spot, indigenous to this locale. Similar phenomena may develop on other worlds that share the Earth's size, composition, and distance from a sun-star of the same size (or corresponding distances from stars of greater or lesser size), but since random distribution of materials and chance events may tilt environmental gradients in idiosyncratic directions, worlds can never be the same. Apparently, it is only the dispersion of peculiar carbon reactions which has given the Earth its present oxygen-rich atmosphere and climate. An identical planet without such life would be an entirely different locale. Venus, which is not dissimilar from our world in its size and distance from the Sun, is a baked desert under a thick acid cloud-cover, apparently lifeless for all of history.

The original Earth was a hot gaseous star plunged into a cold bath; landscape emerged slowly from fire and mists. Even as a crust formed, lava continued to burst from under its surface. As oxygen appeared, hydrogen discovered its natural affinity with its outer shell and shared its electrons with it. Instead of escaping the planet, these molecules were held as water. Water is a unique and talented substance. In a pool the structure of its individual molecules fluctuates chaotically, so that hydrogens are constantly displacing one another and bonding with new oxygens. This internal disorder gives water a peculiar viscosity and a high electrical conductivity. The strong hydrogen-oxygen bonds allow it to absorb heat as the ionized atoms vibrate and trap the energy. Water thus resists easy melting or boiling; in general, it has a moderating influence and provides a gentle environment.

The first water condensed into droplets as vapor and fell as rain, but as soon as the rain hit the hot stones it returned to the steaming atmosphere. The rain continued for millennia, heavily pounding and eroding the ground and filling the basins. The planet pulsated with lightning and shook with thunder. Over the epochs the rocks cooled, the low places filled with pools, and water seeped down beneath the ground.

The Earth was no longer a burning star, and the long interstellar winter took hold on its surface. Discrete seasons evolved, with their own storms and currents, and local zones of climate. The surfaces of the tinier inner planets were baked dry whereas those of the outer planets were frozen. The large Jovian worlds remained stellar and formed no crusts. But on the Earth, internal liquids gradually oozed onto the stones, and there, one chemical in particular displayed remarkable abilities. Denser as a solid than as a liquid, water froze on the surfaces of its own lakes and ponds, insulating its underlying layers and keeping them fluvial. As the ice warmed along its line of contact with the air, the hydrogen bonds between its molecules stretched, and thin layers of it peeled off into vapor. All of these other talents were to take on a special meaning when water ultimately revealed its *tour de force:* it came alive.

In the belly of the planet, molten plates shifted, new basins and islands were formed, and continue to be formed, for the core of the Earth is still hot and active. During the long rains, mountain ranges arose and were swallowed, seas filled with volumes of water which then flooded inland after earthquakes, making new seas and leaving ancient seabeds as dry land which would no doubt be seas again before the dawn of life. This was an epoch of volcanoes, hot springs, gullies, rocky shields, and turbulent oceans. In some regions lava still poured from the crust. Elsewhere chill water accumulated, always oscillating, ever in expanding motion, picking up its own rivulets from geysers and floods, absorbing the salty sediments of the land, and rushing jaggedly over the raw statuary into mineralized bays and world oceans. As the rivers cut their way through the elements, they left similar patterns in wide-ranging areas, giving the planet its characteristic veined and ridged topography. The geosphere became a unity, for the same forces were globally at work.

The spreading waters moderated the climate of the primitive Earth, evening temperatures, cooling off volcanic areas, and spreading tropical currents into regions that were already beginning to feel the cosmic ice. It is estimated that the original flooding of the Earth provided only about a fifth of the water for the present oceans. The rest was squeezed to the surface gradually over millennia, condensed or spat directly into the oceans by submarine volcanoes.

A global map of this Earth would be totally unfamiliar to us now. There was a heap of stone in an amorphous sea. It could have been any planet anywhere. It was just rock, and liquid, and billowing gas.

There were no mosses or shrubbery to cover the bumps and craters, no diatoms or crabs pulsating in the sea. It was a vast desolation, an exotic version of the Moon. The rain may have given intimations of something promising, but it wouldn't have looked inviting and it wouldn't have looked like home.

We might want to imagine that a foreshadowing of life hung over it like a nostalgia that was actually a prescience. Most scientists would deny this. Scientific logic, of course, assumes that at this point things could still have gone either way. It could have looked much like that today but for the accident of life. No railroads and operas—just rocks and rain. Even those scientists who argue that life was inevitable, given the chemistry of the primal Earth, would not propose a visible foreboding anywhere on this ancient world. No wild ghost on the wind, life would bring its own spirits into the land.

As the turbulence of the geosphere retreated to a partial slumber within, something resembling the present terrestrial geography began to form. This long ago the continents were bunched together, with South America and Africa unriven to make up the ancient Gondwanaland, as geologists later named it. In this yolk, a primitive North American rock was joined to the northern reaches of Gondwana, and was in the process of being torn from Eurasia, leaving Greenland in a polar sea. The Tethys Sea, the aboriginal Atlantic, lay to the south of Eurasia, but it was breaking through as the New World and Old World parted. Oceanus lay to the west—the natal Pacific.

Rain and wind washed the cosmic organic debris from the atmosphere, and the simple salt water was sown. The outer membrane of the planet was charged by ultraviolet light and electricity, ion-molecule reactions, and other creationary sparks on the edge of night. Submarine and terrestrial volcanoes also spouted the molecular seed.

Life probably began in the waters of the Earth between three-and-a-half and four billion years ago, only about a billion years after the formation of the planet and long before the oceans had settled to their present levels, and the lands had receded into the continents of modern times. So, early life was subjected to all the turmoil of rearrangement; and subsequent life, even today, must withstand some of the same (and other) environmental forces: floods, volcanoes, earthquakes, waterspouts, and perhaps even pole shifts, cosmic collisions, depletion of resources, and high levels of radiation. (The Earth has never been a peaceful place for any creature to raise a family.)

According to classical Greek scientific theory, the four basic elements, when set in motion by nature, spawned living as well as nonliving substances. Life preceded and permeated matter. Aristotle's description of this process survived over two millennia:

"The hermit crab grows spontaneously out of soil and slime, and finds its way into untenanted shells," he declared in *History of Animals*. "Some insects are not derived from living parentage, but are generated spontaneously; some out of dew falling on leaves, ordinarily in springtime, but often in winter when there has been a stretch of fair weather and southerly winds; others grow in decaying mud or dung; others in timber, green or dry; some in the hair of animals, some in the flesh of animals, some from excrement after it has been voided; and some from excrement yet within the living animal.

"Eels are derived from the so-called 'earth's guts' that grow spontaneously in mud and in humid ground; in fact, eels have at times been seen to emerge out of such earth guts, and on other occasions have been rendered visible when the earth guts were laid open by either scraping or cutting."[1]

The vitalistic paradigm was so compelling that the early seventeenth-century chemist Jean Van Helmont was still able to give a recipe for the creation of mice in twenty-one days—from soiled clothes left in a dark, quiet place and sprinkled with wheat kernels. The discovery of tiny animalcules through the microscope only reinforced the conviction that higher animals were made of indestructible primal germs and could disintegrate into their components.

In 1859, with the publication of *The Origin of Species,* Charles Darwin presented a thoroughly mechanical explanation for the variety of life on the planet. Five years later, on April 1, 1864, the French scientist Louis Pasteur opened a number of test tubes in which he had incubated many of the popular "recipes" for infusoria, maggots, and rodents, including hay and dung. Because he had tightly stoppered the vessels, there was nothing alive in them, nothing at all—no fungus, no bacterium, no infusorian.

Pasteur did not deny the existence of "germs"; he showed that they were abundant in our midst beyond even the wildest fantasy of infestation, but they were not the spontaneous product of the elements. A cubic meter of air in Paris during the summer holds ten thousand viable germs, Pasteur announced. They float freely and invisibly about us and spawn under favorable conditions.[2] But life can arise only from other life; it is neither elemental, nor primal, nor indestructible. The paradigm was changed. "The impossibility of spontaneous generation at any time whatever must be considered as firmly established as the law of universal gravitation," declared a contemporary physicist.[3] Even protozoa are so complicated they could not spring from any chance association of inanimate substances. But if life could not ever arise by spontaneous generation, how did it occur in the first place, before there was any previous life?

One curious nineteenth-century solution proposed that spores ostensibly ascended through the atmospheres of other worlds, travelling dormant and dehydrated across the vast spaces between solar systems, and seeding themselves in the primitive oceans of new planets. These "panspermia" would have to be hardy in order to survive the absolute cold of interstellar space, as well as the airlessness and radiation, and they would have to pass through the atmosphere of their new home without incinerating from friction.

The theory of cosmozoan animalcules was revived quite recently by Francis Crick and Leslie Orgel.[4] Their intuition was that life emerged on the Earth too suddenly to be indigenous and that protoplasm itself varies in key ways from the terrestrial medium in which it might have arisen (cells use the rare element molybdenum in critical enzyme function, for instance). Their improved scenario involved anaerobic (nonoxygenating) cells packaged by aerobic creatures anxious to preserve life in the galaxy. Anaerobic organisms would stand the best chance of surviving the journey and adapting to any of a number of alien environments. Upon arrival they would spin a new life chain, undoubtedly different from the one on their native world.

Panspermia theories are always speculative and based on a fantasy of cosmic history. For instance, Crick and Orgel imagine an original sun-star from the most ancient epoch of the universe, with billions more years than the third-generation Sun, in which to draw from the deck of all possible occurrences—leading (as here) to the origin of life and culminating in the extinction of the star and the launch of interplanetary seeds. We do not even know the history of the European Middle Ages that well, so we can hardly speak for the ancient peoples of the Milky Way. The more serious cosmological problem is that panspermia do not solve the riddle of the origin of life; they simply transpose it to other worlds.

Another kind of solution was first advanced in 1924 through the publication of a paper entitled "The Origin of Life" by the Russian scientist A. I. Oparin.[5] Oparin perceived that the best remaining alternative was to violate the law of the impossibility of spontaneous generation with a single instance that occurred at the dawn of the Earth's history and gave rise to all subsequent plants and animals by natural selection. Oparin described a previously unsuspected planet—the Earth before the biological infestation. On this dawn world carbides shot through the rocky crust from compounds of carbon and heavy metals, and a steady rain of hydrocarbons dropped from the superheated organic steam of the atmosphere into the rising sea. J. B. S.

Haldane later called the ocean of this world "a hot dilute soup."[6]
The "soup" was unique to its epoch and could occur only in the absence of life—a perception Charles Darwin had back in 1871:

"It has often been said that all the conditions for the first production of a living organism are now present which could ever have been present. But if (and oh! what a big if!) we could conceive in some warm little pond, with all sorts of ammonia and phosphoric salts, light, heat, electricity, etc., present, that a protein compound was chemically formed ready to undergo still more complex changes, at the present day such matter would be instantly devoured or absorbed, which would not have been the case before living creatures were formed."[7]

Haldane, Oparin, and others began to reconstruct the meteorology and incipient biochemistry of this unfamiliar pre-Earth. The atmosphere must have contained little or no oxygen but much original hydrogen that had not yet escaped into space or been bound into other compounds. Oxygen would have consumed primitive organic molecules and, as ozone in the upper air, blocked out ultraviolet radiation needed for energy in the absence of both photosynthesis and prior life forms to eat. The most ancient life was thus anaerobic and produced energy in a reducing (hydrogenizing) environment by fermentation (rather than oxidation).

This hypothesis was experimentally tested in 1952 when Stanley L. Miller and Harold C. Urey recreated the primitive "soup" with a Jovian atmosphere in their laboratory. Methane, water, ammonia, and hydrogen molecules were subjected to electrical discharges. The solution spawned many organic molecules, including some of the twenty amino acids used in the protein code. However, many other amino acids and organic molecules not used by terrestrial life also appeared. The addition of oxygen to the original "atmosphere" of the flasks eliminated all molecules found today in living systems.

Obviously, days in a test tube cannot represent tens of thousands of years in ancient seas and tidepools, and no experiment has spawned anything resembling even a poor subvirus; our past will always be actual and historical, not theoretical. But the dawn planet probably shared much with these Earth jars that it does not share with the contemporary planet that evolved from it.

The primitive Earth was the matrix and first abode of life; the modern Earth is the by-product of life. The first chlorophyll molecules transformed sunlight into energy, enabling cells to feed themselves and, ultimately, their predators, from the single proximate energy source. At the same time they took carbon dioxide from the atmosphere and returned oxygen. In the bombardment of ultraviolet radiation, ionospheric oxygen molecules were bonded into ozone, a membrane of protection around the world that absorbed subsequent cosmic rays. The cells generating this membrane lay deep in the biosphere under miles of water. As their own breath shielded them, they subsequently rose to the surface and into the sun. Between twenty-two hundred million years ago and fifteen hundred million years ago, the surge of photosynthetic organisms breathed out a world habitable for a new oxygenating life, and together these organisms stablized the atmosphere into its present equilibrium. If life were suddenly to disappear from the Earth, it is estimated that the weathering and oxidating of ferrous iron would remove all the oxygen from the atmosphere in a million years.

The first plants and plant-animals were the unconscious purveyors of a design so elegant we almost wonder today if the Earth is not a single organism, as homeostatic and preconscious as a polar bear. It has already transformed its own atmosphere, developed a protective skin, adjusted its temperature, and scorched a vast nervous system across its crust. It has used the cells of algae as integratively as it has used the computers and satellites of the higher primates.

3.   The Materials of Life

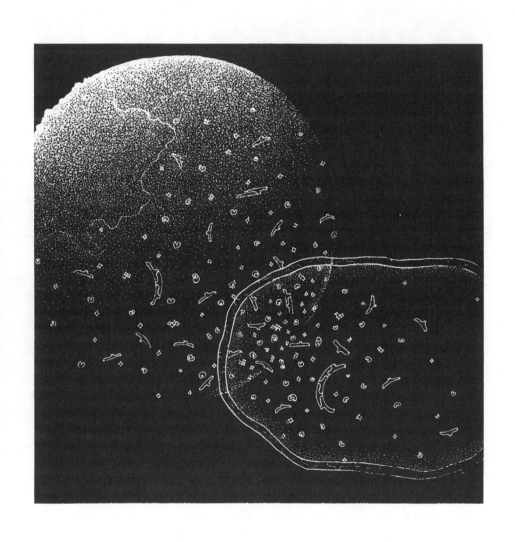

The main building blocks of life are the lighter and more reactive elements, the first musical notes built by stars on the primal hydrogen scale. If we broke the human body down into atoms, we would find that 63% of them are hydrogen, the most basic and abundant element in the universe. Sixty-six percent of seawater is also atomically hydrogen. The next most abundant element in both the human body and seawater is oxygen—25.5% in us and 33% in the oceans. The present-day atmosphere is 21% oxygen, but this is a by-product of photosynthesis and was not the aboriginal condition. The Earth's crust itself is still 47% ancient oxygen atoms trapped in rocks.

The next two most abundant elements in the human body are carbon (9.5%) and nitrogen (1.4%). Carbon atoms make up 3.5% of the Earth's crust, but seawater contains virtually no nitrogen and less than .01% carbon. Since the biosphere, as a whole, is almost 25% carbon, we might assume that raw carbon was appropriated from the primeval waters and atmosphere by nascent organisms. Much of this must have been cosmic carbon, infused in the stellar dust with silicates and metallic iron and nickel, and, later, blown in the solar wind off the Sun's corona into the Earth's field. Thus, living organisms are the crust of the ocean—in the old elemental sense, its "earth." The nitrogen of life originated in the atmosphere which is still over 78% nitrogen gas. A portion of the hydrogen and oxygen atoms of the primeval atmosphere were fused into water vapor which later condensed into the liquid of seas.

Carbon is the smallest atom in the group of the periodic table that rests midway between the chemical elements that give up electrons in their bonds and those that take them up. The carbon atom has equal proclivity to gain or lose the four electrons in its outer shell, so it can form a high variety of stable compounds, including ones using sodium and chlorine—prime ingredients of salt water. In addition, the smallness of the atom allows intimacy with other elements, for carbon can bring its electron veil very close to the nucleus of an atom with a positive charge. Such bonds are strong ones, and, if attraction and *eros* lie at the basis of life, we see them first at an atomic level. In many spiritual systems (such as the one proposed by the Russian

29

mystic scientist G. I. Gurdjieff), the goal of consciousness is said to be expressed in every atom in the universe; elemental particles build toward mind and spirit, from the hearts of suns to the dark oceans of new worlds, expressing aspects of their hidden nature in morphology at every level through which they pass.

Carbon atoms have the same capacity to form bonds with one another, so they attach in extraordinarily long chains; and when the ends of the chains meet, carbon rings occur; many of these rings then collect secondary and tertiary chains, including ones attached to other rings. This carbon choreography (actually performed as a ballet in 1939 at the national meeting of the American Chemical Society in Baltimore) filled the sea with chains and rings made of carbon, hydrogen, nitrogen, oxygen, phosphorus, sulphur, and other occasional elements.

Four of the most abundant and simplest elements constitute 99.4% of the human body (and 99.9% of the whole biosphere). Hydrogen is the unit (1) on the periodic table; carbon, nitrogen, and oxygen are 6, 7, and 8, respectively. The mystery of life begins in these bare mathematics, but it has numerological aspects the integers, as we know them, do not reveal. The quantum numbers of the electrons of the elements and the kinds of multiple bonds they form (especially in the context of water molecules) lead to the complex and varied macromolecules used in organisms. They are the components of amino acids, nucleic acids, proteins, fats, and starch, and they are fundamental to the molecular embodiment of protoplasm.

As matter sifted timelessly through its own mesh, the biosphere gradually differentiated itself from the other zones, incorporating aspects of earth, air, and water. It did not form a separate geographical belt; instead, its chemistry borrowed particles from the other three layers, always to return them later. Jellyfish and whales are literally hydrosphere, but, along with birds and insects, atmosphere too— blubbery balloons. All creatures are lithosphere as well, elaborate gems in but temporary exile from the planet's crust.

Molecular carbon spun the protoplasmic cloth of life, for the bonding propensities of the carbon atom filled the ocean with strange lattices, billions upon billions of novel chemical structures. If carbon was the loom, then the warm sea, which has ever after been a breeding ground, nurtured these fragile membranes. Water's own molecules continuously reorient in relation to one another, so it is a soft, chaotic liquid, flexible enough not to freeze protein in crystals (as the hydrogen bonds of other hydrides like ammonia do). Water forms secondary hydrogen bonds with proteins and thus preserves their complex structure. These oceanic molecules also help develop electrical charge in cytoplasmic material, thus opening a channel for life energy. Water is by far the most abundant molecule yet found in the universe in a liquid state, so its proliferation on Earth is natural.

However, we must not think of the biosphere as developing solely in the sea. Some of the simpler compounds of hydrogen, oxygen, and carbon, including sugars, glycerin, fatty acids, and amino acids, also occur wherever the constituent materials are present, on not only planets but meteors, asteroids, and in the "dust" of outer space. Meteorites, cracked open, have revealed amazing configurations resembling vacuoles, sea urchin spicules, double membranes, and, in one case, even the fossil of a cell in mitosis. The vitalists were not wrong on one point: apparently, form precedes metabolism. Biogenesis is thus physically grounded in cosmogenesis.

The other elements that are used in life have been borrowed from the environment in infinitesimal amounts, not for molding of animate clay but for their specific properties in enzymes and proteins. Both phosphorus (.22% of the human body) and sulphur (.05%) are essential in coenzyme molecules. In addition, phosphorus is a crucial component of adenosine triphosphate, which participates in the basic energy relations of cells and is a unit in the formation of the nucleotides of DNA. Phosphorus is also used for support in creatures with bones, and sulphur is a component of many of the amino acids in the genetic code.

Innumerable other elements are now required by living things, though none of them make up even one-half of one percent of the atoms in the human body or the biosphere, and many of them are present in amounts of less than one-hundredth of a percent. It may be that some elements became part of living systems initially because they were part of the dawn waters and so reacted with existing carbon chains. Once their atoms were present in the protein structure, their characteristics were incorporated and used. Mutations and gene variation regularly established new biological regimes, and no atoms could hide, not even those that just happened to be there. Their atomic properties and the properties of their compounds were integrated either into the structural elements of developing skeletons, the chemical reactions of respiration, digestion, reproduction, or the subtle biochemistry of enzymes, coenzymes, and hormones.

Cells are not electrically neutral; they maintain a charge distribution on both sides of their membranes—negative within and positive without (or the reverse in nerve cells). Positive and negative ions, cations and anions, regulate this charge and also the osmotic pressure within and without the membrane. From the beginning, cells had to keep some ions out while including others, though all were ostensibly part of their environment. It is the process of internalizing elements and maintaining their properties inside membranes that marks the beginning of individual bounded life forms.

Potassium may have adhered to the more clayey parts of the dilute organic broth when it was first forming, for this molecule has difficulty sticking to water. Once potassium is in a cell, it establishes its electrochemical properties as a cation (lacking an electron in its shell). From our oceanic origins, our cells have hoarded potassium and magnesium as cations and excluded sodium and calcium, other cations, maintaining an electrochemical neutrality in bodily fluids. Anions used in this balance include chlorine in the form of chlorides, sulphur in the form of sulphate ions, and phosphorus in the form of phosphate ions. These charged particles also help to regulate blood, lymph, hormones, and other internal liquids.

The metals with their unique traits have become crucial trace elements. Copper is used in hemocyanin for oxygen transport among invertebrates and in various enzymes for photosynthesis and skin pigmentation. Iron is integral for oxygen transport in hemoglobin and is incorporated in a wide variety of other enzymes. Cobalt can replace iron in some functions of blood chemistry and is vital for DNA biosynthesis and amino acid metabolism. Nickel and manganese participate in the formation of red corpuscles. Zinc is indispensable for protein digestion in enzymes, and, as well, in the formation of carbon dioxide, and in alcohol metabolism. Molybdenum is a prerequisite for the metabolism of purines. Selenium is necessary for liver function. Vanadium and niobium are used in the respiratory systems of sea squirts and other invertebrates.

Some original elements are expressed again and again at different levels of structure as creatures evolve. Extracellular calcium provided the morphological basis for the shells and bones of the first fleshy invertebrates. Magnesium has been the crucial component in the photosynthesis of plants and the derivation of numerous enzymes in animals.

"Why blood is red and why the grass is green are mysteries that none can reach unto," wrote Sir Walter Raleigh. But the physical basis, at least, of this reality has been disclosed. The biologist J. D. Bernal reminds us that the greenness of plants originates in the alternating single and double bonds of the magnesium-bearing chlorophyll molecule, and:

"The redness of blood . . . is written into the molecule of haemin; this is to be found not only in the blood of vertebrates, but also in the larvae of some flies, the bloodworms in stagnant pools, and in the nitrogen-fixing nodules in the roots of peas. In all these cases the colour is effectively due to the quantum states of the complex, partially filled electronic shells of ferric iron as modified by the porphyrin groups in which it is placed. Electron shells have existed as such ever since the first iron atoms were built inside a primitive supernova."[1]

Whether quantum theory explains the phenomenological color is another matter, as when the poet Robert Kelly cries, "I have wanted her all these years and for the same reasons I have wanted the color blue . . . The supple musculature of its light."[2] But both begin in the stars.

Just about all of the lighter elements are used in life, with the exceptions being those that are inert, like helium and neon, and those that are unambiguously poisonous like beryllium and arsenic. (Shades of Hecate in the primeval caves.) Fluorine probably has a structural role in bones and teeth. Silicon, similar to carbon, has the capacity to form chains (though not as extensive), and is used structurally as well, for instance, in diatoms. However, both fluorine and silicon, as well as other elements, have unknown and conceivably subtle uses in living organisms. Though highly corrosive, iodine is a component of thyroid hormones; it is the heaviest element known to be essential to life, but its exact role is obscure (flocks of sheep become ragged and diseased without their minim of iodine). If we take into account the possible roles of subdetectable quantities of substances, like the homoeopathic microdoses, and the psychic molecules of Gurdjieff, Rudolf Steiner, and other occult chemists, then no substance naturally occurring on the Earth can be excluded from suspicion of biopoesis. Gold and silver have been used medicinally, as has mercury—why not also before the first pharmacist, before the baboon Thoth? Perhaps the biosphere has developed "herbal" isotopes, microdoses of these elements within coarser substance. About rare potions like ytterbium, cerium, and rhenium, we can only guess.

Life has seemingly formed from the beginning by being poisoned by intruders who, in most cases, were accidentally trapped within membranes. Where the intruders did not fatally disrupt the nascent organism, they got included in such a way that their toxicity became neutralized and then later incorporated in the framing of genetic or enzymatic messages. In a sense, we can think of their incorporation within enzyme systems and vital proteins as their neutralization. Since they had to give rise to new overall properties, it became a matter of whether the organism would creatively integrate these properties or be overwhelmed by them. In many instances, they likely neutralized first, then developed new uses later. Oxygen was no doubt one of the early poisons, but without it, it is doubtful life could have generated enough energy to maintain itself and diversify. No doubt iron and copper were at first toxic too. Our antecedents probably fought off magnesium and phosphorus before they accepted them, and became them. At every level of our emerging complexity we deny ourselves, even chemically, and then use that denial to continue our growth. No wonder on a psychic level we continue to deny ourselves. We were never made whole. We were created, as Heraclitus divined, from strife itself.

The elements of life did not, of course, assemble automatically in an oceanic environment. Before there was an organization or plan, parts came together haphazardly and situations developed from these random associations. Phylogenesis is one long planetary embryogenesis which began without seeds, without organic structure, and without genetic material. Any living creature exists through the confluence of two events so remote from each other in scale that the life of the creature is their only meeting point. One of these events is the continuum of successive generations of replicating mutated life forms over several billion years; and the other is the succession of stages of a developing embryo, lasting anywhere from a few hours to a few thousand hours. Unicellular life becomes multicellular, and simple membranous animals develop tissue structure and a nervous system only through accumulated adaptations within increasingly complex ecosystems made up of other plants and animals. Embryogenesis carries out a brief synthetic phylogenesis. If embryogenesis took even a billionth of the duration of phylogenesis, creatures would not survive their gestation, and, in fact, would not survive at all, for the succession of generations would occur too slowly for adaptation.

Everything about life on Earth suggests that it arose from a unique site and then spread. Life is a singular chemical event. There are not two kinds of life, or three kinds of life; there is only one. All living cells resemble all other living cells; their asymmetrical molecules rotate the plane of polarized light in the same direction; they use the same chemical reactions to metabolize; they reproduce from the same genetic molecule.

The kinship of life is a more fundamental characteristic than is the difference of species. That is why the cells of caterpillars can metabolize the cells of elm and apple leaves, why the anteater's cells draw energy from the cells of the ant. Even viruses and bacteria belong to our protoplasmic lineage; if they did not, they could not read our codes and appropriate them. We could say, then, that all life on this world is the clone of a single primordial cell, whether it was indigenous to the Earth's water or seeded itself from some other world where it arose presumably in a similar fashion.

Bernal tells us: "Life is, indeed, an epiphenomenon of the hydrosphere."[3] This is true even for creatures who have never seen or touched the ocean. Where life exists beyond water it does so only synthetically, by including the sea within its membranes, the perfect alembic of our bodies that keeps the cell cultures alive. Generations of living creatures have captured the aboriginal waters in their tissues through a complex series of embryogenic inversions, and so their insides are a replica of the environment in which life began.

We do not fall as a seed into the ocean; we arise as a seed *from* the ocean, using whatever is already there and exploring its molecular properties within a membrane; all else dissipates back into nature. Two fundamental characteristics distinguish living creatures from the environment around them—their ability to assimilate other substances for their own metabolic requirements without losing their identities in the mutual reactions, and their ability to reproduce themselves precisely. Both abilities are relative and transient. Proteins and genetic molecules are among the most fragile structures in the universe. As even the hardest stone is worn down eventually by air and water, life must be reincorporated into the general chemistry of the planet; however, its metabolizing fabric resists such degradation far longer than it would if it were an inanimate chemical compound.

The survival of single tiny organisms in the great ocean is remarkable enough, but they would be solo acts amidst anarchy were it not for their complementary capacity for replication. If a living creature came about by chance and could not reproduce, then natural mortality would eliminate not only that zooid but its template. We would have to await another such creature to pass on the heritage of having been alive. If any creature finally, by chance, replicated itself, then its death would mean nothing in the evolution of its progeny. They would replace it, and be replaced at their own deaths. Precise reproduction is, of course, distorted ceaselessly by mutations, but life merely uses this crisis to change and evolve. Once the voodoo was established, the sorceror's apprentice was set loose, and no one has found him since.

Looking back on the miracle that spawned him, the biologist Francis Crick wrote: ". . . it is impossible for us to decide whether the origin of life here was a very rare event or one almost certain to have occurred."[4] Conventional Western religion assumes the former but solves the problem with divine intervention—creationism. Conventional Western science invokes the vastness of the ancient seas and the number of possible marine, tidal, and estuarine sites for molecular association and, thus, assumes that life was inevitable.

We can tilt the odds any way we want, but it seems strange, indeed, that all this came from nothing and now sits contemplating its own transcendent event. If we assume that life had outside help, we must still ask how the agency that aided it came itself into being and whether it made the only symmetries and lineages possible on this world or chose them from a latent reserve of potential structures. And who might follow us if we departed forever?

4. The First Beings

Molecules take on properties from the configurations of their atoms, and, from their own configurations, form compounds. At each level of synthesis new properties and potentials come into being. The elemental singularity of carbon is finally overshadowed by the complex variety of traits in chains held together by carbon bonds. Polymers are made up of monomers just as monomers are made up of molecules. Each new aggregate invents its own chemistry. There is no exogenous place for such chemistry to come from and in no way is it predetermined by the atoms and molecules before they arrive at it through their interactions and bonds. That is why we may speak of life as "creating itself." When new characteristics come into being, they invent and define a universe they have now made possible.

Monomers were apparently abundant in the primeval "soup," but in order for these organic molecules to polymerize, two conditions had to coincide: concentration and energy, the former bringing components of a potential system together long enough for some prebiotic associations to occur and the latter catalyzing molecular reactions. We have no idea how rich the prebiological soup was or, for that matter, how rich it had to be. Urey suggested that it consisted of as much as 25% organic material; however, other biologists surmise that as little as .1% would have been sufficient for biogenesis. Orgel actually measured the amount of organic material present in chicken bouillon and accepted it as a rough equivalent for the primeval waters. Life sucked its fiber from these waters by concentrating matter within membranes. Only if all living things and their by-products were dissolved back into the sea could the broth be restored. Through aeons of untended transformation, the biosphere was once and forever projected out of the hydrosphere. The bath was thinned and demystified by the process, but it is now bubbling with the vital gems of its experiment.

The original open sea would have presented two obstacles to polymerization: a cold aqueous environment dissipates energy, while at the same time, active waves tend to break up incipient chains. Oparin felt that the ancient ocean overwhelmed these objections with its sheer quantity of novel molecules. The repeated bombardment of ultraviolet rays and light would stir complex reactions of matter near the surface of this hydrosphere. The earliest life forms were then viscous droplets which separated from the water with colloidal particles. In these hypothetical colonies the colloids would be differently charged but in a state of equilibrium as they floated through such similar, but more dilute molecules. These coacervates (as Oparin named the colonies) were made up of polymers such as primitive proteins, albumin, and gum arabic. Whole regions of the ocean came partially alive, like vast cells.

The main objection to Oparin's coacervates was that they seemed too complex already to precipitate from a raw unbounded sea. The least complicated way around that dilemma was for relatively small portions of seawater to become isolated in fortuitous microenvironments. For instance, thin surface layers of high organic concentration might be blown along as sea foam and deposited on shores. Potential biological water might collect in shoreline cavities or even be closed off temporarily in inland ponds.

Crick hypothesizes that the Moon might have been significantly closer to the Earth once. Its back-and-forth tug ''may have produced continual wetting and drying . . . in pools near the margins of the oceans and seas.''[1] The tides would also have spread the variety of hydrocarbons throughout the seas, creating more opportunities for novel arrangements.

Even with present-day tides, organic material would have been deposited along the estuaries, always in suspension with mineral clay and water forming a kind of ooze. In the turning of the tide these colloidal layers would be superimposed on the mud banks, the adsorption of prebiological molecules creating a charge between different sheets (such as negative silicate layers and positive aluminum ones). The charges would bind molecules to sites and provide free energy for initial protein synthesis and polymerization. Life would arise from the mud like the golem, a heap of clay animated by a spell. The setting is Cytherean: a quiet tidepool on an out-of-the-way Archean beach, a cradle soaked in sun and fed by waves and sea foam.

More recent experimenters have discovered that amino acids synthesize in polymers of 200 or more in hot-spring temperatures between 160° C and 210° C. Small spherical bodies formed artificially in such an environment have been called proteinic microspheres, and they are alternatives to coacervates. If some of the oceanic ''soup'' found its way into a volcanic cinder cone, perhaps in a wave during high tide, the heat might have polymerized it while it was evaporating, and then the microspheres would have been washed back into the sea in the next tide.

Concentrated environments allow different molecules to "explore" their possible relationships. Atoms of similar form (either identical ones such as gold or different ones with similar size, shape, and/or charge) draw together forming crystal beds. Groups of molecules cluster into balls like those of copper inside pipes from which molecules have been worn. Associations of this sort are natural; in the absence of complex proteins serving as enzymes and coenzymes they were probably catalyzed by primitive minerals in the same roles, that is, transferring energy to maintain a repetitive sequence of chemical bonds.

The analogy to crystals is striking. These also repeat and restore themselves. The migrations of stones through the Earth's crust suggest layers of cells composing organs and dressing wounds; we see how ice knits a bandage like skin over a lake. The collagen fibers of bone are similarly regenerative, sliding past one another to fill gaps (with a pressure great enough to generate an electrical current). Twinning, dimensionality, and regulation are properties of biological fields as well as of crystalline fields. Oparin was infatuated with the "ice flowers" that formed on windowpanes; they looked to him like tropical vegetation and suggested the transition between the world of minerals and the world of cells.

Crystals have so long tantalized us with their life-like characteristics that it has become difficult to separate metaphors from morphologies, or transcultural events from subatomic and galactic ones. Aerial photography has revealed cities growing outward like stones, axis by axis. Houses are also crystals, as are cars, clocks and cyclotrons. Computers have now exposed human symbol systems as vast quasisymmetrical lattices arising from protoplasmic computers, living symmetries of mind. There is an invisible active crystal behind laws, languages, and concertos. The mathematics describing a crystal must also, in some sense, *be* a crystal.

It is no wonder, then, that early polymers should have this form and that DNA helices reveal series of spirally displaced axes. When we observe a crystalline transition between matter and life we are looking out through ourselves into an atomic wave pattern that apparently permeates the universe and gives structure to a variety of phenomena prone to "snap back" into stable patterns—from proteins to swarms of bees, from myth cycles to supergalaxies.

The dividing line between a mere chemical polymer and a living creature is impossible to mark hypothetically. Many chemical phenomena (fire, for instance) behave like living beings, and some creatures seem as inert as stones (the tiny tardigrades in diapause can remain dormant for decades and viruses can remain latent seemingly forever). Less metaphysical but equally imponderable is the question of which came first: the cell metabolizing substances, or the naked gene copying itself without beginning or end? And how did they come together? Did a freewheeling chemical reaction become surrounded by a membrane, or did a macromolecule begin reproducing itself from scratch? Either possibility presents the insoluble riddle: How did the metabolism within the membrane survive the aeons without the ability to replicate itself, or how did the naked gene copy itself accurately without catalysts, or sustain itself without hunger? There is only one possibility: They developed in concert, coevolutionarily, though we can hardly guess why or how.

After the miracle of their emergence the original creatures must have been anaerobic, producing energy by fermentation like yeast. Anaerobic metabolism would be sluggish and inefficient by present standards, but time was irrelevant in the Archean world. Rudimentary life forms would not rot, for there would be no microorganisms to disturb them. Some contemporary bacteria use hydrogen to produce methane, others metabolize by sulphur. Like creatures could have evolved in total darkness in subterranean rivers, energized partially by radioactivity; they could occur even in the damp cores of meteors and asteroids. Without air, without light, without water, they would be torpid quasi-living forms. (Perhaps such anaerobic beings on Mars greedily devoured the food sent to them from Earth by spaceship in order to test if they existed, or perhaps "they" were only the crystalline sands.) Not only do we always ask the same question, but we always get the same answer.

A new millennial chemistry tied metabolism directly to the Sun, and, with that burst of energy, novel creatures arose and spread across the planet, absorbing the beings of the anaerobic age. Photosynthesis is simply the ability of certain molecules (resonating lattices of carbon rings around central magnesium atoms) to split apart hydrogenous substances (usually water), freeing the hydrogen to reduce any carbon in various compounds, including carbon dioxide. The chlorophyll molecule is able to fluctuate between configurations, and its oscillating grids trap, store, and translate the energy of the light quanta passing through them. Groups of chlorophyll molecules distribute light quanta among themselves so that energy continues to be transferred (instead of being degraded into heat) until it is used in metabolism. Although receptive to light, chlorophyll is inert structurally so does not steal for itself; at the same time it is able to catalyze the bonding of hydrogen to carbon in the splitting of carbon dioxide.

It is pointless to try to guess how carbon and magnesium discovered these intricate and fragile properties, but, once imbedded in biological systems, chlorophyll reversed the balance of energy in the microcosm. The carbon-magnesium grid produces far more than the metabolism which it supports can use, so that energy is primed for other reactions, and has been ever since. Proto-enzymes ensured that currents generated through one metabolism were transferred to another, or, to state it in neo-Darwinian terms, substances that simplified bonding became included within membranes in an organized way. Some of the early polymers may themselves have served a catalytic function for other polymers, hence the birth of true protein enzymes and the type of active chemistry that can assemble macromolecules like proteins and nucleic acids. The energy of the Sun has now been stored and tapped for billions of years, and is located in the bodies and lives of plants and animals, and in the civilizations of this planet.

There is nothing in principle requiring living entities to be constructed of cells, but this is the manner in which plants and animals on the Earth evolved. The only units smaller than cells that can maintain life are parasitic upon cells. These entities, called, viruses, are extraordinarily simple organisms consisting only of a coat of protein around a strand of DNA, but they are not apparently ancient or primitive. They know too much about advanced organisms to have evolved before cells. Their entire life mechanism involves placing their own chromosomes in one of the cells of a plant or animal (or even a bacterium) and inducing it to make viruses instead of themselves. As their invasion spreads within an organism, more and more cells are destroyed and more and more viruses are synthesized, the collective effect often being diseases such as polio or rabies. In between such flurries viruses are imperceptible inert stones without metabolism. They are a separate kingdom of nature that has developed in response to cellular life.

The most primitive type of cell still in existence is the prokaryote, with no distinct nucleus and a single DNA molecule. Nowadays, prokaryotes are represented only by the bacteria (including the blue-green bacteria once thought of as algae because of their ability to use sunlight to split carbon from carbon dioxide and nitrogen from nitrogen gas). Although bacteria are contemporary organisms, not aboriginal remnants, they suggest what a primitive simplified cell might have been. They lack basic structure and organelle differentiation;

they do not have significant internal membranes or even the molecules to assemble a mitotic spindle, so they do not divide by mitosis. They can regulate metabolism and transmit genetic information in only one type of cell, which they already are, so they cannot differentiate as tissue or develop specialized cell functions. Incapable of synthesis, modern prokaryotes are limited to a unidimensional world of organization.

The cells that constitute plants and animals are called eukaryotes, and they cohere through the integration of a variety of subcellular structures called organelles (in a sense, the organs of cells). These organelles are quite different from each other; they include: bodies that are like cells within a cell (mitochondria, chloroplasts, and the nucleus); active specialized sub-animalcules like cilia, microtubules, and microfilaments; and primary structural components, including the membrane, intracellular membranes, lysosomes, ribosomes, Golgi bodies, etc.

Life on this planet has spread from zone to zone and microcosm to macroscosm by one dominant mode: Self-complete units are integrated into larger animate systems in which they lose a portion of their identity—cells into tissues, tissues into organs, organs into organisms, organisms into societies, and societies into civilizations. Cells are made of molecules, molecules of atoms, and atoms of particles. Since all matter, life, and consciousness can be reduced to these particles and their ostensible components, we can explain the transcendence of particle nature only by the introduction of utterly novel characteristics at each new level of synthesis. The tendency to combine and coalesce is a property of raw molecules, and so is passed on to primitive life-stuff, to minute sea creatures, large animals, and to thought itself which continuously sifts, associates, and links, and cannot stop this activity even in sleep. The first animals were compelled to associate by inherited nature, and their descendants continue their associations as jellyfish colonies, termite nests, flocks of birds, tribes of hunters, and, of course, parasites and hosts, and mating pairs.

The most obvious origin for the differentiated cell is in a gradual commensalism of subcellular organisms whose life cycles became so trapped in one another that they unintentionally merged. This is the only way we can explain the mind-boggling complexity of the simplest remaining animal, the protozoan, which could *not* have been assembled, *ex nihilo,* in the primal soup. The extremely rudimentary creatures that emerged from tidepools or volcanic cones were most closely akin to organelles. If parasitic bacteria can serve as actual flagella in protozoan parasites of Australian termites, and nitrogen-fixing bacteria can regulate the digestion and protein synthesis of leguminous plants by attaching themselves to root hairs, then life forms, ancestral to both organelles and bacteria, surely could have combined to form the first one-celled plants and animals.

In 1970, the biologist Lynn Margulis hypothesized that eukaryote cells originated in the endosymbiotic fusion of prokaryotes and their transformation over time into organelles. Thus, the mitochondria, the nucleus, and even the membranes of cells would all be the descendants of different free-living organisms.

The likely first stage in this fusion was the capturing of the mitochondria. These enigmatic intracellular animalcules even today have their own membranes, their own DNA and RNA of a distinctly different nature from that in the cell nucleus, small ribosomes for carrying their hereditary material, and a separate metabolism. The tiny rings of mitochondrial DNA synthesize some twenty proteins, and the mitochondria continuously extend, fuse, and split in two, just like cells themselves but at a higher speed and energized by their own intermembranous enzymes.

Mitochondria breathe for modern cells. Whereas anaerobic metabolism occurs throughout the eukaryote, oxygen exchange for the whole cell is localized in its mitochondria. If the mitochondria were prokaryote parasites to begin with, they fused with the precellular unit in such a way that the nature of both was transformed.

The chloroplasts of plant cells also have their own genetic molecules, ribosomes, and membranes, and likely share an ancestor with the blue-green algae they resemble. Whereas mitochondria probably infected subbiological clusters, the chloroplasts were cell invaders of a much later aeon. The original chlorophyll organisms might well have been forerunners of both plant and animal cells, the entire zoological kingdom arising later from those early bionts able to develop new life-styles out of the oxygen and sulphur produced by their anaerobic chemistry.

The intracellular membrane (known as the endoplasmic reticulum) may represent a complex association of creatures, some of which became vestigial in all but one aspect (like the organism that forms the bright blue float of the Portuguese man-of-war). Within the endoplasmic reticulum, Golgi-vesicle "creatures" synthesize cellular secretions; "predators" in the form of pinched-off membranes around vesicles, known as lysosomes, circumscribe digestible materials. Both organelles are present in all cells but are more prominent in cells that carry out their respective specialized functions.

Microtubules are animalcules made of two distinct proteins stacked in parallel circumferences around a hollow center. They are, literally, tubes. In combination with lysosomes, the microtubules form phagocytes, which engulf and transport food material and even small organisms. It is compelling to imagine these proto-organisms feeding only themselves before moving into the polymorphic colony that will become a cell.

The little tubes are not part of the reticulum but appear to be assembled within other organelles, the centrioles and basal bodies. The main role of microtubules in the modern cell is structural: maintaining and changing overall shape. In such processes they may collectively adjoin the membrane or line up parallel to the cell walls. Microfilaments are related creatures which act like the contractile fibers of muscles. They regularly run parallel directly beneath the membrane, appearing suddenly before cell contraction and disappearing afterward. Like microtubules, microfilaments play a fundamental role in cell differentiation and, thus, are integral during mitosis and tissue formation.

From their mutual (and perhaps spontaneous) synthesis of proteins and lipids, all of these separate ''creatures'' were enveloped by a great crystal, an outer membrane composed of sheets of cells of uniform thickness. The membrane gave their endosymbiosis a bounded environment—an enriched metabolized pool within the larger sea. Its osmotic system selectively allowed molecules into the pool while removing others. The organelles were trapped, but they were also protected and fed. The inside of our own bodies, hence, the inside of space and time, began here.

The nucleus, like the cell itself, may be a gradually fused association of free-living prokaryotes. The ribosomes have a particularly independent role within the nucleus as they travel back and forth between messenger RNA and amino acids. Perhaps they invaded other ancient organelles like viruses, and, while attempting to steal their protoplasm, were instead impressed by the emerging cell into using their artistic talent to copy *it*. A natural selection over many generations would then preserve the moment of subterfuge in regularized scripts.

Whereas all of the other organelles are concerned with immediate metabolism, the nucleus is the center of genetic transmission: the storage site of hereditary information, and the assembly plant for its reproduction. It participates in the continuum of life and the conveyance of the chemical identity of the cell through time and space. Of course the nucleus has no such high ideal. It has been trapped in the cell by an unknown prehistoric symbiosis and must enact the process that its nucleotides require. As it expends its intrinsic energy, new proteins are formed on the blueprint of its body. In the suitable medium a nucleus can program the formation of trillions of new cells, not only modelling them on prior cells but differentiating them from one another so that they create intricate sentient masses in three dimensions.

The true effect of nuclear activity is expressed only in a superorganism transcending time and space, dwarfing both the original cell and the genes. As tiny and prescribed as it is, the nucleus is a link between two totally different types of universes, each of which exists in terms of the other. Imprinting the linear scrolls through the narrow corridor of space and time, it is metaphorically the "temple" of the cell. It represents the connection between transient secular life and eternal life.

The nucleus has its own membrane within which the nucleic acids and associated proteins are maintained in a nucleoplasm. The nucleic acids form in distinct structures—the DNA (deoxyribonucleic acid) in the chromosomes and the RNA (ribonucleic acid) in the nucleoli in association with protein. The nucleoplasm operates by analogy with the cytoplasm of the cell itself, and it contains a variety of regulatory proteins, enzymes, and other associated bodies while providing a matrix for the long chromosome chains and dense nucleolus bodies. The nuclear sap carries out anaerobic metabolism, and it maintains a different ionic state from the cytoplasm. Thus the nuclear membrane, like the membrane of the cell, is a selective barrier preserving a chemical and electrical differential between the nucleus and the cytoplasm.

The replication of cells is a difficult event to imagine. Crystals repeat structure, but they are inert and, in a sense, two-dimensional. They give rise to three-dimensional entities but only by operating repetitively on a sequence of surfaces. In biological replication, the genetic molecule must be able to "feel" what a preexisting crystal is like in all its three dimensions and reassemble it. This holographic capacity is very old, perhaps as old as any part of the cell, for, as Bernal points out, "without definite molecular reproduction it is very difficult to see what an organism means: if it is merely a piece cut out of an undetermined extension of metabolically active material, it has no *raison d'être* of its own."[2]

As the cellular cluster cohered and "accepted" its unity, a new hierarchy emerged; the daily business of the cell's internal milieu took priority over any lingering independence in the individual organelles. The nucleus claimed genetic control of them and rebuilt them in generations, like robots, always including its own imago at the heart. Nevertheless, the organelles continued to carry out aspects of the distinctive behavior that brought them into a community in the first place. Something in them remains forever out of the full reach of the cell "mind."

Because of its unique function, the nucleus is the one organelle that seems to revert to a wild state, shedding its membrane during cell division and enacting in a round dance its ancient flight. Mitosis is the last throes of a trapped creature to escape the colony, to break the trance that holds it. Instead of kicking free, the nucleus turns in place, replicating its own nightmare to infinity and casting forth the living creatures of subsequent evolution. Its gyrations win it a sort of freedom, if not *out of* the cell then *through* the cell into the multiplicity of plants and animals around it.

So, where is our unity? From where does the existence we experience arise? For most of recorded history human beings have thought of themselves as completed entities with rational goals. They might be possessed by other entities or temporarily lose consciousness, regularly, in fact, in sleep, but such episodes were deviations, and any invaders were also whole and intentional. We could lose ourselves in the forest or the stars but not in our own profundity.

The cell was a serious breach of our unity. Historically, it was the forerunner of the unconscious mind—its prophecy and now its replica (many scientists still hope to prove that the biological substratum *is* the unexperienced mind). Through an analysis of the simplest behavior, Sigmund Freud demonstrated an unconscious vortex at the identity and core of every organism. He did not mean it to be necessarily componential and subcellular, but this is the direction in which we have looked for just about everything since his time. Post-Freudian philosophers acknowledge the multiplicity and fragmentation of their own personalities, the primacy of a vast interior realm of material they will never know. They accept that they will be in the dark and that their work cannot be completed. Where Platonists once hoped to find qualitative geometries regulating being, neurologists discovered only equally indeterminate animals in search also of a paternity.

We now know that our imagined wholeness is but an association of cells in a body, our brain a collection of nonthinking entities conducting thought. Each of these cells is a mosaic of individual bacteria archetypally stalking their own food and breathing for themselves in the shadows. This is the basis of the illusion of identity felt by the cart-drawing ox, the sleeping baby, the giant whale. Associations of bacterial creatures make up the oak leaf and the lizard. It was established in the protozoa, and there was nothing the snails or octopi could do but live it and pass it on. By the time proto-human apes formed tribes and stalked the first dreams of the Earth, they were already doomed to discover in some twentieth century after twentieth century that minute predators filled every cavity of their being down to the end of the world and back to the beginning of time. Self was our invention, not our heritage.

5. The Cell

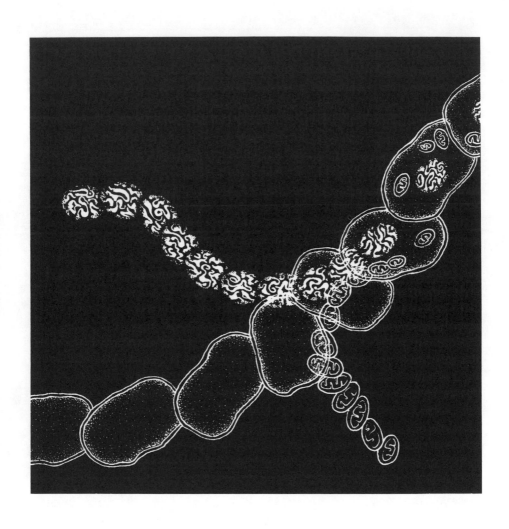

Free-living cells were discovered in 1674 through a microscope that was little more than a bead of glass mounted in metal and held up to the eye by a focusing pin. When the Dutch amateur naturalist and lens-grinder Anton van Leeuwenhoek examined standing pond water under a magnification greater than a hundred diameters, he saw, to his astonishment, tiny luminous creatures swimming about. Some spun like wheels, running into each other and retreating. Others crawled like shapeless snails, engulfing their unfortunate neighbors. In ocean water he watched little "fleas" hopping great distances (for their size) within "the compass of a coarse sand grain." In rainwater he found himself looking at a vorticella whose eye-stalks he described as "two little horns." He called it "the most wretched creature I've ever seen for when, with its tail, it touched any particle it stuck entangled in it, then pulled itself into an oval and did struggle by strongly stretching itself to free its tail, whereupon the whole body snapped together again leaving the tail coiled up serpent-wise . . ."[1]

It did not take long for van Leeuwenhoek to realize that these creatures appeared also in infusions made from hay, dust, dried mud, dirt from roof gutters, and moss. Since they arose spontaneously from infusions and in such staggering numbers, they were first named "infusoria." It took a full generation for people to get used to the disclosure that the inner spaces were far more densely inhabited than the visible forest or sea. The variety and numbers were shocking. In his own words, van Leeuwenhoek "dug some stuff out of the roots of one of my teeth . . . and in it I found an unbelievably great company of living animalcules, amoving more nimbly than any I had seen up to now. . . . Indeed all the people living in our United Netherlands are not as many as the living animals I carry in my own mouth this very day."[2]

Eighteenth-century biologists came to classify the infusoria as elemental seeds out of which "real" animals are made. The renowned taxonomist Count Georges Buffon described them as a kind of organic molecule that did not have the vitality to reproduce as plants and animals do but which contributed to their formation. Albeit

53

"through a glass darkly" and without realizing the implications, Buffon had perceived an essential aspect of the animalcules; they *were* the building blocks of plants and animals, not in their present but their ancestral form as the forerunners of dependent cells.

The modern world actually contains two distinct orders of cells. One is composed of these minute organisms that prey, carry out chemical reactions, and reproduce by dividing. The other contains the individual atoms of living tissue which also have membranes, derive their energy from phosphate reactions, and breed by mitosis or meiosis. Early magnifying-glass biologists saw at least the fuzzy outlines of these constituent cells when they looked at ant eggs, fly brains, larvae, and human blood. In 1665, Robert Hooke viewed faint divisions in cork and named them after the "cells" in a monastery. Their actual character was discerned in 1838 by Matthias Jakob Schleiden, a German lawyer and naturalist. After close examination of botanical material under a microscope Schleiden declared that "plants consist of an aggregate of individual, self-contained, organic molecules—cells. The cell is the elementary organ, the one essential constituent element of all plants, without which a plant does not exist."[3]

With better magnification observers began to see that there was a fundamental similarity between cells and infusoria: they each had distinct membranes and nuclei; and they divided into pairs from a crisp star-like pattern suggesting a spider's web or a crystal of frost. The same honeycombed structures with nuclei existed in animals as well. An analysis of the cell jelly showed it to be a kind of protein, one of the complex compounds of carbon.

By the mid-nineteenth century, the cell had become the unchallenged first principle of life, and biology was a mere extension of cytology—life was the consequence of the physical characteristics and relationships of cells. Medical materialists were explaining that "the brain secretes thought as the kidney secretes urine," and that "genius is a question of phosphorus."[4] In an industrial age the cell was reborn as a machine, and, although this machine was transformed into a computer in the twentieth century, we would have trouble today explaining what "thought" is other than the secretion of the brain.

Named after cubicles with stony boundaries, the cells of the seventeenth and eighteenth century were discrete formations. However, the entity which modern biologists inherit is hardly a cell at all in the seventeenth-century sense; it is a physical field through which a dynamic transformation of polymers proceeds. Hooke's image merely casts a cell-like shadow from a prior century.

* * *

Since we precede protozoa in our own unravelling of natural history, we call them (anachronistically) "cells functioning as organisms." In the context of our own embryogenesis they are zygotes in which the potential biological energy for cell multiplication and differentiation is translated instead into motility and predation. In a number of species free-living cells form colonies which are ecologically, if not morphologically, the forerunners of multicelled plants and animals. The Volvox is a spherical colony of flagellates in which each cell feeds on its own but is synchronized to its siblings. Despite its resemblance to a blastula (noted frequently by early recapitulationist biologists), Volvox is a collaboration of separate cells not the differentiation of a genetic unit.

Dependent cells have substantial functional connections that must have evolved at the beginning of their lineage. These include protein molecules in their cell coats that recognize one another, filaments and seals between membranes, and gap junctions through which minute channels ions and other materials pass. The latter generate electrical coupling and metabolic symbiosis. These are not external connections like those of the Volvox; they are internal polarities by which the cells make contact, eat and breathe in a unity, and send one another chemical messages.

Junctions, although they originate from the deep morphology of cells, are not fixed or permanent structures. They disassemble instantly with treatment by certain chemicals and are quickly restored when cells are put in contact with one another, even if the cells are from different regions of tissue or from different animals of widely separated species. In cell cultures made of human cancer tissue and chicken embryo, the cells recognize one another as cells despite the gap of history and content among them. These flexible connections are crucial in embryogenesis, because groups of cells often cluster together suddenly and move as a single field, whereas other cells that do not participate in the field must be able to decouple themselves.

In the widening gulf between ontogeny and phylogeny, the cells of multicellular organisms lost a large degree of their independence but not their discreteness. Cells do not melt and blend to create organisms despite the seeming homogeneity of most plants and animals; they continue to act very much like little beasts, each concerned with its own survival and doing the kinds of things animals of its nature do. Yet it is because the cell is capable of independent and integrative activity that it can become the building block of life.

If cells develop a functional coherence in tissue (and tissues constitute organisms), these are still incidental matters to the lives of cells. Of course, once cells have been organized and their collective properties subsumed in larger entities, they are effectively symbionts and cannot survive outside of the environment they have created. Still, their symbiotic condition is never a replacement for basic cell function. Symbiosis is a variant of conventional cell life, and, if (as has happened to all dependent cells, except the germ cells in special brief episodes) their capacity for separated survival has been lost, it is only because such capacity became an expensive luxury. Parts of their physiology have become vestigial or degraded as they were no longer required. Other aspects have been transformed into mechanisms of collaboration (like the gap junctions between cells).

Although it is merely a human metaphor, we might say that cells remain cells to themselves even as they form organisms; they are not "aware" of giving up anything or of participating in a group effort. It simply happened that their mode of survival was transferred to a new environment; they went on metabolizing, growing, and dividing like their free siblings, as they do today, providing our metabolism through their own. It would be a surprise to them indeed to find out that in conjunction with one another they had such elaborate and exotic by-products. We ourselves could hardly be interesting to the cells that embody us, for they are still independent animals in a primeval sea. They are barely large enough or wise enough to see that the surrounding waters have moved and changed. And then again, in their terms, perhaps it *is* still the same sea, and its curious plasma packaging is little more than another current or tidepool. (Would *we* know if our galaxy had become part of the fabric of a gigantic being?)

Protozoa lack the cytoplasmic substance to exist as anything but a kind of crystalline by-product of water. They do not have separate organs or specialized parts under the control of their own nuclei. The early twentieth-century mathematician D'Arcy Thompson pointed out that almost all the protozoan shapes, from oblong tubes to floating bells and pears to swimming ciliate balls and assemblages of balls, can be derived from the surface tension between fluids. These animals are virtually independent of gravity, and their high surface-to-mass ratio makes specialized respiration unnecessary; each one breathes throughout itself. Protozoans are, equally, lung cells and gut cells. Their energy comes from the universal process involving the breakdown of adenosine triphosphate to its diphosphate, though it is unclear how that energy is translated all the way to remote filamentous protein molecules in the locomotory fibrils.

Although protozoans are entirely aquatic, they have adapted to a diversity of moist environments over millions of years. Some species live in damp soils; the little bits of water clinging to the dirt particles serve as their small lakes. Others live as parasites in the organs of animals and plants; they have adapted to interior ponds even as the cells of those organs have. The protozoans have differentiated in order to be whole creatures. They are the primeval monads of not only life but individuation.

Paramecia are highly coordinated one-celled animals, able to retreat quickly, for example, after striking objects, as their cilia reverse. Threads, shot out from their basal bodies, can capture prey or hold the creature to a spot like an anchor. They show clear "hunting" strategies and apparently "learn" from their errors, which is remarkable in a creature of this size. It is not just that we are seeing the dawn of mind; we are seeing a hidden power like gravity that casts an image on every stone and in every skeleton and nerve organ, a unity whose shadow comes into being even before there is apparently enough surface for it. Then again, we may be watching simply our own projection of mind into matter.

The locomotory organelles of the protozoans are critical to their survival as predatory animals. Simple flagellates like *Euglena* lash their single protein whip back and forth so that their body is thrown to one side and another in an overall spiral progression. The plane of motion lies at an angle to the animal's angle so *Euglena* rotates as it swims. The flagellum also strokes food into the gullet. Ciliates, like the *Paramecium,* have a more controlled gait generated by the waves of tiny cilia. Amoebas are more like mobile guts. The outer part of their endoplasm is a gel and the inner part a sol which becomes a gel along the advancing lobes of the pseudopod. Vacuoles secrete enzymes and absorb the food in the protoplasm.

These methods of locomotion have also evolved in specialized cells within organisms. Ciliate and flagellate cells move particles in digestive and excretory tracts, and amoeboid lymph cells and antibodies behave like independent pond-dwellers as they circumscribe trespassers. It would appear that the morphological repertoire developed by aboriginal zooids has been transmitted through multicellular as well as free-living nuclei.

Within the limits of their size and dimension unicellular organisms do complexify. Protozoans form membrane-bounded vesicles, some of which are excretory; other structures contract to squeeze out water. Ciliates develop cavities, funnels, and cell mouths—elaborations of compound cilia fibrils. Although the organelles are functionally the forerunners of the organs of animals, it is impossible to see how they could share any structural continuity with analogous forms made *of* cells. Somehow, when the protozoan transcended its atomicity, it passed on the template of its organismic coherence, and its incipient anima.

If single-celled creatures grew much larger, they would not provide enough protoplasmic surface for absorption of either oxygen or food. Their evolving descendants would not have been able to pack in their new cells like crystals in an expanding snowball without choking and starving the original core. Only if their "one-dimensionality" were extended in a line so that no cell were more than a few microns from the surface could they occupy exponentially increasing space. This linear extension is the basis of a flatworm.

A more substantial way to fill space with cells is to increase the complexity of surfaces through the layering of tissue and the branching and folding of organs. All large animals do this. They breathe and eat through the devious topology of their lungs and gut. Our bronchial pleats and twisted esophagus record the long and irregularly internalized passage of events through which we evolved and shaped ourselves surface-to-surface with our planet's atmosphere and surface.

We can explain multicellularity as the animal discovery and occupation of space and time. Increase in physical size and depth leads protoplasm to new habitats and scales of mortality and to an organic dependence of parts. Worms creep across the torso of the Earth. Connection, communication, and coordination establish themselves again from within. Synchronization of cells creates internal environments—central cavities through which organs are molded and their functions integrated. Ultimately, a skeletal underpinning is needed for this blubber of swelling cytoplasm, a girth by which it can thrust itself out of the droplet onto the hollow planet among stars.

Single-celled animals are continuously reborn in their offspring, dying only to recur. This is not the fate of the metazoan zygote, which will differentiate into many diverse cells and form an organism; only its germ cells, those few resembling the protozoa, will live past it, and then only insofar as they generate whole organisms that will also mature and die, each of *them* spawning a small number of germ cells to carry on their life. It is no wonder that we have the intuition of life as both ancient and new, both mortal and eternal.

The cells of our bodies maintain their individuality by a membrane, but that membrane is as thin, in a relative sense, as the film of life around the Earth's crust. Since the membrane develops within the cell and combines with adjacent extracellular membranes, there is no concrete boundary between the cell's internality and externality. The membrane is a subtle combination of a gate and a barrier, sensitive enough to transmit information about substances in its environs without letting those substances in. Certain hormones (like insulin) influence the cytoplasm at a distance.

Cells are internally composed of membranous material separating the cytoplasm into regions and resulting in vesicles, vacuoles, and other substructures. The internal cell membrane is one continuous ribbon of interconnected structures that arise from each other and lead one to another. It is a soft maze through which a minute particle could wander for miles without coming to an end or crossing its own path. The various endoplasmic membranes are one membrane, and that membrane *is* the cell structure. The cell as a whole is a functioning series of boundaries not inside or outside one another but arranged as layers of atmosphere are, by the chemistry that takes place among them.

The environment within a cell is the result and also the source of its chemistry; the particular internal milieu that made cells possible continues to flourish and spread because cells must continue to synthesize their environment, or, more precisely, come into being as it is synthesized.

Intracellular polymers include proteins, nucleic acids, fats, polysaccharides, and adenosine phosphates. By the machine analogy, the adenosine phosphates are the batteries of cells. Adenosines are members of a class of sugars bonded to nitrogen bases in units called nucleosides. Specifically, they are ribose (C_5) sugars with nitrogen-carbon bases called pyrimidines. When inorganic phosphorus chains (phosphates) are in their environment, adenosines can form mono-, di-, or tri-phosphates depending on the lengths of their chains. When adenosine diphosphate (ADP) already exists in a locale containing phosphates, adenosine triphosphate (ATP) will be formed if there is energy to attach an additional phosphate to the chain. But this is no small undertaking, and the amount of energy required is not generally available; however (and this is the key to the movement of living things from columns of ants to jetting squids), the necessary energy can be provided by the decomposition of organic stuff, for instance in digestion of proteins or sugars. The energy of the digestive reaction is stored in ATP bonds and released, along with ADP, when they are broken.

We eat for energy. The cells experience the chemical reaction or they die. Life on a cellular level seeks to survive only as cells, and, likewise, dies as cells. When the wolf eats the rabbit, the intimacy of that relationship lies in the fact that their cells are virtually identical, and the carbons and hydrogens move from one chain to another.

Polysaccharides and fats are structurally part of the cell and mem-

brane and participate in energy transfer. Polysaccharides are polymers of sugar molecules, and fats are formed by single glycerin
molecules bonded to three fatty acid molecules. By the standards of
inorganic chemistry these are complex substances, but they are
eclipsed by proteins and nucleic acids which are polymerized in macromolecules of unprecedented size and intricacy and with molecular
weights often in the millions.

Proteins are elaborate crystals whose different parts bend around
and through one another to form multiple bonds; their bodies twist
and attach and grow until they fill and even seem to distort the three
dimensions of ordinary space. They dwarf the molecules that assemble them, but their tangled heaps are chemically innovative and provide the materials for life.

The proteins are polymers of the otherwise common and unpromis-
ing amino acids. There are often one hundred thousand or more
amino acids in a single protein, and these can be arranged in trillions
of ways, so the possible diversity is astronomical, enough to express
our unique multiplicity. There are more *possible* proteins than atoms
in the known universe.

Proteins, as the first modern biologists discovered, are the stuff of
cells. Their reactions are catalyzed by substances that are also
proteins—enzymes and coenzymes, which evidently replaced minerals in the early life chains. Enzymes change molecular structure tran-
siently to make certain things happen more easily. They form
intermediate structures that facilitate more enduring structures. For
instance, if the energy to form a particular bond is not regionally
available, the enzyme provides sites for each of the molecules of the
potential bond, and it brings them to a place where it will take less energy to complete their combination. Enzymes work not by chemically
reacting with substances but by changing the way they lie in space;
they use the natural shapes and spaces in molecules (which fit their
own bodies) to twist the molecules into new positions.

Essential life reactions, that now happen almost instantaneously or
in minutes, would take years or even centuries without such help.
Yet, enzymes are not predesigned for their roles; their occasions are
accidental until they happen. Within the ancient prebiotic material
their properties changed relations among other molecules without the
enzymes being altered themselves, so they were necessarily incorporated within compounds. Since enzymes do not burn with chemical
activity, they are available again and again for the same reaction.
There is nothing to stop them and nothing to wear them out. Once
they are integrated into the embryogenic chemistry they must go on
catalyzing, for every time they release their partner a new dancer will

grab either hand. Without this intruder life would wait forever to happen, and by then all its other preparations would have fallen into rubble, into the anonymous waters, to occur again, if at all, at random isolated sites. The whole possibility for highly integrated animate systems lies in seizing the fragile moment when the pieces are present. Volcanic action might bond proteins at high speed, but it would also destroy their delicate configurations. Enzymes allow the living stuff to congeal by providing the necessary energy without extreme heat and pressure.

Proteins were such creative and promising molecules that it was assumed, until 1944, they were also the basis for heredity. In an experiment that year, biologists destroyed selective macromolecules in treated bacteria cells and discovered that DNA, a nucleic acid, was the only molecule required for a new generation of the microorganisms. DNA was pictured then as an extremely long and dense molecule arranged in such a way that sugar and phosphate groups alternate on a backbone, with nitrogen bases attached to each sugar. The sugars and nitrogen bases form nucleosides which combine with the phosphorus units in compounds called nucleotides. Nucleic acids are enormous chains of polymerized nucleotides, regularly hundreds and often thousands of units joined in diverse and intricate ways. Their stability and the consistency of their arrangement give them a potential reproductive function. By making replicas of themselves they pass on information about their own chemical structure divulging how they are made. So basic was this blind transmission that it lodged as code in the heart of the cell and became the first language spoken on the Earth.

6. The Genetic Code

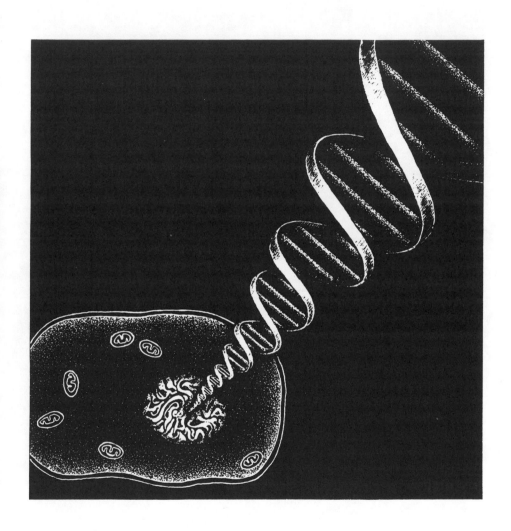

Reproduction has a molecular explanation in the double-helix model of DNA (deoxyribonucleic acid) proposed in 1953 by Francis Crick and James Watson. Their immediate predecessors had already discovered that four different nitrogen bases were attached to the nucleic-acid sugars. Two bases (cytosine and thymine) were of the pyrimidine type, bonding carbon and nitrogen in a single hexagon; and two were double-hexagon purines (guanine and adenine). Chemical analysis had shown that adenine content equalled thymine content, and that guanines and cytosines were in similar balance. Using this hidden geometry and X-ray diffraction studies, Crick and Watson arrived at their icon: two chains twisted about a shared axis. The sugar and phosphate backbones were on the outside, and the bases were turned in toward their common axis and periodically joined—adenine to thymine, guanine to cytosine. With each chain tracing a right-handed helix but running in an opposite direction from the other, DNA resembled a winding ladder: two parallel ribbons joined by purine and pyrimidine steps.

The purines and pyrimidines are poles of a complementarity. Their hydrogen bonds express the creationary force at the heart of the Sun that also holds together water molecules. The molecular biologist Harvey Bialy compares them to the *yin* and *yang* of Chinese physics, the opposing primal forces producing a whirlpool through which the universe manifests itself.[1]

The spiral staircase is an ideal hologram for heredity. The genetic molecule must bear its realms of information in a reproducible state. If it were solid it could not be read; if it were a plane it would give parallel images. That it was essentially a line, a linear strand, was suspected before Crick and Watson. The surprise was that it was a bent line, a helix, with another helix, a twin, wound around it. This is now as much a seminal image for the latter half of the twentieth century as the mushroom of the split atom and the blue-white mandala of whole Earth against black lunar night. The twin helix suggests the mystery of structure within each of us, an interminable winding dance of growth and form, its molecules spinning ribbon out of ribbon, half of them always upside-down.

Because DNA is actually a flowing transformation we can portray it statically only through a kaleidoscope of different appearances—twisted ribbons, parallel lines, a continuous ladder, or the wave phenomenon created by the polymerization of nucleotides. If we look at the ladder from the steps to the backbone, then the nitrogen bases combine with pentose sugars to form nucleosides, which are held in the backbone as nucleotides by the phosphates. Each nucleotide actually starts with a sugar and a base joined to three phosphates in a row, but two of them are consumed in the polymerization of the chain.

The ribbons themselves are long and predictable: they go: phosphate-sugar, phosphate-sugar, phosphate-sugar millions of times, with the nitrogen bases strung on them as close as natural forces will allow and each ribbon making a complete circuit of the axis every ten bases.

If the DNA in any one of us were disentangled from the helices inside the nuclei of our cells and then stretched, the cable would run a thousand million miles. Unwound, each strand can transmit its information to an assembling sugar-phosphate chain. New double helices occur spontaneously—one old strand, one new strand—matched in antiparallel fashion and separable too at the proper impulse from the environment. The genetic messages, the genes, are incorporated in sequences of nucleotides formed by the seriality of nitrogen bases: they have no other physical reality. Crick describes the transient bonding of the strands: "The two chains of DNA are like two lovers, held tightly in an intimate embrace, but separable because however closely they fit together each has a unity which is stronger than the bonds which unite them."[2]

The nucleotide sequences, the genes, program the assemblage of twenty amino acids; from this seemingly small number, arranged in different combinations, 10^{23} different proteins are possible, each an average of a hundred amino acids long. Apparently, this is enough to turn clams into octopi, and worms into blue jays—and to accomplish the remarkable transformation of protozoa into human beings. However, without precise timing, synchronization, and the capacity for pause and termination, the raw code would be merely a prescription for organic chaos. Given the exquisite subtlety of carbon's bonding ability, such muck may abound in swamps bigger than the Earth on Jupiter, Saturn, or Uranus—raw organic material without life, sludge and slag lacking a genetic molecule.

The precise strands ensure that cells are reproduced and differentiated according to their ancestral plan. There is no other way for them to form, and there is also no way for them not to form, so they continue to aggregate and interact out of their intrinsic chemistry and the sequential fields they themselves bring into being. In ontogeny as well as phylogeny life can be created only from life itself.

The same anonymous loom that polymerized other macromolecules spun the antecedents of nucleic acids. Phosphates were already abundant in the primal "soup"; the ribose sugars were cosmic debris. But some ancient Spider God would have had difficulty sticking together bases in space-time and then knitting sugars both to them and phosphates. Perhaps sections of the code assembled themselves separately, a babble of chance words that came together when their strings proved meaningful afterward. No doubt innumerable "wrong" chains appeared in the same environment, were able to contribute nothing to the emerging message, and so were obliterated. Once fully replicating sequences developed, their assemblage was likely regularized by a mineral or a primitive enzyme. This kept meaningless units out of a potentially functional code.

Such is RNA (ribonucleic acid), likely the first genetic molecule, an accidental polymer assembled from random sections of loose material with the capacity to transmit repeated chemical instructions. Alone it was not able to do much, but if such a chain were to associate with other organelles, perhaps ribosomes and primitive mitochondria, then rudimentary protein synthesis could begin.

DNA was a later refinement, a cumulative library of genetic information set in the nuclei of cells. Old DNA never leaves the chromosomes and does not participate in protein synthesis. Only the modern forms of RNA continue to transfer primordial warrants from the nucleus to the cytoplasm where the proteins are manufactured. There are slight chemical differences between the two, like different dialects of the same mother tongue (RNA has simple ribose instead of deoxyribose sugar, and uracil instead of thymine), but the complementarity of the message does not change.

RNA is usually found in single strands, though it may form double helices with itself or a strand of DNA. Whereas once it served as the actual genetic material (and still does in some small viruses), its function has changed in the cells of most contemporary organisms. In combination with a complex group of protein molecules RNA is used structurally in the formation of ribosomes; but, most fundamentally, RNA is the working copy of the DNA in the nucleus; in this form, as messenger RNA, it carries the genetic information to the ribosomes. Another structural form of RNA, transfer RNA, leads the reinless amino acids to the correct codons at the ribosomes where they are assembled into proteins. The possible forerunner of DNA, transfer RNA is both structural and informational, and it folds easily into a structure resembling the double helix.

The proteins are synthesized (the genetic information transferred) in a series of events that suspiciously resembles an assembly line. Our simple machines foreshadow, after the fact only, the assemblage of proteins, but this does not mean the cell is merely a factory. Rather the cell has an aspect which suggests a factory in a historical sequence of formative paradigms. The process of protein synthesis is far more complex than any machine metaphor. When we portray individual effects in linear chronological order, we are transcribing the shadow of nonlinear synchronous events, forces creating a field out of which they themselves emerge.

DNA composes RNA from the nuclear sap. RNA enzymes locate the precise points of contact on DNA molecules for transcription; nucleotide by nucleotide they proceed through the genetic message as they hold the chain in place. The RNA molecules are of course complementary, not identical to their DNA molds; they are anti-codons—but then the cell knows of the existence of DNA only through the mirror structure of RNA. In fact, RNA is the dynamic reality of the genetic molecule. Its strands come into being only as the nucleoside triphosphates copy the DNA chain, uracil aligning with adenine, guanine with cytosine.

DNA replication requires dozens of discrete proteins to catalyze the unwinding of the helices as well as the rearrangement of the chains in space, their rotation about one another, and the legible marking of the backbones. All of this must occur while the helices are separating at one end and being copied at the other so that transcription is going backward on one as it is going forward on the other.

Any one transcriptional event must bear different though consistent relationships to others in time and space—to those within the same cell but also to those in other cells in the same tissue and ultimately to those in other parts of the same organism. Each environment is transformed by and transforms the events it requires. New environments emerge seamlessly from existing ones.

The messenger RNA molecules are translated into amino acids with the aid of enzymes which recognize the shape of only one amino acid and so maintain the precision of the message. They are the active bond between the static nucleotide sequence and the emerging amino acid polymers. The transfer RNA molecules weld these sequences together. One end of each kind of transfer RNA brings the codon for which it carries the complementary message; meanwhile the molecule is folding such that its other end can be fastened to the ribosome at the node of protein synthesis.

The ribosome is like a lock in a zipper. Its RNA component allows it to attach the sequences of codons from the messenger RNA to the amino acids of transfer RNA. This zipper travels the length of the codons picking up successive transfer molecules with their amino acids. Numerous ribosomes work sequentially along a messenger RNA strand at the same time, catalyzed at the sites of protein synthesis. Ribosome by ribosome, growing peptide chains form as a result of the mutual reactivities of the amino acids and the emerging configuration of the protein. When the last amino acid has been attached at the ribosome, another RNA molecule arrives at the site, and the process recurs. When the full polypeptide chain has been formed, the protein molecule is released and a new one begun.

Proteins are made in straight lines—the lines of their amino acid sequences—but they take on much more intricate three-dimensional shapes when amino acids interact and bend their chain around to join itself in loops and twists. Some of these are actual chemical bonds; some are electrostatic interactions; others are hydrophobic groupings. The latter form when those amino acids that do not associate with water gather in the dry center of the protein. Then, those amino acids attracted to water form a ring around the core and fold together with electrostatic and hydrogen bonds. Covalent bonds anchor the lattice.

The proteins are not "life," but they are able to mold complex figures from their own intrinsic shapes and the dynamic chemistry of their components. Through the contours and constellations of the polypeptide bonds they improvise a startling array of differentiated structures: tissue, skin, blood, silk, feathers, nerves, stinging cysts, phosphorescent bodies, bone, scales, horns, and so on. The approximate rate of translation is one amino acid per half-second. Hemoglobin molecules, with 150 amino acids, take one-and-a-half minutes to synthesize. This same process occurs at different sites throughout the organism but with no external synchronization of scales or regions; the assemblage is self-referential and derives "meaning" internally.

The transition from cell to organism is immense, but it happens only as discrete interrelated steps. The genetic code is not a set of instructions simply played back; it is a self-made language that continues to be translated. The metamorphosis of creature-assembly is orchestrated in the highly adaptable cells which must continually change their interpretations and uses of the same information.

Except in rare cases a cell contains the DNA for the construction of its whole organism, but it will transcribe only those portions of the hereditary code that pertain to its position in the developing embryo; that is, the role of a cell is determined by selective activation of its genes. The relative contributions of different genes rest also upon their differential catalysis by enzymes and the length of time their messenger RNA survives in the cytoplasm before it is degraded. The more stable it is, the more protein chains a gene contributes.

The code is degenerate.

Each three consecutive sites on a strand of RNA constitute a codon for a particular amino acid. Since uracil, cytosine, adenine, and guanine can form sixty-four triplets, there are sixty-four possible amino acids in the code. However, three (and possibly four) of the triplets simply punctuate the end of a polypeptide chain. Adenine-uracil-guanine begin chains but are always excised before protein synthesis. Many of the other triplets signal identical amino acids. There are finally only twenty different amino acids transcribed by the nucleotide triplets. There is no fundamental relationship between these triplets and their amino acids; utterly different nucleotides can carry the same message. It is not a code made by a computer genius; it is a code made by the ocean.

The code requires blind intermediaries between itself and living structure.

Genes speak through amino acids; that is the extent of their "intelligence." There is no relationship between the code and physical characteristics in living beings. Genes do not directly program eye color, sex, blood, or instincts. They do not even choose between mammal and amphibian, or starfish and sunflower. Amino acids are their sole voice. The creativity of the genes lies only in the potential of the complex structures assembled from amino acids. Regardless of their differences, Darwin and Lamarck both intuited that morphology orchestrates heredity by remote feedback; certainly there is little evidence that the genes have the intrinsic capacity for making structure or responding to function.

The code is mutable.

If a perfect transcription of genetic material simply recurred, life would remain static. There would be no parade of living creatures, only identical crystalline forms. The transcriptive process is not only fallible but subject to mutagenic revision. It can be changed by motiveless cosmic forces, and changed it is: whole creatures are lost, stage by stage, and replaced by other creatures. The fact that the code is written in amino acids, rather than in precise blueprints for organs, protects life. Because the code does not contain final-stage information, it can be altered randomly without being totally degraded. A mutagen rearranges amino acid sequences and relationships, but the creature bearing this "error" will spawn no lineage unless the newly selected proteins can be organized functionally during embryogenesis *and* the altered offspring are able to survive in a competitive ecosphere.

The code is arbitrary.

The genetic code has no regular feature that could not be attributed to accident. The most important proteins are not coded by the simplest sequences. There is *no* generic relationship between proteins and nucleotide triplets.

If we fear that such a code is too simple and random for such intelligent creatures as roam the planet, we must remember the other equally arbitrary code—human language. The fact that the phonemes and morphemes of this code have no intrinsic meaning does not stop their chains from carrying philosophical systems, laws, poems, and sacred concepts.

Both human languages and amino acid chains begin with units that bear no signification in and of themselves. Organisms transcend their genetic code in much the same way science transcends the nonsense syllables in which it is written. The initial randomness becomes irrelevant once the systems are operational. Not only is this randomness not a hindrance, it is a source of creativity. The arbitrary base allows novel properties and radical structures to originate from haphazard changes. Known words turn into unknown words, ideas into their opposites and then into whole different ideas; exotic species of plants and animals arise. How else could we explain Apache and French as dialects of the same language? Walruses and orchids are likewise idioms of the same original "speech." No logical nonarbitrary system could invent such divergences.

If there is a primary order to such things, it lies deep in the universe beyond our ken. If we could penetrate the nature of number and the meaning of space-time, we might be able to go deeper into the world of phenomena. Until then, we must watch cells and creatures and stars and ourselves polarize out of gibberish, establish meaning for existence, wither, and die.

The genetic code may be arbitrary, but it is unique to our planet.

Life on other worlds, if formed in our manner, will use carbon or
other elements in entirely different reproductive languages.

The full message is unexpressed.

Scientists have uncovered some of the workings of the genetic process, but, as we shall examine in more detail in subsequent chapters,
they have also discovered that biological manifestation (like consciousness) is a matter of hierarchical selectivity: What the cells collectively do not *suppress* is what the organism finally becomes. In the
most immediate instance, each somatic nucleus must withhold its potential replica of the whole organism (every cell of our bodies,
whether skin, hair, or liver, was at one time able to make another one
of us; billions of our twins lie asleep in our flesh). The content of the
dormant aspects of the code as a whole has become almost a theologi-
cal concern.

On the simplest mechanical level suppression can be as effective
an evolutionary force as excision or replacement. If a mutation is able
to muffle the expression of a particular gene, either by changes in enzymes or more subtle transpositions of the entire embryogenic field,
previously unknown beasts come forth as certainly as from whole
new codons. Over vast generations global morphogenesis succeeds—
the dormant loci sinking beneath like Sleeping Beauty, to be awakened in a subsequent aeon by a kiss from a pagan knight.

We, as well as the other creatures on this world, contain vast documents of information about life, but most of it has been absorbed in
the intracellular unconsciousness of our soma. The futurist and biologist John Todd told me recently of a discussion he had with another
futurist, Lyall Watson. They were talking about the way that we have
been reinvading each other through viruses for millennia, so parts of
plants and animals are continuously transferred back and forth between one another and stored in viral nucleic acids. Watson's theory,
in Todd's words, is

". . . that the silence in you represents the genetic imprint of all
other beings. The silence in the oak is the genetic imprint of beings
other than the oak. Even extinct creatures continue to exist. They're
carried in some way in other creatures. Can you imagine! We could
dance back that which is gone. What a project for civilization! We
could bring back the pterodactyl or some ancient armored fish.
Watson thought it might be easier to recreate a species that left recently. The animal he would elect to dance back is Stellar's sea cow
which was last seen about 1886 off Alaska. It's not science-fiction

stuff. It swings back directly on who we are and what our place is in the universe. Maybe it will make us more careful stewards. I haven't said much about love but I can't help but feel that's what this is all about."[3]

Whether or not the biological universe exists hologrammatically in each of us, we can see that we are in the bare infancy of understanding the genetic code and who we are. If so much information lies buried within, life is a unity to a depth not believed possible since the mechanical universe replaced the Gnostic one. But we should not be overly optimistic, for much work lies ahead. If we are to recover our destiny, we must soften to what we are and allow ourselves to reach and reconcile those aspects of our nature (and nature itself) that we have destroyed in becoming, or have embodied silently. We also contain frightful malignancies, tumors and shapeless growths that destroy both form and memory. We have yet to discover whether we are the stewards or the destroyers, the radiant ovum of life or the malignant horde. There is no doubt we intuit the battle within, and it is not a simple matter of light and darkness, consciousness and unconsciousness, for sometimes the way to go is into darkness, where we begin.

We are gradually coming to accept a disturbing script. We are the collective imprint of trillions of individual creatures each of which has had its full manifestation suppressed. If we hear voices, some of them may be very ancient indeed and (in fact) inseparable from what we are. It takes some effort to summon unity from our many origins, and it is not because we are made of parts. We are made of other unities. Splintered, we become first dozens, then hundreds, then thousands of distinct creatures, who would no doubt stir to life if awakened, and would manifest as we do.

The first punctuation in our realm falls not on mind but on the ribosome, for without the premental translation of code there would be no mind. We are a phase of nucleic acid transcription going back to the dawn of time. If we try to imagine our life from *its* point of view, whoever and wherever it is, we are a brief florescence in a rigorous exercise, as the peaches of a single tree in a single summer are to the Primordial Root of all trees. The genetic material goes on fissioning and copying itself as though it were always in the same organism in the same environment forever. Our phenomenal world is severed from its representation in the transhistorical nucleic thread. Alienation may be a twentieth-century symptom, but the identity crisis that it portends has existed from the dawn of our species. Because we can never become whole, because we can never speak the original language, our lives become a yearning toward wholeness, an attempt to reconstruct sacred speech. This is the Dreamtime that aboriginal peoples intuit through the mythic beings who occupy Earth and Sky. The Ancestors are as present as they are absent. We have never been anything other than their transposition through time and space.

7.   Sperm and Egg

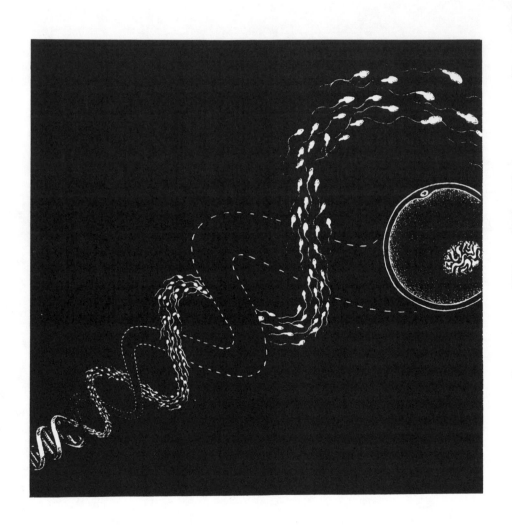

The genetic material of multicellular creatures is transmitted between generations by primordial germ cells. Almost all cells retain the nucleic blueprint of the organism, but early in embryogenesis certain cells become sexualized and their role henceforth is the union of their polar material with each other.

These germ cells appear to arise independently of the structural layers of tissue. They migrate to the gonads not with the blood or lymph system but by amoeboid crawling. The large human germ cells are first visible in the endodermal epithelium of the yolk sac, in the region where the placenta and umbilical cord will later form. From there they travel inward like a Stone Age tribe crossing a subcontinent, arriving in a mass of other undifferentiated cells which will form genital ridges.

The British neurologist Sir Charles Sherrington refers to them as "old ancestral cells . . . one narrow derivative line of descendants, nested in the rest of the specialized collateral progeny."[1] They are ancient parasites that continue to be reborn in our germinal liquid.

When Adam squeezed onto his fingers his own seminal fluid from the dream of Lilith, it was truly the mercurial waters of the alchemists he held, the protean white seed of all future consciousness and humanity. Each of us touches that seed again—billions of ne'er-to-be-born offspring, beyond individuality, beyond death too. These cells are drawn from plasma inside of tissue and will die as their semen is exposed to the air. Each one of them is an *imago mundi,* a complete universe in miniature, but it must be sacrificed to this one corporealized world. Masturbation is a link between the life of the individual and the eternal history of cells; it is also the revenge of time against timelessness. Surely that is why psyche reaches blindly for *eros* at the center of sleep, as an affirmation of the sun. Sexual fantasy, as the old poet-priests divined, is the antidote to the portent of the corpse.

"But at my back I alwaies hear Times winged Charriot hurrying near/" wrote Andrew Marvell in the mid-seventeenth century. "And yonder all before us lye/Desarts of vast Eternity."[2]

The primordial germ cells have a basic sexuality that transcends our temporal attractions and seductivities. They are pagan creatures with their own sacraments and rites, a pale representation of which appears in pornographic centerfolds and erotic art. The sperm is far too old to have human tastes. When the male releases this animal in lovemaking with a woman he experiences the accumulated desires built up in layers of tissue around an original deliquescence of cells. Through his own *eros* he completes their migration (although for most of them it is back to the abyss). "As a sperm . . . you and millions of your companions, all of you brightly shining there in that river of elixir, were drawn out into the ocean it seemed, or suddenly scattered like flying ants released from the mud hive to chase the queen. All fifty million or more of your companions died that night. And you—how is it possible?—you out of all those millions were permitted to be metamorphosed so that you could contemplate the Mystery as we are presently doing it."[3]

So the Master Da Free John told his disciples. Death and birth, extinction and desire are indissolubly joined, and though some cells perish before life, all perish eventually, or are changed into another thing:

"What happened to all those millions of others that night? How terrible that every last one of them died! How terrible that there are billions of us here now flying toward who knows what, and we are all going to die! This existence is a hopeless flight in the night toward death, the urge of Nature to repeat the cycle."[4]

The woman's *eros* has a far subtler and more profound relationship to primordial substance. She is continuously preparing her eggs for syzygy and then sacrificing them (when they go unfertilized) in a cycle tidal membranes have tied to the Moon. Her sexuality involves a deep spasm, a round dance or gyre about an egg. She is always on the verge of contacting the ovum's nature, of communicating with it psychosomatically and moving from her time to its time. But part of her is also phallus, and in lovemaking that phallus meets the male's spurting, while her lunar receptivity completes the ring of desire, the life cycle of generations.

Even the "wasted" germ cells have meaning in the flesh; desire is a translation of cytoplasm into tissue and image. Fish did not invent the experience of quivering fins, and bees did not create the hive; we did not fashion our own swelling copulatory organ nor the orgasmic spasms. These precede us and precede our rituals; they create time somatically and thus are timeless. Life is not long, and there is no way to perfect these organs or contemplate them before acting. Men and women must be creatures like all other creatures and for the same reasons.

Sexuality is built again and again upon asexuality, so the latency of gay sexuality in our species should be no surprise. A man feels in another man his own phallus; his germs are liberated not into a polar receptivity but a mirror of his own desire. Even though this may frustrate the seeming intention of the sperms, cells do not act out intentions anyway, only desire, which they translate to the organism. The life cycle of the species is broken in gay sex, but this does not foil nature. The cycle is always broken in the evolution of species; sexuality and desire are reinvented again and again, from crabs to squids, from frogs to monkeys. New customs may not (in a simple Lamarckian sense) change the biology of procreation, but they create their own experiences on the level of myth and symbol (once the biological act is internalized and cast back through society). This is what we see on Castro Street in San Francisco—not a new species yet but the iconography and impersonation of one.

Additionally, male and female are ever fused; they are aspects of each other. Even biologically, woman and man emerge from homologous ducts in their gonads; in the woman they become oviducts, and in the man ducts of the testes. So in any form of sex the man can never not be a woman too and the woman can never not be partially a man.

We are not passive recipients of life; both psychically and somatically we seek the meaning of our organs and the riddle of our germinal surge. Now that the cells have made conscious beings, their own long sexual history must be unravelled. We cannot become them ever again in their naiveté, but we also cannot deny them by our complexity. So we will carry out each other's ceremonies until nature itself changes.

From the moment it is born from another cell, a cell knows where it is and either divides again or specializes. It may specialize permanently like nerve cells or red blood cells and never divide again, or it may reproduce its specialization.

General cell division, known as mitosis, is initiated by the nucleus and foreshadowed by the splitting of cytoplasmic centrioles just outside opposite edges of the nuclear membrane. The centrioles are like cracks in the mirror of time in the way that they initiate fission by shooting across the nucleoplasm, but they are also mirrors through the crack in time, for they reflect interminable future generations as they pull apart cell after cell to reveal only new centrioles pulling apart cells. When the nucleolus dissolves, the cell's components melt and merge. The chromosomes, which have been copying themselves in the raw material of the nucleoplasm, gather in pairs of new and old strands and coil in the nuclear core, their diffuse units now tightly bunched and the single complementary strands joined like shocks of wheat at the centromere. The daughter centrioles cleave, and then migrate around the nucleus to the opposite sides of the cell, fine gel microtubules flying behind them. Some of these threads traverse the nucleus—a mammoth span for such tiny entities—and connect the centrioles.

These events become semivisible to us as three-dimensional representations of things in another time and space. Their delicate coordination suggests the invisible complexity behind them. A current rips upward through the monad and twins it. The nuclear membrane deteriorates, and the spindle fibers penetrate the nucleoplasm to snare the chromosomes at the centromeres and pull them into alignment at the midpoint of a plane perpendicular to the spindle axis. The chromosomes break, and the pairs migrate poleward as if repelled by one another. The spindles do not do the splitting. The cleavage is in the centromeres; the fibers merely pull the halves apart, and a gel fiber seals the division. Then a thin furrow forms in the surface of the cell and encircles it, cutting through the spindle, tightening like a knot, and squeezing out two separate cells with nuclear membranes and nucleoli reconstituted.

A population of cells divides like this, synthesizes fresh DNA, and divides again. This is its breathing. The rate of fission is highest in the embryo and slackens with age. But age itself does not render cells decrepit: wounds still heal in elderly creatures, and cell cultures can survive indefinitely. The mere fact that the cells of an Afro-American woman's tumor, preserved in the 1950s and adapted to plastic dishes, have continued to spread through the world in laboratories indicates the primitive potential that is suppressed in all but extraordinary circumstances. Aging is a crisis of the organism, not (for the most part) of its individual cells.

In some plants and animals simple cell division breeds full-fledged gametes. In order for human cells to function as gametes another division of chromosomes must occur—a resynchronization of mitosis such that the cell divides twice while its chromosomes are replicated only the first time. Fissioning with halving is called meiosis. When the chromosomes are densely packed during meiosis, the homologous strands come together along their length as if drawn to each other from head to toe. During this first phase they actually break and rejoin across their centromeres so that homologous sections of genetic material are exchanged between original maternal and paternal helices. "Crossing-over" causes recombination of traits so that a virtually unlimited variety of gametes can arise from a single organism. In a human being, with twenty-three separate chromosomes and forty-six pairs, there are already ten million different kinds of sperms and eggs possible, *before* crossing-over.

Spermatogonia and oogonia carry the chromosome number of their species. A cell spawned by their merger would have double chromosomes, and its offspring would have four times the functional number. The meiotic variation sexualizes the cells by chromosome reduction, and, in species in which propagation was established this way historically, it must happen again each generation.

Chromosome reduction is an aboriginal event. The colonial *Volvox* divides meiotically to produce large egg-like cells and clusters of sperm-like gametes. Bisexual paramecia exchange nuclei from mouth to mouth in an act foreshadowing copulation. In *Euglena* a new basal body arises from the centriole and generates a flagellum. As the nucleus splits, a global severing proceeds lengthwise down the body. Among paramecia the gullet is the first organelle to divide, the entire parent then splitting transversely and sharing its vacuoles with its fission-self. The cytoplasm always restores missing organelles after mitosis, so the protozoa continue to splinter, dissolving their nuclei and reconstituting.

The halving of its chromosomes is a form of crisis for a normal cell. Suddenly, while carrying out standard mitosis, it is jolted and its parts become entangled, its breath speeds up, it divides quickly, and then finds itself half an animal. Most germ cells die without becoming whole again. The few that replace the missing chromosomes do so by fusing with their benefactors, i.e., with each other, turning their brief tryst into a multiplicity that swallows them.

This intimacy is so much greater and more final than the sexual intimacy of creatures. Like Plato's original males and females the germ cells have been truly riven. So when they find their Other they require it forever. Gametes give themselves up without regard for race, beauty, or any actual love exchanged between partners in the larger multicellular world that juggles their fates (apparently they are just as passionate for union in cases of rape and unwilling fertilization). To some degree the emotions of the sexual partners encompass the feeling of the egg for the sperm and the sperm for the egg, and this primordial attraction may overcome other repulsions and give even an undesiring partner a sensation of *eros*. The line between love and hate is thin in most of creation. The intimacy of sex is strong enough to overcome, at least temporarily, the natural enmity between male and female in some species of spiders, octopi, birds, and cats. They may sting, bite, growl, and hiss, but they cannot keep themselves from mating.

"Let us roll all our Strength, and all/Our sweetness, up into one Ball/"concluded Marvell, "And tear our Pleasures with rough strife Thorough the Iron gates of Life."[5]

Sexuality is mitosis internalized and writ large. We experience a glimmering of the fundamental act of the biosphere, making sentient what cells merely do. Of course if they could speak they might not judge the lugubrious gyrations of massive creatures as anything like their own graceful fission-dance.

The two ancient animals that meet as sperm and egg in the syzygy of complex creatures are similar but not identical. If they were not similar their blueprints would not cohere in protein assembly (obviously, there is no compromise between a fin and an arm, despite mermaids). They are not identical because they have accumulated different histories since their one point of absolute identity in the common ancestor of their line. Actually the individual genes share many points of identity, some on the path to their common ancestor (for instance, the basic amino acids present in all multicellular life) and others in shared ancestors within their species (millions of present-day humans can claim Alfred the Great as their ancestor, even many Africans and Asians, and there is nothing special about a ninth-century King of Wessex in this regard). Where common ancestors have left a trace, there is simply identity of chromosomes. Where there is difference, then some characteristics from each lineage must be selected and some eliminated in the surviving reproductive cell. We are related not only to blood kin; strangers may share our identical traits from duplicate mutations or remote ancestors, yet our own siblings may diverge.

If we were to try to locate ourselves just before the moment of our conception we would find that, at least on the physical plane, we were two creatures. This is almost as disturbing as being three-dimensional lattices of fifty trillion changing cells. Whereas the lattices were always fused in an expression of unity, the sperm and egg come from different organisms and are distinct animals right up to the instant of their merger.

We would experience this identity problem in another way if phylogeny had dealt sentient creatures a different hand and we were to reproduce by splitting into exact twins. Which of our fission products would we then be? Would each of the twins have our memories as well as our nuclear material? If not, how would a clone without memories know who or what it was? Or how would clones that began with our memories individuate their separate existences?

Frank Herbert asks these questions implicitly in the latter volumes of his *Dune* sequence. The warrior Duncan Idaho is reembodied thousands of times from a single patch of skin preserved from the corpse of the original who died in battle. Even though his memories up to death are preserved identically in each successive clone they become independent personalities with their own idiosyncrasies, ignorant of one another's inner lives and memories. The original Duncan Idaho is never awakened.[6] Thus, the history of the universe gradually changes despite the fixity of the genes and the struggles of kingdoms to make themselves immortal through advanced weaponry and biological technology. However prescribed by the force of a previous generation, new cells are always free and rebellious. Herbert has given a hypothetical solution to a hypothetical problem, but beneath it lies the old dilemma of who we are, if anyone yet, when we awake (again) at the beginning of time.

As the roles of the germ cells became known during the seventeenth century, some scientists surmised that the species of plants and animals were imprinted in the cell stuff and that successive generations already existed, nested one within another within an egg. As each egg developed into a mature organism the next largest egg within *it* became visible, but it still held in reserve an infinite hierarchy of ever-smaller eggs infinitesimally encased in one another until the end of time. Scientists and philosophers could not otherwise explain the ready development of embryos; thus, they must already have been there, preformed, waiting to flower.

Sperm was ostensibly discovered in 1677 when a student at the University of Leyden brought van Leeuwenhoek a bottle of semen from a man who complained of too frequent nocturnal emissions. The student had a preliminary diagnosis: After viewing the semen through a microscope and seeing thousands of small animals swimming with tails, he concluded that the man was genitally infested with some infusorian. Van Leeuwenhoek confirmed the existence of these small animals under a more powerful microscope, but he found them likewise in the sperm of healthy men, men suffering from a variety of ailments, and animals ranging from rabbits and dogs to pike and cod. He wrote:

"I have seen so excessively great a quantity of living animalcules that I am astonished by it. I can say without exaggeration that in a bit of matter no longer than a grain of sand more than fifty thousand animalcules were present, whose shape I can compare with naught better than our river eel. These animalcules move about with uncommon vigor and in some places clustered so thickly together that they formed a single dark mass."[7] He added that there were ten times as many animalcules in the milt of a large male cod than there were human beings on the Earth.

A new theory of preformation now arose: animalculism. Although van Leeuwenhoek did not advance it himself, others used his published findings to argue that future generations were preformed in sperms and that the womb served only as an incubator, a matrix for this seed. One Dutch scientist even drew the angelic human homunculi he "observed" in spermatozoa, their umbilical cords wound into their tails. So the argument was joined between the proto-feminists and early male supremacists over whether Adam or Eve carried the whole future human race in their loins. Both sides shared the belief that the germ-plasm was fixed from the beginning to the end of time.

"This precaution of Nature," wrote Immanuel Kant, "to equip all her creatures for all kinds of future conditions by means of hidden inner predispositions, by the help of which they may . . . be adapted to diversities of climate or soil, is truly marvelous. It gives rise, in the course of the migration and change of environment of animals and plants, to what seem to be new species; but these are nothing more than races of the same species, the germs and natural predispositions for which have developed themselves in different ways as occasion arose in the course of long ages."[8]

The preformationists were opposed by the epigenesists who contended that the sperm and egg each contained raw unorganized material, particles which came together dynamically to form the individual. In his book *Vénus Physique*, the French natural philosopher Maupertuis wrote: "The elements suitable for forming the foetus swim in the semens of the father and mother animals."[9] The chemical attractions between the particles lead to the formation of heart, intestines, limbs, etc.

Each school intuited a different aspect of biological reality: The gametes *are* preformed, but the organism is not; organization begins from scratch each generation; the only design that passes between parents and offspring is a code for the lineage that must be retranslated and reenacted from its elemental proteins to the finished organs of creatures. Sperm and egg are not different kingdoms; they are merely different packaging for the neuter gametes, their sexuality arising from slight variations in the amino acid sequences of some genes.

The decision to form germ cells has already been made from the moment prior germ cells have merged. Spermatogenesis occurs cyclically in the male gonads throughout a creature's life. New germ cells (called spermatogonia) are formed by mitosis and remain dormant in humans from the foetal period until puberty at roughly thirteen to sixteen years of age when they begin to multiply. Human spermatogonia undergo an average of seven divisions before growing in volume and, as primary spermatocytes, entering the meiotic phase. The first division during meiosis yields two secondary spermatocytes, and the second splits these into four spermatids, the nuclear material of the four sex cells.

The cores of these spermatids become dehydrated, losing their RNA and almost all of their reserves of protein. Their single strands of DNA are packed into the head by the idiosyncratic protamines. Only the fragile primary plan is protected, imbedded in a "spaceship" for the intergalactic flight. In our fantasy of microcosmic sentience, unconscious entities (or their shadows) are preparing a projectile for a journey they cannot undertake.

A granule sprouts in an enlarged Golgi vacuole of the sperm head, and it swells with a pointed tip called an acrosome. The vacuole gradually dehydrates and spreads as a double-walled sheath over the body. Virtually all the living substance of the cell is exhausted in making this little animal. It has only enough matter to keep it alive while it swims toward the egg. If it does not reach the egg it has no mode of feeding and dies. If it does reach the egg it fuses with that gigantic cell and its contents are preserved in its sanctuary.

In man the acrosome is ovoid and flattened at the sides, and a long flagellum-tail grows out of the filaments of two axial centrioles. Rodents and frogs have bent "sword" sperms with points, birds have corkscrew heads. Sperms are remarkably similar throughout the animal kingdom; even plants have sperm-like male germ cells. It is astonishing that a fern and a monkey could arise from such similar mites; it shows that the sperm is a very ancient creature.

Eggs carry the same genetic information as sperms, but since they do not have to transport it, they are not streamlined or motile. By contrast, they are plump storehouses of proteins and fats, for once the sperm arrives, the ovum is the abode of the newborn. It must provide the energy for the embryo's metabolism and growth, and it must also have on hand all the material needed to assemble a creature after its kind.

An egg is bipolar. The seed is wrapped in a nutritive liquid with particles of protein and fat suspended in it. The sphere of the egg that has the least of this yolk is called the animal pole, and where the yolk accumulates is called the vegetal pole. The material for the yolk comes from a variety of maternal tissues; in birds and humans the liver supplies a large portion, but in insect eggs it is the blood that supplies the material. Even before it is fertilized, the egg is organizing itself into an embryo and beginning to foreshadow organs. Brilliant reds and yellows in some species are a natural staining of the otherwise invisible texture of presumptive muscles and intestines. Although yolk is not itself structural, its relationship to the cytoplasm determines the contexts of layers, tissues, and organs which require it. The surface of the cytoplasm gradually gels and forms a superficial but fixed membrane, a cortex which serves as a reference point for the emerging structure in the liquid. The shell of the robin and the jelly of the newt form atop this membrane. In species where the egg develops sealed off from maternal tissue, the full complement of nutrients must be stored from the beginning. So the frog's egg grows by twenty-seven thousand times its original size over three years before maturation whereas the human egg, which can draw on the mother till birth, remains small. Most eggs are round even if their cortex is oval or tubular (but insect eggs are elongated).

To become eggs the original germ cells acquire a simple asymmetry during meiosis. The spindle of the oogonium tilts sharply into the animal region with the result that only one fully formed oocyte survives division, and it receives the potential larder of four. Ripe with extra nuclear sap, its chromosomes bristle out like lampbrushes and become despiralized. The egg forms in a rich envelope of protective follicle cells which also nourish it. Although the source of this residual tissue differs by species, some of it comes from the aborted eggs. Insects, mollusks, and annelid worms all form nurse cells from unused oogonia. As many as fifteen siblings may contribute to the growth of one functional egg.

In most of the invertebrates the yolk reserves are small and evenly distributed through the cytoplasm. Yolk takes up a relatively much larger portion of the amphibian egg and is regionalized toward the lower vegetal zone, its protein located in large oval granules. Birds and reptiles have enormous eggs because of their stores of vegetalized yolk; the cytoplasm is but a thin layer around the yolk with a thicker cap at the animal pole where the nucleic material resides. A chicken egg cracked open reveals a single giant cell bloated with yolk, so much yolk that the functional part of the cell is not immediately apparent. The animal pole and genetic substance are both squeezed into a cloudy film that looks at first like mucus in the yolk. The white (albumin) is simply a protective membrane secreted around the egg as it passes down the oviduct.

Arthropods, notably insects, have evolved through eggs with yolk in their interior surrounded by a thin layer of cytoplasm and surrounding another pool of cytoplasm. So the emerging ant or spider must develop both interior and exterior to the vegetal pole.

The consumables in the yolk determine the possible duration of embryogenesis except where the egg can draw on a mother's organs or the external environment for food. The simpler eggs of water-dwelling invertebrates are neuter gametes, not unlike sperms. The seas, lakes, and rivers are their wombs. The planula larva of the jelly-fish is shot from the mouth of the adult soon after fertilization. The connection between the tissues of the egg and the tissues of the "womb" is brief and minimal where the mother has few layers of complicated cells. The development of maternal tissue in the more advanced multicellular animals represents a gradual cumulative internalization of watery habitats, as land animals become mobile ponds in which their cells live.

But evolution does not proceed along simple linear parameters or unidirectionally. For instance, the mammals are not the first creatures to have developed the placental relationship. Some species of fish and the sea squirts nourish their unborn babies in this way (although in mammalian embryos a new kind of intimacy between organisms is established in the womb). The yolk reduction of the mammals is a reversal laid down atop a prior system of yolk manufacture (unlike the original invertebrate yolk sparsity which was an unadorned consequence of the first multicellular animals). Yolk was required in greater amounts as embryos became more complicated, and then in *exponentially* greater amounts as a particular lineage transferred its embryogenesis to external eggs on land, then in drastically smaller amounts as that same lineage developed an internal marine environment for its young, "reverting" to the simpler invertebrate condition. It is possible that the complexity of extra-embryonic organs developed out of the very yolk it replaced.

During human embryogenesis thousands of eggs are formed in the ovary and remain dormant, but only about 200 of these will ripen in the span between female puberty and menopause. The sperms are newly made by the hour and discarded easily; the eggs are finite in number and may be fifty years old when they are fertilized.

Just as an artist does not produce partial works in his youth because he knows masterpieces are forthcoming, so the differentiating egg cannot postpone its survival until it has more complex organs. An artist who does not work on each symphony or canvas as if it were his last does not progress or discover new stages. And if he is killed suddenly (as was John Keats), his last works must be considered as the final stages of his growth. The same is true for the starfish embryo swallowed by a clam.

The egg like the sperm is a living animal. The illusion that it is only one-half of a yet-to-be-formed organism has nothing to do with its moment-to-moment crisis of survival. It cannot postpone its responsibility. Nature requires that the egg live by the same rules as other animals; it must be able to metabolize organic substances to generate energy. It is just as hungry, just as anxious to breathe as any mole or lizard. There is no special dispensation for being embryonic; the ovum is one more fish in the sea.

For the egg itself, there is no embryo, there is only an egg, forever. As it passes through the stages of ontogeny it adds organs, but not as a means of assembling (one by one) the functions of its adult destiny. At every stage it is another animal, and its tissues are added gradually and only in relationship to each other. Each transitional form is the basis of a subsequent form. The embryo cannot give up its unity to accommodate even the most evolutionarily promising modification. If a structure has become critical for the adult form, then it will have been achieved through a historical sequence of living phases. The blueprint must already contain even the most radical deviation, and then, only from a seamless transition of mutated creatures any one of which is a possible terminus of development. Even an animal parasitic upon its yolk or its mother's tissue is effectively competing for space on the Earth.

The embryo, said Ross Harrison, the pioneer American embryologist, is not just "a developing organism, in which the parts are important potentially, but also an organism in which each stage of development has functions to perform that are important for that particular stage. . . . Organic form is a product of protoplasmic activity and must, therefore, find its explanation in the dynamics of living matter. . . ."[10]

8. Fertilization

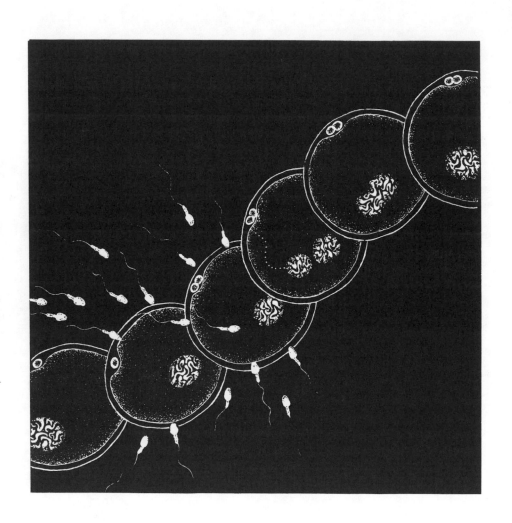

Fertilization occurs in an aquatic environment, whether externally in a body of water or internally in the tissue ducts of animals. Most water-dwelling species, especially simpler invertebrates, spread sperms and eggs through the hydrosphere in the same way that plants cast seeds into the lower atmosphere. Random dispersal would seem doomed without a species attraction between a sperm and an egg; yet some scientists believe that sperms are capable only of undirected swimming and that fertile collisions are, in essence, accidental. Scientists who challenge this assessment claim to have observed spermatozoa speeding up as they approach eggs of their own species or species with which they are interfertile. Selective fertilization occurs because eggs respond only to their complementary sperms (from the thousands with which they might come into contact in an average aquatic zone); only such a sperm can normally penetrate the "egg water" around the jelly coat; foreign sperms slipping through may also be sterilized by indigenous protein. Some eggs periodically attack sperms from their own species, a vigilance perhaps necessary to prevent abortions from multiple fertilizations.

In human beings fertilization occurs in the female genitals when ovulation and sexual intercourse coincide. As it is undergoing the final meiotic division the egg is swept out of the ovary into the uterine tube where repeated passes of the tube's feathery fingers propel it along. Meanwhile, the male genital sends five hundred million spermatozoa into the cervical canal. With wriggling tails they beat their way along, aided by the muscular contractions of the uterus. Egg and sperm meet in the tube within an hour of our time, though it is for each of them an eternity.

The egg is surrounded by spermatozoa, and although this is the whole of life for all but the one that penetrates it, the end of their individualities has come anyway. In rare cases sperms can live for hours in the female ducts, and, in even rarer cases, longer. So women can have one egg fertilized by two men and conceive twins with different fathers. Even before the era of artificial insemination, live sperms got transported unsuspectingly into uterine tubes. In one reported case a Caucasian woman gave birth to a part-African child, and her husband sued for divorce on grounds of infidelity. Since the woman had not had intercourse with anyone else, she hired a detective to solve the mystery. It turned out that the husband had visited a prostitute hours before having intercourse with his wife, and he came away with the sperm of a previous client adhering to his penis. This hardy sperm was then used by him unintentionally to impregnate his wife.

Even a ripe egg is not easily penetrable. The wriggling sperms would be rebuffed at the outer cell membrane if they did not bear acrosomal enzymes which break down the proteins in their path. Aquatic eggs bond by proteins to the head of an eligible sperm. In the disturbed vitelline jelly the acrosome dissolves and releases its granule, and the components of the sperm are delivered into the interior of the egg. The plasma membranes of the sperm and egg make contact with each other, fuse, and dissolve into minute vesicles. The continuous membranous structures internal to both become one membrane, and the male nucleus continues into the cytoplasm of the cell. Its journey will end at the center of the oocyte where its separate identity is annihilated in the Other.

The German Oscar Hertwig was the first scientific recorder of this event in 1875. After fertilizing a sea-urchin egg, he turned the glass of Galileo the other way and saw the single large moon in the center of a meteor shower. One of the meteors then burst through the atmosphere, and a shell formed keeping out the rest. The sky changed color and became stormy and irregular. The point of contact bulged. The tail disintegrated but the head collided with the cell center and there formed "an extraordinarily characteristic starlike figure."[1]

With the fusion of the membranes the sperm and egg have become a single cell, a zygote. They now function chemically as a unity; their unique identities disappear, and they embark upon a journey that will transform the contents of their egg into an organism. The female pronucleus must migrate to the center of the egg where it contacts the male pronucleus. When the pronuclei approach, they appear to respond to the nearness of each other. In some species they send out protuberances which embrace and pull each other together. The male pronucleus provides a centriole for the ensuing mitotic divisions. The nuclear membranes break down and the male and female chromosomes mingle freely. and rejoin in a new pattern

The penetration of the sperm does many things. It unmasks enzyme activity so that messenger RNA replication and protein synthesis are intensified, and it contributes the specific hereditary traits of the male lineage to the fused pronuclei, but it does not contribute mechanically or structurally to the new organism. Eggs in nature have regularly been activated by sperms from other species. We know from experiments that in some cases of related species (as the toad and the frog) the nuclei may fuse and cleavage begin, but the embryo is aborted soon after because of the unmatchable genes. When the species are from different classes or even different phyla, the sperm may activate the egg and then die without damaging the female pronucleus, in which case a normal offspring of the mother will occur. Sea urchins have been fertilized in this way by sperms from sea lilies and mollusks.

Unfertilized eggs can also self-activate without any sperm. This process (known as parthenogenesis) has been regularly observed in rotifers, nematodes, and various species of insects. Pure sunlight can begin development in turkey eggs. When such virgin reproduction was first observed in plant lice and myriapods in the years after the discovery of the sperm's role, it was an unexpected triumph for the ovulists who had been all but defeated. If the egg could develop without the male, then the newborn must exist already in the egg. Throughout the eighteenth century a woman might conveniently claim that her pregnancy had occurred from a day in the sun and (certainly) without copulation.

Laboratory experiments have even caused parthenogenic offspring in mammals. Rabbit eggs were activated by an embryo extract after suppression of the second meiotic division, and then inserted in the womb of a virgin female rabbit where early stages of the embryo regularly formed, and, in one instance, a complete rabbit (which stands fat and healthy beside its rather diminutive mother on the pages of most embryology textbooks).[2] Parthenogenesis is also part of the natural life cycle of many species, alternating with generations of sperm activation to produce offspring of different sexes. The sperm that the queen bee collects on her mating flight can be used to fertilize her eggs, with only females being born from this process. If she lets them develop on their own, they will become males. Although there is no known case of human parthenogenesis, it is theoretically possible. Our eggs are not fundamentally different from those of rabbits.

At the nucleic level the egg already has what the sperm provides: a set of genes. Meiotic halving (in some species) has made the sperm necessary, but meiosis is only the germ cells' secondary adaptation to sexuality. If this heritage were artificially negated by a suppression of the second polar body the egg would then carry the full (diploid) number of chromosomes through oogenesis. Even a human egg thus treated could produce a child without a father.

The halved (haploid) egg must extract its lost genetic material from a sperm if it is to develop. The sperm body is absorbed back into the egg's cytoplasm, but its genetic component has been incorporated in the nucleic "brain." The oocyte accommodates the full heredity of the sperm—not just its male characteristics but all the allotted chromosomes of the lineages of males and females through which it is incarnated; likewise, the egg contributes its male and female lineages both.

Before syzygy the egg was lethargic, but the zygote is engulfed by motion and change. Cytoplasm begins to flow, electrical potentials of membranes fluctuate, and (in many species) great gulps of the vital *prana,* oxygen, are suddenly swallowed throughout. The new cell may seem to throb before it divides. The combination of changes prevents other sperms from entering the egg, or, if they are already within, it causes them to decompose.

The entry point of the sperm is one of the orientation sites for the embryo. In frog zygotes a gray crescent develops on the side of the egg opposite penetration (like a recoil). This crescent will become ectoderm, and it is an organizing center for other tissue layers. In sea squirts yellow cytoplasm flows down to the vegetal pole and a crescent forms just below the equator of the egg in a region that is to become muscles and mesodermal tissue.

According to our present mythochronology sexuality appeared as a system of chromosome variation and gamete polarization some twelve hundred million years ago, possibly in association with heterotrophy, at the dawn of the Cambrian era. Sexuality and heterotrophy are different modes of cellular attraction between organisms. In heterotrophy creatures feed on the remains of other creatures (life comes to steal its energy from other life). Sexuality causes cells to draw together and exchange external products (germs) which become internal (intracellular) to the emerging zygote. Sexuality, followed by embryogenesis, is a form of eating, cannibalism as differentiation; however, sexuality also establishes contrarieties only to overcome them by unity. It is no wonder that creatures ever since have had difficulty keeping food and seduction separate.

The mouth, the genitals, the womb, and the stomach regularly take one another's places in evolution. Mouth-breeding fish carry eggs and young, both, in their jaws, a potentially deadly ambiguity. In many invertebrates, eating and mating overlap. One species of bristle worm is fertilized as the female eats the sexual apparatus of the male which darts and twists provocatively, like a piece of food being offered, off the anterior portion of his body. The sperm reaches her eggs through the abdominal cavity. The octopus uses a copulatory arm (the hectocotylus) to reach into his own breathing funnel for a packet of semen at his mantle cavity which he then stuffs into the female's cavity. She is partially choked by the long invading arm in her gill chamber but aroused by the erotic red that lights the male's whole body while his arms also titillate her. In a few cephalopod species the arm breaks off and lodges in the female's cavity where it dwells as a semi-independent animal. Live sperm-bearing hectocotyli have been observed swimming at sea. In another instance, female grasshoppers chew up the remains of the semen tubule left at the base of their ovipositors.

Sexuality is a complex association of germ cells, layers of tissues, nerves, fixed rituals, and symbols; it develops through the animal kingdom as many different organs, life cycles, and levels of cognition and behavior. There is no clear lineage for most of the human aspects of sexuality. For instance, an organ for introducing sperm directly into the female's body from the male's gonads exists among some species of worms, snails, and insects but not among others. Crabs and spiders, like octopi, have "arms" with "hands" that detach sperm from their own body and plant it in the female. Simple amphibians have penises; lizards and snakes have spiny, hooked double penises so strong that the mating animals are often temporarily unable to pull apart. Yet salamanders and frogs have no organ of copulation (though they make contact in the seeding of gametes). *current*

This dichotomy is present in the simplest and apparently most ancient creatures. Protozoans copulate in a fashion, but many invertebrates (for instance, ocean-dwelling annelid worms) merely shed sex *as are snails* cells into the water. Flatworms are hermaphroditic. At the back of *see below* their ventral surface is a genital atrium opening into a vaginal duct. The testes, scattered through the lateral margins of the body and developing from mesoderm, drain through a series of sperm ducts which fuse into a genital muscle. Eggs and sperm ripen at the same time and sperms are exchanged in copulation which appears to be ejaculative. Other worm phyla have such variations as long hooked penises, vaginal structures that swell up to a greater size than the rest of the animal during fertilization, and permanent marriage in which the mating worms grow together at the genitals and remain united for the rest of their lives. Flatworms also reproduce asexually by fissioning, and, in fact, a section down to a tenth the size of the whole worm can produce a healthy complete offspring.

Snails rise to press up against one another, and, in rubbing their bellies together, give huge smacking kisses and dance like couples in the late hours of a party. They are also hermaphroditic, and each con- *Snails* tains a quiver with which it pierces its partner, visibly wounding it (often seriously by puncturing the lung or abdomen). Yet the strikes are also pleasurable. An enormous swelling tube functions as a penis, and each snail inserts it deeply in the female genital of the other. They cast sperm cells simultaneously and then separate permanently, each crawling off in a different direction.

Male spiders fertilize females with sperm-bearing appendages (palps). The male arouses his lover in some cases by pulling on her web and tapping a few stylized dance steps. As she comes to his call he may hold her at a distance with his forelegs. He is not sure whether he will be considered mate or prey. One species of hunting spider clarifies the difference between himself and "it" very carefully, by courting with a fly wrapped in his silk. During his presentation of this gift his whole body shakes and his palps quiver and stretch out to the female. It is a haunting mixture of two ambivalent rituals—one of courtship, the other of cannibalism—caught in a biological trap. There is no way for spiders to choose between these as they dance together on separate threads, so they go on in seeming contradiction, a contradiction we must also inherit from nature if not directly from spiders.

Many species of fish engage in intense kissing. The gurani suck their lips together for as long as twenty-five minutes before mating. Carp-like fish rub and writhe prior to simultaneous expulsion of gametes. The male lamprey sucks the neck of the female while wrapping his body around her and pressing against her abdomen. The stickleback draws the female into his nest, and then, when she is thoroughly buried therein, stimulates her caudal region with his fins causing her to lay thousands of eggs. After she has left he is compelled to burrow through his tunnel, spreading his sperm on the eggs. The erotic component is separated in time from actual fertilization.

The male Surinam toad squeezes the female so hard that his thumbs may penetrate her abdomen; their joint spasm leads to the simultaneous discharge of germinal cells. The male, hunched over the female's back, presses the eggs out of her belly. The male newt writhes on the ground as he leaves his sperm wrapped in jelly. The female passes over this packet and fertilizes it by pressing her cloaca against it. She is present because the male has gotten her attention with a display of bright colors, a dance, and an erotic waving tail.

Turtles and crocodiles have swelling copulatory organs like penises. Turtles seduce each other with a slow head-to-head swim. The male extends his organ into the female's cloaca, and then he may ride atop her shell for days, titillating her genital with whips of his spiny tail, which causes her to push the hind end of her body as far as she can out of the shell.

Bats mate while hanging, the male pushing his bent tubular penis from behind between the female's hind legs into her vagina. Whales must leap out of the water belly to belly and, while they hang there for a moment, the male's long and leathery penis enters the female and ejects semen.

Nature, however, is a whirlpool of unconscious violence and insatiable desires, and in some instances *eros* is resolved only in death. Male bees have organs which break off in the queen's vagina, plugging the sperm duct so that the eggs are fertilized before the seed runs out, yet the loss of the organ causes the male to bleed to death. Females of other species of insects devour their lovers during copulation. The gentle turn-of-the-century entomologist Jean Henri Fabre was shocked by this sexual cannibalism, but he was also the first to describe it in detail.

"What should we say," he wrote, "when the saddle grasshopper, before laying her eggs, slits her mate open and eats as much of him as she can hold? And when the gentle cricket becomes a hyena and mercilessly pulls out the wings of her beloved who performed so magnificent a serenade for her, smashes his harp and shows her thanks by partially devouring him?"[3]

Fabre further describes the golden beetle's marriage: "A vain struggle to break away—that is all the male undertakes toward his salvation. Otherwise, he accepts his fate. Finally his skin bursts, the wound gapes wide, the inner substance is devoured by his worthy spouse. Her head burrowing inside the body of her husband, she hollows out his back. A shudder that runs through the poor fellow's limbs announces his approaching end. The female butcher ignores this; she gropes into the narrowest passages and windings in the thoracic cavity. Soon only the well-known little boat of the wing sheaths and the thorax with legs attached are left of the dead male. The husk, sucked dry, is abandoned."[4]

In his lecture on the fate of the sperm Da Free John addresses our imperiled situation: "A death that occurs in the moment of reproduction has a terrible aspect. It causes us to consider whether we are born to eat or to be eaten, to consider whether we are born to be eaten in this cycle of reproduction or whether we are born to be. To be or to be eaten—it looks like 'to be eaten' is it! Even reproducing is eating and dying."[5]

The phylogenesis of gametes in the microcosm trans-signifies the evolutionary history of sexuality in the macrocosm. From a sheer biomechanical point of view the purpose of sexuality is to bring together the germ cells, so the efficiency of different gonadal tissues and their adaptability to changing environmental niches would appear to be the baseline of sexual morphology and innovation (in these terms, cloned germ cells are very efficient, requiring only themselves). However, neo-Darwinian biologists have long suspected that the sexual differentiation of plants and animals is a secondarily favorable adaptation despite the perilous adventures sex cells must complete in order for embryogenesis to occur at all. Through meiosis and syzygy sexuality upsets the deep-rooted and repetitive fixity of the genotype, and this means that the combination of traits available for innovation through natural selection are continuously shuffled through breeding populations, from males to females and from females to males, leading to unusual creatures. If originally maladaptive, sexuality became the source of the myriad variety of living beings.

In any case, we cannot begin to know whether sexuality arose from accidental mutations on Earth or is an expression of a basic force in the universe. Although the disjunction into opposite sexes occurred transhistorically (through successive creatures), the union of male and female is an explicit temporal event, rhythmic at all levels and expressed in a single synchrony of sperm and egg. Generally, neo-Darwinian biologists have not considered that the reunion of germ cells might have a significant metabolic effect on the conjugating creatures; its power lies more obviously in the transfer of life energy to zygote. However, if sexual energy expresses a cosmic force, then the embodiment of that force on Earth would be inevitable. The demographic role of sexualizing mutations might be subsidiary to their energetic quanta—subtle waves of nourishment between mating creatures. (Some aspirants have developed exercises to cultivate this erotic charge in medicines and as a trigger of religious trance, for instance, in tantra and taoist yoga.) If sexual differentiation represents a primary and universal archetype, then gamete fertilization (and ensuing embryogenesis) would be the terrestrial manifestation of its cosmogonic force.

The psychologist Wilhelm Reich formalized this set of beliefs in his twentieth-century proposal of an orgone energy—a current exchanged between lovers but also between galaxies, planets and suns, and organic and inorganic material—in other words, a fundamental stream of particles impelling all other physical motion. *Eros* then becomes the expression of that energy in the human psyche and complements its reproductive function. In Reichian terms sex cells form organisms because they translate sexual energy into the zygote, but the energy itself is cosmic and ancient and is realized in plants and animals only because it was already present in stardust.[6]

* * *

Embryogenesis is carried out entirely by female tissue, which fashions a male as naturally as it does a female. The gut originates in an internal polarity of the ovum; the brain is crystallized "egg stuff." Even the penis and scrotum are made from female substance. Behind every male (including the misogynist rapist, the macho cowboy, the chief of the military) there stands a woman, the Great Mother, beyond simple gender.

[margin handwriting: not true the original zygote was ½ male]

No one can replace this original Lady of Fate. In her breath arises not only the oxygen to drive the heart and feed the cells but the chambers of the heart itself. Even the emerging consciousness of the male is grounded in the matter of her body.

Human life may drift far from the womb in the independence of its separated existence, but we are only secondarily adapted to the land, and the Great Mother has a lasting attraction. She is not any one woman but the chain of maternal tissue enveloping all embryos back to the dawn of our species and, on a cellular level, the beginning of life. She is never fully seen; she is sight itself (which comes from the proteins of her cells that spin iris and lens). Her flesh may not be touched, but it *is* touch.

The Great Mother is an archetype, a series of images illuminated from within. Her countenance appears through men and women and their symbols; she is parent, goddess, lover, movie star, matriarch. She is the fat hag-like medium we flee from in a nightmare, but (as we might suspect) she carries the grail that will restore our courage and end *all* nightmares. She will accompany us as a female astronaut on a journey to Saturn in a dream, and when someone has to descend and capture a fragment of its incandescent surface, she will have the ability to shoot down like a spider on a thread, bringing back not only the soft, gaseous metal but a sprig of alien life. She is the hero of our own embryonic possibility.

[margin handwriting: very confused, misleading]
[margin handwriting: heroine]

Each individual creature experiences the Mother and the anima as an aspect of its own particularization. The newborn is first attracted to the archetypal feminine as the projection of its own libido and its source of nourishment. As it becomes more fully conscious, the ego realizes that it has barely escaped oblivion. The womb may have been life-giving to the body, but it is also a place of death—the death of consciousness in preconsciousness and the death of all cells that will never become conscious.

Trillions of creatures are heartlessly sacrificed for every one that escapes, a legacy we on Earth can never forget. Consciousness wants to live; it does not want to be snuffed out in the allure of the Mother. Even the self of women seeks a male principle, a warrior-priest or shaman. Through him we may disarm violent and unconscious forces that will steal our images and fantasies if we cannot so reclaim them.

Although the attraction of the Great Mother is directionless and internal, it occurs as exogenous desire; she is a siren and seductress of both men and women through their fantasy figures, their Heroes. Many men cannot even clearly see Earth women, for they are blinded by phantasms of the extraterrestrial Woman within. So women chronically become sex objects, i.e., ways of contacting an eternal being. But behind every woman there also stands a man.

By making the Great Mother "visible" in horrid figurines and grim myths we experience the pain of her more directly and so are in less actual danger. How many times this emanation will have to be played out in the wars and genocide of nations we hardly know. A terrible nostalgia for the Great Mother can afflict whole cultures, setting loose her dark and bloody myrmidons. The Nazis resurrected her caricature as an Aryan queen and delivered more than six million corpses to her altar. She was a dark and bloody place to begin with, long before the blood of the womb. (The psychologist Erich Neumann names her three transformation-mysteries: the blood of menstruation, the life-blood of pregnancy, and the blood of the breast converted to shining milk.)

Historically and mythologically, beasts of prey lurk behind the beatific faces of the Mother. The flesh of the young princess peels away to reveal a rotting skeleton (presently, in horror movies); the belly dancer becomes a rabid wolf. The Japanese Ninja female warrior displays her breasts in order to startle her opponent, and while he is distracted, she unleashes a concealed dagger. *Kunoichi* approach their victims seductively with a flower which then explodes with blinding powder in their faces.[7] There are actual statues of the Goddess with vaginal teeth. The Great Mother as a carnivorous witch is portrayed by Heinrich Zimmer:

"In her 'hideous aspect' the goddess, as Kali, the 'dark one,' raises the skull full of seething blood to her lips; her devotional image shows her dressed in blood red, standing in a boat floating on a sea of blood."[8] At the foot of her altar flows the blood of beheaded sacrificial animals, their heads piled in mud. She is stroking her cobras while gorging on entrails, and her own innards are connected by an umbilicus to the innards of the corpse of a beast.

This is not a representation of visible nature, but as much as the double helix and the Golgi body, it is an image of invisible nature experienced through our symbols. The Great Mother is not a whim of surrealism or an idol of pagan culture; she is a force, impelling us toward change, and the transformation she embodies is part of our own being and part of the natural world of cells and tissues. The sexual desire of men for women and women for men is inextricably wedded to the place and manner of birth and is fused again with the mysteries of the natural world. The archetypes arise *with* us, perhaps even suddenly within our materiality. They are basic properties of the field of the planet, as much as water and light and gravity. They express a unity that imbues every developing embryo as the Moon imbues the seas.

In his later pessimistic years Sigmund Freud proposed an all-embracing death impulse which he named *thanatos,* the opposite of *eros* (and *orgone*). His Hungarian disciple Sandor Ferenczi elevated this drive to the central force of nature, not only of psyche, not only of life, but of matter down to its very atoms. Through this instinct he proposed a primordial psychological link between embryogenesis and phylogenesis, a mythic ritual played out by cells and resulting in the construction of a genital.

Unless aroused by a trauma, substance (according to Ferenczi) could never have come alive. Nature cannot generate life randomly and neutrally out of inanimate chemicals; it requires a psychic catalyst. In true homage to Freud, Ferenczi chose *thanatos* over *eros.* Creatures are literally shocked into existence. Evolution is merely a succession of failed suicides, the efforts of animals to erase the pain of prior traumas, desperate actions which give rise to new organs and life-styles in classic Lamarckian fashion: Desire is channelled through nerves and unconscious tributaries into germ cells where it is somaticized by the genes—not only inheritance of acquired characteristics but embodiment of the Wish. In particular lineages these attempts incorporate the old traumas in new ones which mask and displace them organ by organ. In some manner the desires created at the levels of cytoplasm and tissue make possible new tissues and organs through speciation. This psychosomatic aspect of evolution leaves no fossils and exists only in holisms of mind and body. It allows the libidinal crises of early creatures to be reexperienced by their descendants (even those of different species) through the ontogenesis of their very organs.

The formation of life from dark lifeless waters was the first trauma, forcing something to awake unwillingly from the untroubled womb of eternal unconsciousness. The denial of this event is imprinted onto matter at the level of the protoplasm which it brought into being. The "catastrophe" (as Ferenczi defined it) recurs ontogenetically in each new organism with the maturation of its sex cells. The second "catastrophe," spawning individual unicellular organisms (that darted about in wonder and dismay) left its permanent recurrent scar in the mature germ cells formed from the gonads at the onset of each new life. Flesh was simply trauma materialized and shaped around its own wound. As many-celled marine animals began to propagate sexually, they introverted their inherited death wish into an organ which was finally recapitulated as the mammalian uterus.

The recession of the ocean and the subsequent adaptation to land was highly catastrophic, but once again pure *thanatos* failed, and these events were somaticized in the protective covering of the uterus and the development of sex organs. This "primacy of the genital zone" (as Freud named it) transfers desire from the whole body into libidinalized nodes of flesh. For Ferenczi this desire is merely the shadow of the wish to return to the Great Sea. Humankind then came into being only after another global holocaust: the exterminating glaciers at whose heart lay the seed of consciousness. Trillions of creatures whose ancestors were drawn into lush tropical lands and who adapted to the temperate seasons were now forced to undergo an epochal winter. They sought death again, and although many of their races were extinguished, *eros* prevailed: the human child was born (along with birds and other mammals). The Ice Ages were somaticized in these creatures as the sexual latency of their childhoods.

Each regression is more intricate and deceptive than the one before it and shifts tissue further from the ancient sleep. The death wish becomes profoundly unconscious, but its latency somaticizes germ cells and organs through which it is then replaced by sexual latency; from this further sublimation come the symbolic realms of language and culture. Each transfer of libidinalized energy takes the organism further from the original desire, which is to go back to sleep forever. *No!*

"Yet we have gained much from civilization," the Freudian and Ferenczian disciple Géza Róheim reassures us. "We have learned to conserve fore-pleasure, and to prolong youth and life itself."

Then he quotes the master directly:

" 'At one time or another, by some operation of force which still baffles conjecture, the properties of life were awakened in lifeless matter. Perhaps the process was a prototype resembling that other one which later in a certain stratum of living matter gave rise to consciousness. The tension then aroused in the previously inanimate matter strove to attain an equilibrium; the first instinct was present, that to return to lifelessness. The living substance at that time had death within easy reach, there was probably only a short course of life to run, the direction of which was determined by the chemical structure of the young organism. So through a long period of time the living substance may have been constantly created anew, and easily extinguished, until decisive influences altered in such a way as to compel the still surviving substance to ever greater deviations (retardation) from the original path of life, and to ever more complicated and circuitous routes to the attainment of the goal of death.' ''[9]

Sexual organs are literally the introversion of the battle between *eros* and *thanatos*. Our insides are a psychosomaticization of the salty flux in which our ancestors were spawned, fed, and into which they discharged billions of generations of sperms and eggs. When the sea is lost, primal desire creates internal waters. First the sea becomes tissue, then the reptilian egg (used also by birds), and finally the mammalian womb.

From the Greek name for "sea," Ferenczi called his theory "Thalassa"; he defined sexuality and primary development as hopeless flights to the ancient ocean, contemporized as regression to the birth waters. The grasping rush of the infant to get to the breast is an oceanic craving which matures into the sexual consummation of adults. In general, thalassan *eros* drives male creatures to penetrate the females of their species, to get back into the womb. In amphibians, which still have access to external breeding pools, Ferenczi saw a foreshadowing of coitus, the male frog clasping the female with the pads on his front legs. The excited discharge of urethral materials in salamanders shapes the tissue for the partial erections of crocodiles.

The reptilian forerunner of the mammal struggled to get the waters back around him, to burst into the salty moist interior of a woman. Its embryo was a lungfish in the "mud," breathing by osmosis from the water. The unconscious memory of that existence was so intense that several lines of ancient mammals went against the trend of latency and returned to thalassa as sea cows, dolphins, seals, and whales.

The penis and vagina are scars of the combat between creatures trying to penetrate each other, pieces of flesh folded around germ cells. The attempt at regression succeeds, but only to a degree. Penetration is accomplished and rudimentary organs make it a lasting and inheritable experience, but those organs also trap the thalassan drive at a sexual level. The desire to retain the prior, more primitive, and more peaceful sea (and through it, the original sleep) is frustrated. Like a field of energy this desire polarizes evolution in an opposite direction from itself and leads to new organs, new creatures, fur and lungs and land habitats, and, finally, the greatest disturber of its craving for slumber, consciousness. Once these organs and beings developed, their psychosexuality and ecology followed, and, in our aeon, their sociology and symbolic representation. As death was translated into love, the subcellular realm became multicellular and superorganic. Sublimation was successful because it not only brought sperm and egg into being but genitalized them and compelled them to seek each other, and thus ensured its own inheritance and ultimate globalization. Eroticism is an intimation of embryogenesis, its disturbed dream. It is the closest we can come to experiencing the yoga and pain that lie within our formation. Normally we suppress this agony, but sexuality allows us to be thrown headlong upon its mystery, and to be reminded, as Søren Kierkegaard reminded his readers, that life is not a riddle to be solved but a reality to be experienced.

9. The Blastula

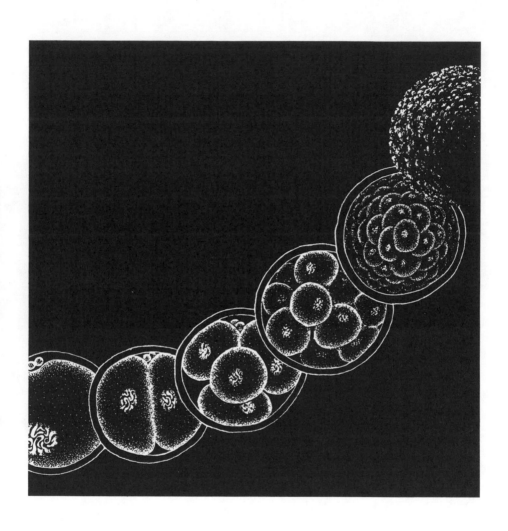

The blastula is a heap of cells, the first stage in the formation of a new organism. Its general appearance is spherical (or ovoid) because the free surfaces of the constituent cells continue to be globular (like bubbles) while the pressure of the other surfaces against each other compresses them inward. The blastula starts as a single round cell, the egg, and then fizzes outward into many cells, but always keeping the basic topology of a ball.

Soon after fertilization the zygote divides by mitosis. Its daughters each split, and their daughters split. The cells go on fissioning, yielding two for every one, until there are enough cells for the embryo to differentiate according to its ancestral plan. This initial phase of cell multiplication is essentially similar throughout the animal kingdom.

With each division the cytoplasm is distributed among more and more cells. Every succeeding generation has less cytoplasm and more nuclear material, so the cells become smaller and smaller until they dwindle beneath the size of the cells of normal tissue (each generation is roughly one half the size of the cells from which it came). At the same time the cells become more deeply joined and develop communication junctures across their minute gaps. In this form they are known as "blastomeres." They are a unity of separate cells, so their divisions are perfectly timed and coordinated.

The nuclear material, spindle, and outer membrane of the blastula are inherited from oogenesis; the sperm has fused with the egg, so its image is imprinted on every blastomere, but its contribution is withheld. In one sense, the egg was already building toward cleavage at the time it was fertilized. Even an enucleated cell will develop a pattern of serial divisions if stimulated, for the rhythmicity of the blastomeres is solely under the influence of the cytoplasm.

During oogenesis, mostly RNA is synthesized, but this cadence is arrested by the entry of the sperm. The zygote is briefly dormant; then it picks up a new signal: The transcription of hereditary material takes priority over the assembly of proteins. It is as though a ship were given an entirely different mission before even a twentieth of its voyage had been completed, the new instructions being locked away in secret logs inherited from a prior ship.

The mission of the blastula is panspecific and affects all eggs, from invertebrates to mammals, in much the same way. The cells assemble a primordial semihollow organism, a mold from which particular animals can be shaped. Even the fusion of unfertilized eggs with fully differentiated cells from mixed species has activated DNA synthesis in the resulting heap. However, in healthy blastulation the periodicity and duration of cleavage must synchronize with the full transcription of the DNA molecules of the specific animal; otherwise the daughter cells will not receive a complete set of genes. This hive of sibling balls is not much of a creature. It grows as though a Zeiss projector were continuing to fill dome after dome of nested planetaria with the sparkling imprint of a single night sky, yet all the planetaria were phosphorescent jellyfish stuck to each other as they swim apart. This hall of mirrors picks up the minute two-dimensional engraving of the genes and radiates it exponentially outward—first, by expanding its actual space, and second, by creating an enormous variety of contexts for its potential read-out.

The plane of the first cleavage is usually vertical, cutting through the animal-vegetal axis; the second division is also vertical but at a right angle to the first so that the four hemispheres stretch from the top of the embryo to the base, each crossing the equator. The third division then slices the equator, separating the animal and vegetal spheres. The upper four blastomeres may rest perfectly atop the lower four (in the case of radial cleavage), or they may be tilted slightly to the left or right into a spiral cleavage pattern. Because the cells originate in mitosis, their off-center rotation reflects oblique arrangements of the spindle fibers twisting the axes of their cleavage furrows; spirality is projected from the heart of their fissioning and is basic to their self-organization.

If we view the embryo as a ball in space, the cleavage furrows can be seen as a ring developing simultaneously around a globe and pressing inward at all points. The sphere breathes, pauses, and then breathes again; at each breath the number of chambers has doubled, visibly and within. As with mitosis, we see only the surface of an internal event. The pauses between divisions can range from several minutes (as in a sea urchin for instance) to several hours in a mammal.

Insect eggs are partitioned between yolk and cellular material such that cytoplasm does not initially divide with nuclei. After the nuclei fission they cluster in a central mass which migrates toward the embryo's surface, each nucleus bearing a section of cytoplasm. As these fuse with the animal sphere, furrows roll inward and blastomeres are formed. The segmented individuality of the insects may thus be fore-shadowed early in their development.

In animals with a great deal of yolk (or segregated yolk) the rich protein and fats of this material hinder surrounding cell division. The upper hemisphere divides rapidly into tiny blastomeres (called micro-meres) whereas the vegetal zone remains smooth and yolky, the em-bryo having the overall appearance of small crystals on a stone. The "crystals" are continuously displaced upward, away from whatever margin they share with the uncleaved "stone." In general, where the sizes of blastomeres differ, micromeres form at the animal pole with macromeres at the vegetal end.

A scientized version of preformationism persisted into the later nineteenth century as a belief that the body parts of any future organ-ism were precisely dictated by granules already in the egg—discrete particles that were qualitatively different from one another and could transmit those differences into biological traits. The first significant experiment in the laboratory science of embryology was a test of pre-formation.

In 1888, Wilhelm Roux used a red-hot needle to destroy one of the first two cleavage cells in a frog embryo. The surviving cell went on to develop a half-embryo, apparently lacking just those parts that would have been provided by the eliminated blastomere. This was seemingly a dramatic confirmation of preformationism. Expecting the same results, Hans Driesch mutilated a sea-urchin egg in a similar way three years later. To his astonishment he found not a half-organization in his dish the next morning but a whole gastrula. The embryo had mysteriously resupplied its own missing half.

At the turn of the century, the contradictory results of these experi-ments were still unresolved, and many prominent embryologists and geneticists continued to argue that genetic determinants were passed on discretely to individual blastomeres. In 1928, Hans Spemann un-dertook the definitive experiment: he split the newly fertilized egg of a newt in such a way that the nucleus was isolated in one sphere. A thin bridge of cytoplasm was left between the spheres. Up to the stage of, roughly, sixteen blastomeres, only the nucleated half devel-oped, but at that point a single nucleus slipped through the bridge and then the other half began cleaving. Both produced normal newts.

By not removing the dead blastomere forty years earlier, Roux had unintentionally placed the living one in a situation it interpreted as: "My other half's still there, so I'll go about my usual business." Without realizing its implications, Driesch had then made a clean di-vision and so given each blastomere autonomy to become whole.

With the reinterpretation and synthesis of these experiments, the presentiment of a biological field eclipsed the linear particularism of the genes.

More sophisticated variations of Spemann's experiment were undertaken during the ensuing decades. Eggs given unequal divisions of genetic material restored themselves as well (their advanced nuclei were able to backtrack and start embryogenesis from the beginning). As techniques for transplantation improved, genetic material could be taken from later and later stages of organisms. By the 1950s, one cell nucleus from a blastula with sixteen thousand cells had organized a healthy animal. In 1968, nuclei from cells of the neural plate of a tadpole—cells well on their way to becoming a nervous system— were transferred back to an enucleated blastomere and assembled a mature frog. Gut, lung, and kidney cells also worked on rare occasions, though all of these experiments produced monstrosities and nonfunctional blastulas far more often than healthy creatures; but the fact that they succeeded at all demonstrated the nucleic potential that could be realized in ideal circumstances. Even the nucleus from the cell of a tumor growing in a frog, when isolated from the malignancy and placed in an enucleated blastomere, yielded normal tadpoles. Cancer may be uncontrolled growth, but its cells, at their nuclear hearts, are not alien to the creature in which they form. Their DNA contains the imago of complete and healthy creatures, each one able to come alive if cloned. Something in the organismic context of tumors has severed their tissue from the coordination around them, but the individual cells are "human."

These experiments do not prove that kidney cells, for instance, are the same as blastomeres, but in an emergency created by intelligent-species interference a kidney cell can stop being kidney and call up the sequential properties of the embryo of which it is part. Where such experiments fail, the problem, likely, is in the tempo. Relocated nuclei often must grow as much as thirty times their size upon finding themselves in a new zygote. Endodermal and neural nuclei have difficulty responding to the swiftness of early cytoplasmic division. Embryos are also extraordinarily complex entities that defy consistent experimental results, sometimes behaving like machines with interchangeable parts and sometimes contradicting their components with an overall unity that seems to arise from nowhere. A machine cannot suddenly change itself, but the early embryo has such underlying malleability that tissues and organs not only appear but disappear, and sometimes reappear. There are no simple rules in systems with this many factors. The same method of blastomere isolation produces one result with sea urchins, another result with mollusks and sea squirts.

The terms "mosaic" and "regulative" have been applied by embryologists to the ostensible extremes of determination in development. Spirally cleaving eggs exhibit mosaic development; that is, each of the cells formed by cleavage furrows in the blastula has a determined fate in the adult organism. Nematodes are typically mosaic; their early blastomeres already correspond inalterably to endoderm, alimentary tract, reproductive organs, etc. The homologous traits and organs of many flatworms, segmented worms, mollusks, and insects develop along the same mosaic pattern, blastomere by blastomere, despite the fact that these groups have been separated evolutionarily for half a billion years.

In contrast, the cells formed by the divisions of regulative mammalian blastulas retain equal potential until well into organogenesis. The difference between mosaic (fixed) and regulative (equipotential) development is a subtle one, however, and it is still unclear whether the early determination of the blastomeres is a primitive trait, diagnostic of simpler and older species, or an idiosyncratic pattern maintained conservatively in diverse lineages. If all simple creatures were formed determinatively, then development by equipotential blastomeres would be a kind of higher system involving more complex transpositions of information, but this is not the case. Sea urchins, for example, are regulative. In addition, changed experimental procedures have sometimes revealed greater potential than thought possible in the blastomeres of mosaic species. It is difficult to explain the difference between dragonflies, whose eggs can be divided to form two equivalent organisms, and houseflies, in which final development of the oocyte occurs regionally even before cleavage. Are these real and fundamental differences, or experimentally induced divergences?

Houseflies show the importance of context even in a determinative framework. The discs of the adult fly are completely suppressed at the larval stage. Only in the slumber of the worm-like pupa does the soma read its own genes in such a drastically different manner that, from the chrysalis, a new winged animal lifts itself into a previously unknown habitat. Metamorphosis is a second embryogenesis. However, if these potentiating discs are transplanted generation after generation, from fly to fly but in different segments from those in which they originated, the cells arising from them will eventually change too and form the "wrong" organs—though only in consistent sequences. A genital disc can become a head or a leg, but not a wing; however, a head or a leg disc, including one that was formerly a genital disc, can become a wing, and a wing disc can give rise to thorax.

This shows us again the latent potentiality in each organism and how, millennially, structures may have been transformed. New organs are invented by a series of changed contexts, leading eventually to new species. Homologous discs are transpositioned along continua over epochs and generations. The unadorned worm becomes millions of idiosyncratically sculpted insects and crabs, their various wings, claws, and feelers generated by mutations translating segments into organs and organs into systems. But this is also how, in a single summer, a worm-like grub becomes a lady bug.

In a sense, biologists identify animals from their embryos not because they can see the forerunners of specific organs but because they have learned which embryos give rise to which creatures. The earlier stages of very diverse phyla overlap. In the beginning human beings and sea urchins are not that different, but ancestral patterns interweave in each in slightly different ways, and divergences accumulate until whole new structures appear and creatures utterly foreign to one another emerge. Once upon a time mammals and octopi diverged from the same primordial blastula without a single violation of growth and form and with a subtlety that does not reveal the precise origin of their differences.

Even in cases of regulative development gradients form in the blastulated embryo (and sometimes even earlier in the unfertilized egg). The raw blastula is a mediation of two fundamental polarities: the vegetal pole provides gut material and endoderm; and the more externalizing animal pole propagates skin, feathers, fur, hair, and claws (depending on the species). So the two poles of the blastula embody the opposite extremes of deep gut and surface elaboration.

If too much cytoplasm is removed from the animal hemisphere the blastula will become vegetalized, i.e., interiorized, only a stomach; whereas a dominant animal pole may lead to a hollow ciliated ball. The two anterior blastomeres of the sea squirt, isolated at the four-cell stage, will form only ectoderm, nervous system, notochord, and a trace of endoderm, but will not form muscle tissue or cells for the circulatory system.

The textures of multiple organs originate in the gradient of cytoplasm between the poles. Prior to their actual formation these body parts preexist as potentialities in the blastula's field. The unformed liver emerges closer to the vegetal pole. The unformed brain and eyes appear in the animal plasm. Functional tissue arises almost as if from interference patterns in the currents between opposing poles—a biogenesis foreshadowed by the most ancient Chinese theories of embryology and incorporated in the *Yellow Emperor's Classic of Internal Medicine*. From prehistoric times herbs and acupuncture needles have been directed not at organs but at the current of energy through the channels from which organs emerged and in terms of which they continue to function. The dynamic force of these remedies is ostensibly synchronized to germinal paths, and to modulations of vital flow. (Western pharmacy by contrast targets treatments only by their specific chemical effects on fully formed organs.)

The blastula is a universal animal formed by simple cell division. Yet it is little more than a hollow or partially hollow globe. In many species the fissioning cells remain at the outer wall. Their inner edges (as we have seen) are pushed in by the surface tension of their membranes against each other, so a cavernous region forms at their collective interior, a hollow cavity called a blastocoele. The surface clusters of tiny blastomeres, perhaps of different sizes, maintain the spherical illusion of the original zygote, but the cells are deeply packed and hide their hollow center. In some species the contact between the outer cell membranes seals into a permanent layer, an epithelium or skin.

The surface layer of cells is called the blastoderm, in nomenclattural preparation for the three layers of body tissue, i.e., ectoderm, mesoderm, and endoderm. Cell size determines regional blastoderm thickness. For instance, the yolk near the vegetal pole of the frog embryo makes the blastoderm a puffy mass of jelly, whereas its animal layer is a mere lining of cells. Although the blastula of the frog remains everywhere a hollow sphere the cavity reaches almost to the surface at the animal pole. On the other hand, insect yolk fills the entire blastocoele, and bony fish have convex disc blastulas. But the underlying topology and meaning of the stage remain the same.

In humans it takes about three days to form a solid ball of sixteen blastomeres—called a morula from the Latin name for the mulberry it resembles. As with all mammals the cells of the human morula are packed right through to the interior. For blastulation to occur in this kind of an embryo, the ball must be penetrated. Fluid passes from the uterine cavity into the morula and fills its intercellular spaces. The cells are separated into an outer layer, or trophoblast (which will participate in the formation of the placenta), and a distinct inner cell mass, or embryoblast (which will provide the material for the organism). When the fluid-filled spaces contact each other and run together, a central cavity (called a blastocyst) occurs. Placental attachment requires its own specialized anatomy. It is as if the potential density of the (nonexistent) yolk were transformed into the trophoblast. Perhaps such a mutation lies at the basis of mammalian lineage. In any case, the whole meaning of mammalian life emerges from this subtle divergence.

The assembly of blastula and morula does not create true interior and exterior zones. Cells merely occupy different positions in a uniform grid. The blastula is like a volvox or sponge; it is not an adult multicellular organism. If we think of our outside as communication, protection, and feeding, and our inside as digestion, transformation, and assimilation, then skin and intestines, nerves and gut must be polarized, respectively, away from each other. In the blastula these all lie in linear cotangency. There is no layer of muscles, no skeleton; there is not even a real central cavity (the blastocoele and blastocyst are circumstantial not structural). In order to become an organism the blastula must develop a functional interior and make use of its three-dimensional space.

Dreamed while studying blastulas:
I feel strong waves of gravity. I awake suddenly. I have been in hibernation with the others as our spaceship passed inertially from galaxy to galaxy. Now our body has been perturbed by a sun-star and we cannot escape this system. We are sucked down onto a huge planet. I see light flying out beneath me and widening as I sit up staring through the rear viewport. I am materializing through the blue sky onto this world. Giants, crooked and bent, stand on hills and look up at us as we fly past them. We are the UFO, but we are moving so fast they do not understand what we are. The strange topography looks familiar. Our ship is caught in the branches of a tree. I wait, hanging there, hoping that we will be able to take off again. The ship becomes smaller until it is only me perched there with a blanket wrapped around me. I throw off the blanket and climb down out of the tree. Now I am on this planet for good, and its inhabitants are dwindling to normal size and shape. I join a crowd in a village on the edge of a forest, and I walk with them until we blend in with all the others on the city streets. I go back to the apartment building of my childhood and take the elevator upstairs.

The body's pulse, like the microwave background of the cosmos, is the record of creation, and it continues to time the spatialization of body/mind. In Oriental and Western systems which presume a relationship between cosmic and life energies, the gap between universality and individuality is superficial and illusory. Changes in the climate and the heavens are reflected in the movements of cells and tissues. The Japanese philosopher and physician Michio Kushi describes the fertilized egg as "a replica of the earth, which is rotating, therefore producing electromagnetic belts around itself and periodically undergoing axis shifts."[1]

Cells and molecules are simply planets spinning in deeply impacted zones of the heavens; stars and galaxies are eggs and blastulas strung out across (from our perspective) extensive fields of gravitational energy. The meridian cycle is multidimensional; it does not begin in the skin as such or drain in the gut, or vice versa; it emerges from all directions at once with differentiating tissue, even as giant nebulae and interstellar fields spring from nowhere and create space and time by their convergence. An obscure but indelible unity binds each cell to the birth of the cosmos. This is the hidden astronomy of the blastula.

We cannot hope to recall what it felt like to become. The sense of being born emerges anew from our life itself, again and again. We awake each morning on a planet with blue skies, we feel the ocean waves against us on a hot summer day, we experience (all the time) tendrils of sorrow and joy, parts of our own identities and others'. At the moment of seduction we see ourselves reflected a billion times through the universe, each one about to happen. We also sit calmly on the grass, breathing, doing nothing. The yogi sinks down to the roots of that and finds a whirlpool beginning, the more twisty the deeper he sinks.

". . . living is one constant and perpetual instant when the arras-veil before what-is-to-be hangs docile and even glad to the lightest naked thrust if we had dared:"[2] the words of William Faulkner in the voice of the mind of Rosa in *Absalom, Absalom!*

Embryogenesis is the background of being. Even if it is not remembered as an image, its wholeness is remembered. Our sense of our own completeness is tethered to a fissioning egg and a hollow ball, ". . . the globy and complete instant of its freedom . . . a fragile evanescent iridescent sphere."[3]

We cannot return to a time before wholeness, for wholeness precedes mind, and for mind to exist at all it must be whole from the beginning; otherwise it would have to be everything and everywhere and would disperse into the universe without ever congealing. Psychic unity happens once, at the beginning, and from there shines through the entirety of life.

10. Gastrulation

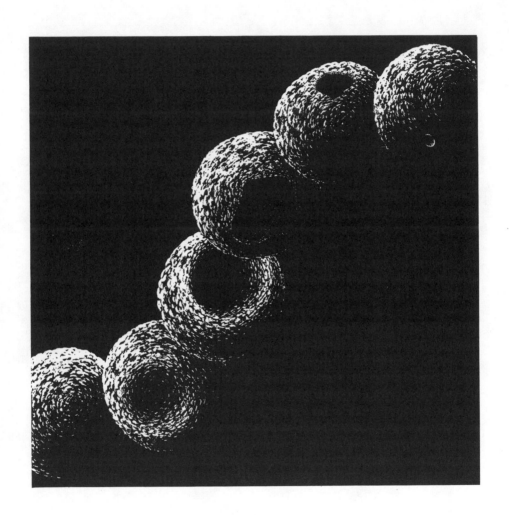

The life of the blastula is brief: its fission gradually slows down and its chemistry changes; its cells start synthesizing exotic proteins distinctive to the animal in their matrix. Whereas blastulation is primarily the breath cycle of the cytoplasm, gastrulation is the inevitable result of RNA transcription. The nuclei become active, and the maternal and paternal chromosomes begin to transmit their hologram. A remarkable transition occurs: substance that was formerly exterior turns inward, and a mass of tissue stretches over its surface to cover it.

The blastula is an undifferentiated animal not unlike a simplified jellyfish; the gastrula is a living sculpture that will mold and carve itself from the inside out. The germinal cells of the notochord, the gut, and the brain form on the surface of the amphibian blastula with some of the eventual skin (most of which is in the interior). If we project a newt back onto its blastula, the epidermis and nervous system flow into the upper animal hemisphere—an irregular island marking parts of the nose, ears, and the skin of the mouth; the eyes diverge to its opposite sides. The notochord bunches up at the equator, strands of its connective tissue falling toward the vegetal pole. As the mouth comes apart, its endoderm flows down into the vegetal zone with the gills and pharynx. Segments of internal organs pull apart like yarn and twist indistinguishably about each other in the yolk of the vegetal pole. The mesoderm of the body cavity and the kidneys unravels onto the other side of the embryo with the rear gut.

In gastrulation the surface of the hollow sphere folds inward and back around itself; the interior of the adult body is sucked inside. Tissue seems to rush toward itself, to reconsider its nature. The only reason gastrulation does not look like a finger stuck in a soft orange is that the blastula is under no pressure from the outside. It folds upon itself with unimaginable grace, putting the organs in their historically proper relationship to one another and obliterating the blastocoele (which is physiologically meaningless) in order to create the archenteron, the forerunner of the alimentary system.

Small fully-cleaving eggs such as those of starfish and sea urchins develop cilia after their tenth division, and a day later the vegetal pole of the free-swimming creature flattens and cells migrate inward to fill

119

the blastocoele and form a cylindrical larva. Amphibians gastrulate between the fourteenth and fifteenth division when they have ten to twenty thousand cells. The vegetal pole suddenly flattens and creases inward. The sphere hollows into a cup, its external wall still facing out while another wall lines the new cavity. The opening into the archenteron, called the blastopore, remains throughout adult life. In species simpler than starfish and sea urchins the blastopore usually becomes the mouth; in species from the echinoderms on, the opening becomes the anus and the mouth must develop from a secondary perforation. This distinction between the more primitive *Spiralia* (or protostomes) and the deuterostomes is one of the most basic in the animal kingdom, though it is overlaid and often obscured by adaptive cleavage patterns of individual species.

In embryos in which the vegetal wall is yolky enough to resist collapse, the cells of the marginal zone (between the animal and vegetal hemispheres) initiate inversion by migrating inside and sinking. As the initial cells disappear, new cells follow them over the lip; this "inrolling" through the blastopore creates the archenteron wall. The migration of course has a mechanical basis: the cells in the region of the lip elongate in the shape of upside-down bottles. This narrowing causes them to cave inward collectively, forming a thin surface groove which splits the vegetal from the marginal zone and continues to spread until it meets itself on the ventral side of the embryo. Ultimately the dorsal, ventral, and lateral lips of the blastopore almost converge (as when a caterpillar brings mouth and hind tip together). Most of the infolded material comes from the dorsal region where gastrulation began; some subsequent tissue moves over the ventral and lateral lips. Even though the bottle necks stretch out, the cells remain connected at their surfaces and move in concert. The "inrolling" ring spreads through the vegetal hemisphere, and the yolky endoderm is sucked into the interior.

Once over the lips, the vegetalized cytoplasm moves in a direction opposite to its migration on the surface, so, by the end of gastrulation, the vegetal region, which entered the archenteron at the dorsal lip, has rotated ventrally through the embryo. While the prospective endoderm and mesoderm are involuting, the animal hemisphere must expand to cover the deformed sphere. The part that is to become the nervous system moves toward the blastopore, and the rest of the ectodermal sheet spreads (as its cells divide) and covers the entire surface of the embryo except for the yolk plug, which sticks out as the underbelly of an interior sphere. The covering of the surface with animal tissue is called epiboly, literally a "throwing, or willing forward."

The different internal organs will ultimately emerge along a gradient of invagination (or distance of migration). In the amphibian embryo the notochord enters the interior from the marginal zone and becomes concentrated along the dorsal side of the archenteron roof in the context of prospective nervous tissue. The mesoderm originates as a sheet of cells moving forward between the ectoderm and endoderm. While one edge of it rolls freely, the other edge remains connected to the notochord. The archenteron lengthens, and a minute invagination of the ectoderm fuses with the cavity to form the stomodeum (the mouth rudiment). Contiguous endoderm becomes the pharynx. Since the mesoderm does not quite reach the anterior end of the gastrula, the mouth forms as a knitting of ectoderm and endoderm at the front of the gut.

The deepest endodermal cells become duodenal and contact the alimentary canal. Surface endoderm becomes pharyngeal and oral. Intermediate sections merge into the foregut.

In insect embryos a median groove forms from the thickening of blastoderm along the ventral side of the gastrula. As the groove advances inward toward the yolk it spreads to form mesoderm and endoderm. Furrows develop across the germinal band, dividing the organism into metameres (segments) and enclosing it in an extraembryonic membrane.

The embryos of birds and reptiles (among others) feed and breathe through organs which will not be part of their adult body. Early cleavage in such eggs is relatively superficial and, as we have seen, limited to a germinal disc at the animal pole. Even before the chicken egg is laid, cells with greater amounts of yolk begin falling into a subgerminal cavity beneath the blastoderm. The actual body cells are located in the upper transparent tissue (the *area pellucida);* yolkier cells surround these in an *area opaca.* Gastrulation begins as a thin layer separates from the lower *area pellucida;* this hypoblast is the source of the yolk sac whereas the upper layer, the epiblast, will generate all three germinal tissues of the body and parts of all the tissues of the extraembryonic organs. As the hypoblast sinks beneath the epiblast, the subgerminal cavity is displaced between the hypoblast and the yolk; and a new cavity, the blastocoele, forms between the blastoderm layers.

Once the egg has been laid, the blastoderm margin advances to enclose the yolk in a sac; meanwhile a strip of blastoderm at the mid rear of the *area pellucida* thickens and stretches forward, pushing out a groove behind itself. The whole area is progressively sucked into this strip, the posterior part to become mesoderm, most of it extraembryonic, and the anterior zone to become both mesoderm and endoderm.

The "primitive streak" (as the strip is called) is not so much a structure as a disturbed region through which the surface of the blastoderm migrates into the interior without infolding. As the cells near the streak bunch together, other cells converge behind them creating multidirectional motion. Tissue sinks down the center and then shifts sideways and forward. The thickened knot at the head of the primitive streak (Hensen's node) contains a single plate of prospective notochord and mesoderm which extends between the migrating ectoderm and endoderm. The material for the heart and kidneys remains connected to the chordamesoderm and follows it in a spiral around the endoderm to the mid-section of the organism. As the hypoblast develops into the yolk sac it is covered by the continued displacement of cells. When there are not enough transparent cells left to feed the primitive streak, it shrinks and moves backward, disappearing into the cloaca and tailbud.

The recession of Hensen's node leaves in its wake the germinal neural plate, which is stretched backward (having occupied the anterior portion of the streak) and drawn in toward the center. The front of the plate then folds around itself to form the neural tube which is subsequently enclosed by ectodermal folds knitting together in a smooth sheet. The rear section of the mesoderm breaks into prospective skeletal units, called somites. The alimentary canal of birds and reptiles does not form directly by invagination as in amphibians; rather folds of the epiblast roll downward and curve in along a narrow median strip of endoderm, the mid-section of this gut remaining open to the tissue forming the yolk sac.

Reptiles and birds develop extraembryonic organs from upward folds of the *area pellucida.* One fold bends backward over the head and then along the body, its free edges fusing in an anterioposterior direction. The fold is originally ectodermal but incorporates mesoderm as it moves backward. The embryo is now enclosed in a cavity within a membrane. The inner surface of fused folds, the lining of the cavity, is called the amnion, and the continuous membrane of the outer surface is the chorion. As this amniotic cavity is filled with secretion the unborn animal floats freely in fluid, buffered from shocks and without friction from its shell. Between the embryo and these "external" organs only a compact endodermal stalk of tissue remains, the umbilical cord. This stalk, rich in its own blood vessels, gradually encloses the neck of another outgrowth, an extraembryonic bladder (the allantois) which is formed from the endoderm of the hindgut and a mesodermal lining. As the storage chamber for urine the allantois spreads out beneath the chorion, developing its own blood network to the dorsal aorta. It too surrounds the organism. Until the animal breaks into daylight, its embryo breathes the air of an unknown planet through external membranes it experiences as itself. It is a complete ecosphere in an egg, a creature functioning as a world.

Mammals have yolkless cytoplasm, so the young blastula (five days old in the human) stays alive like a parasite or tumor by implanting its tissues in the uterine lining. Placental organs are formed by a combination of maternal and foetal tissue, the latter originating vestigially in the allantois, chorion, and yolk sac. In hominids the blastocyst sinks beneath the maternal tissue, destroying areas of the uterine wall with trophoblast cells and replacing them with a complex network of tissues and blood vessels. The mammalian embryo voids waste material directly into the bloodstream of the mother, so the endoderm of the allantois is stunted (in hominids it does not even reach the placenta); however, the mesodermal blood system of the allantois is part of the placenta and carries sustenance from the mother to the implanted foetus. A metabolic exchange is established between the two organisms even though the blood of each circulates independently. Water and oxygen pass into the foetal blood by diffusion; carbon dioxide, urea, and various salts and proteins are eliminated likewise. The mother's antibodies may immunize the offspring against certain diseases, but others, like rubella and chicken pox, can infect the embryo through the placenta. Congenital syphilis and deep psoric miasms are passed from generation to generation through tissues weakened by drugs and suppressed illnesses—a heritage of Ibsenian ghosts.

Though we each build new lives we contain the residues of past generations: their immunities, poisons, and even perhaps their desires and aversions. Some say we also experience the world first through our mother's impressions, that these pass into our tissues as archetypes and feelings. The child thus understands the family and civilization into which he or she is to be born, experiencing it psychosomatically as an aspect of the tissue from which it is being made. Powerful antibiotics, tranquilizers, hallucinogens, and other drugs also reach the foetus. Thalidomide produced thousands of modern industrial tragedies in a world Ibsen and Dickens were already pointing to—infants born with weakened and irregular hearts and without full limbs because pharmaceutical firms chose to promote sedatives. The corruptions of cultures are often incarnated, though not in the way the people expect.

* * *

The human embryo begins gastrulation while it is grafting itself into maternal tissue. The primitive streak originates at the hind part of the dorsal hemisphere in its approximate middle, and pushes forward as a ridge of epiblast with the addition of cells at its rear. The streak thickens in front and a primitive knot develops. When the foetus is sixteen days old, a group of cells migrates beyond the primitive knot and travels sideways between the ectoderm and endoderm until it contacts the prechordal plate of the endoderm, which is attached to the ectoderm. Some of these cells begin forming the notochord and others enter the trophoblast (which is still implanting) and participate in placentation. Subsequent cells migrating forward of the primitive streak flow around the notochordal process and prechordal plate and meet on the other side to begin the construction of the heart. Through the fourth week of embryogenesis the primitive streak continues to invaginate mesoderm in the human foetus; then it gradually disappears into the base of the tailbone. In rare cases where it is not fully absorbed, tumors of partially formed and unintegrated tissue occur. The difference between coordinated growth and malignancy here is a small one: the primitive streak is a global tumor under precise systemic control, whereas cancer is decentralized cell growth and migration.

The notochordal process works its way down through the endoderm, fusing with it to form a flattened plate which rolls from the cranial to caudal end to mold the notochord, then folds up into the mesoderm as the area of fused tissue deteriorates. The notochord contacts tissue which is thickening into horizontal columns; twenty days into the embryo's life the columns begin to split into somites which are pushed up like mountains within the mesodermal plane. Thirty-eight of these paired bodies arise in a craniocaudal sequence along the germinal spine during the next ten days, and two to four pairs develop later in embryogenesis. These will generate the axial skeleton, its musculature, and some deep layers of skin.

The mesoderm separates into two layers: an external body wall integrated with the ectoderm and a primitive gut wall connecting to the endoderm; the space between them becomes the coelom, or the central body cavity of the creature. Initially, the mesoderm is solid, with only small isolated discontinuities, but these coalesce into a horseshoe-shaped cavity. Above and below this coelom, the mesodermal layers remain continuous with the extraembryonic mesoderm.

It is impossible to draw clear lines between stages like oogenesis, blastulation, gastrulation, and subsequent organ development. The blastula is already beginning to form within the egg, and blastulation does not end during gastrulation: the cells continue to divide in phase, but their "breathing" is incorporated in the series of disjunctions. The simple unity of the zygote is translated into a complex harmony of zones.

Although gastrulation differs in detail from order to order even within closely related groups, there are fundamental similarities which confirm that it is an ancient event shared by every terrestrial animal. Relative amounts of yolk, larval and foetal adaptations, and specializations that work their way back to earlier phases of ontogeny camouflage to one degree or another the universal organization of creatures which must all eat, void, breathe, swim, and reproduce. The blastula is a primordial association of one-celled animals. The gastrula transforms that association into three germinal layers (ectoderm, mesoderm, and endoderm) and a central cavity (the coelom) which interiorize multicellular animal life. The only exceptions are those creatures which make up the transition from one-celled life to three-layered coelomate animals: the sponges, jellyfish, and some simple worms and related species.

The original living tissue, before any layering or introversion, was all ectoderm—blastula. Protozoans are pure surface. The Coelenterate phylum includes spheroids of dilated ectoderm around semi-amorphous endodermal swellings. These are the diploblastic (two-layered) jellyfish, sea anemones, corals, and hydroids. Their ectoderm is lined with nerve fibers and their inside is a gastrovascular cavity blind at one end. Between the exterior cytoplasm and the endodermal lining of the gut is a gel layer (called the mesogloea) made up mostly of water and inorganic salts. Undifferentiated cells lying close to the mesogloea give rise to amoeboid cnidoblasts whose Golgi bodies become stinging capsules, harpoons laden with protein poisons. These nematocysts discharge barbed threads with unfolding pleats which hone in on prey at high speed and pull it in.

To a certain degree, the jellyfish are enormous endodermalized protozoa, thickened along only one axis. Like one-celled animals they digest their food in the vacuoles of their cells, though some enzymatic breakdown of large morsels occurs in their interior cavity. Epidermis and gastrodermis absorb oxygen directly and diffuse it to their tissues, excreting carbon dioxide and nitrogenous wastes. Certain of the more complex jellyfish of this phylum have branching guts with canals and pockets, but no anal opening; amoeboid cells carry some of the larger particles of indigestible material out of the mouth.

The diversity of creatures at this level represents millions of years of adaptive radiation of a basic animal type without further complexification. Jellyfish and their kin are polymorphic gastrulas, with polyps that grow from the seafloor like plants and swimming medusae that bud from these strands like seeds from flowers. Some species grow connective tissue that binds different zooids in a superorganism like the Portuguese man-of-war, which includes swimming bells that spend their existences propelling the colony through the water, gonad buds producing medusae, and individuals that become oily bags and fill with gases secreted by ectoderm. These bright blue floats are exposed to the winds, and the colonies are blown along like sailboats.

While ancestors of such creatures explored the exotic varieties of diploblastic organization, expanding into available niches in the ancient seas, others were affected by mutations and continued to "gastrulate," to add on tissues, and their descendants became worms, fishes, crustaceans, insects, starfish, and terrestrial vertebrates.

Worms are a basic expression of the gastrula stage. They are so abundant and diverse because they embody a distinct level of morphology, like living "primitive streaks" that have adapted over aeons to exquisitely discrete niches. Large numbers of worms in all phlya have become parasitic and have lost the ability to move about or even feed. Their guts as well as their limbs wither, and their lives are imbedded in the life cycles of various mollusks, fishes, birds, and mammals. The original predisposition of cells to specialize and merge crosses the boundaries of discrete creatures through both coembryology and coevolution.

The planula, a larval coelenterate, is closely related to the free-swimming larvae of all worms, hence all multilayered animals. In ancient times some of these ciliated embryos (or equivalent forms) adopted seafloor crawling and matured sexually in a foetal state. Subsequent differentiation established their separate tissue surfaces, including contractile layers for propulsion. The larvae of modern parasitic forms all begin as such free-swimming embryos before their organs degenerate—a vestige of their historical state and an indication of shared phylogeny.

Endoderm is interior; by definition it must have developed in polarity to a surface, deepening into muscles and gut. The mesoderm is an intermediate supporting layer between ectoderm and nerves on the outside and deep interiorized organs. Flatworms are triploblastic (three-layered) but lack a true coelom; mesodermal tissue arises between the epithelium and the endoderm of the gut lining. The opening in the gut is embryologically a mouth, but it must serve also as the anus. Simple folding and tubing of tissue has produced muscle protrusions in the pharyngeal region. The flatworm moves as peristaltic waves cross its body in loose coordination with the contractions of mesodermal tissue. The internal flow of liquid is controlled by clumps of flagella (flame cells) which arise along with penis, ovary, and excretory organs from fleshy bulbs in the mesoderm. Some anterior concentrations of cells suggest cephalization, but there is nothing like a real head or seat of identity.

In most phyletic lines, growing cell colonies can reach only a certain size and complexity without exponential increase in volume and layering of internal surfaces and tissues. In one exception, ribbon worms may grow to two meters in length without gut walls, muscles, or segments simply by adding spirally crisscrossing fibers around a tubular gut, but these tidal animals are an evolutionary dead end on this planet. In truly complex living systems, internal organs must be supported by folds or mesenteries in the mesoderm; diverse texture occurs as a polarization away from surface in multiple dimensions. The segmented worms, the mollusks, the echinoderms, the insects, though all simple creatures, represent fully gastrulated three-layered stages with central cavities and intricately devised organs, including ectodermalized sensoria, nephridia (for removing nitrogenous waste), and mesodermal muscles and vessels (for movement and circulation).

Cells are basically single-dimensional forms; when they fission and stick together they are still discrete points. However, when layers of cells move as units and undulate, they have become tissues and are bidimensional, we no longer see the single "cell points." As this tissue assembles itself around cavities in the brain, the gut, the eyes, and the heart, a creature of three dimensions emerges. A thicker animal can exert greater force per unit volume of its body. Internal three-dimensionality is a prerequisite for entering external three-dimensionality. Protozoans, rotifers, and even ribbon worms and flatworms are essentially "lens" animals—cellular and linear. (Of course, cells and tissues are topologically three-dimensional too, but the three dimensions of a cell are confined functionally to one of our dimensions, and the three dimensions of tissue operate in flat vectors in our bodies.)

It is impossible in all but the most metaphorical sense to assign a jellyfish or flatworm stage to the embryos of other phyla. There may be remnants of such a stage in the genes and tissues, but the physiology of a three-layered gastrula is already more complex and less specialized than a jellyfish, and it has no potential for anything as exotic as a stinging polyp. If ontogeny were purely linear, discrete animals might be distinguishable as phases, but complexity is never a matter of terminal addition; once it is implicit in a system it is integrated backward and forward through every stage of development.

Gastrulation is the most radical of all metamorphoses. It rearranges whole groups of cells and turns structures inside out within other structures being turned inside out. So many histories are reenacted and synchronized that they disappear into each other and it seems as though the gastrula could accomplish any animate shape if given enough time. Phylogenetically, this is what has happened; repeated gastrulations of prior gastrulas have derived worms from jellyfish, snails from worms, and monkeys from lemurs. Of course, these precise animals did not proceed from each other, but gastrulas at their approximate stages of tissue configuration and layering gave rise to one another's topologies.

There may be other paths to the same point, but by now we are stuck with our one method of getting our guts inside our skin and our organs in functional relation. Although we do not recognize them, our most deeply held convictions of shape and propriety originate in gastrulation. "Inside" and "outside" are fundamental notions—not only inside and outside our bodies but inside our minds, in our memory, and within our lives. We awaken to find ourselves in bodies in families in cultures in history in time and space. Our senses and experiences continue to flow inside us, to be internalized. Symbolically and mechanically we live among enveloping structures. We erect hive-like apartment complexes and take elevators up through the midlines of buildings to rooms where people live and work, and where information is filed. We ride in cars and trains and are packed into stretched metal and jetted from one city down onto the membrane of another. We watch flattened filmstrips through which light is projected in animate replicas of three-dimensional shapes.

Our whole science has become an invention of "degastrulation," not only of the bodies of living creatures but of their minds and behavior patterns, and of the field of stars and matter wrapped like a membrane in waves of gravitation and light around our world. The surgery carried out by physicians from primitive times has been a scant and surficial unravelling of a concrete solidification of organs formed by waves of introverting cells. We now plan to trace "the entire skein of causality back to the origin of the universe."[1] Our conviction has always been that only by "undoing" what was done to us can we know who or what we are.

Palaeolithic peoples also tried to get inside sacred space. Their star maps were embroidered with spirits, and their caves painted with wild animals which danced in firelight. The later plank houses and rowboats of Northwest American coastal peoples were designed as introverted bodies of totem whales and ravens, the masks their sorcerers wore in ceremonies likewise an experience of self beyond birth.

We have managed to get outside the Earth, first in our imagination of the three-dimensionality of the heavens and then in our design of engines and fuels and sealed vessels on candles to put our body/minds there. Incarnation never was abstract, and the fearful screech of metals against gravity is as necessary as the bloody walls of tissue. The astronaut Joseph Allen described floating outside and then returning to this colossal membranous world:

"You know the Earth is round because you see the roundness and then you realize there's another dimension because you see layers of things as you look down. You see clouds towering up and you see their shadows on sunlit plains and you see a ship's wake in the Indian Ocean and brushfires in Africa and a lightning storm walking its way 1000 miles across Australia. It's like a stereoscopic view of all of nature. . . ."

(And then reentry):

"All of a sudden you begin to hear the sound. Until now, Columbia has been a very silent ship, but now there is a roar or a rush that builds and builds and builds. It's the rush of air.

"The next thing you're aware of is a color on the windows that starts out with just a faint tint of rose red that gets brighter and brighter, then changes to a whiter red, then an orange pink and ultimately a white that flickers around the windows and is the fiery heat of reentry. It's like being inside a neon light bulb."[2]

In viewing gastrulation from the outside we are watching time and scale being created out of themselves—the same time and scale we project from within gastrulation into the multidimensional universe. Mathematics emerges embryonically from fields of cells shifting in relation to one another. The wall of flesh becomes wrapped around the universe, and its interstices become lines of time and space, stretching once across seas and mountain ranges to new lands and, now, in the mind's eye, across galaxies into cosmogenesis. The way in which we were incarnated remains our sidereal limitation. In the gastrula we are seeing this original matter moving out of the universal into the temporal—the lotus mind embodying itself in soundless waves.

"There are some insects," writes the novelist Janet Frame, "that carry a bulge of seed outside their body as the intelligence of the universe carries its planets and stars. A spider has its milky house strung *fragilely* between two stalks of grass; and so God has pitched his worlds; and we who are replicas and live in the house of replicas cannot exist until we have shaped what we have discovered within the manifold; and know in the repeated shaping that we are not Gods, and not avoid knowing that we ourselves have been shaped and patterned not by a shadow of light or a twin intelligence but an original, the sum of all equals and unequals and cubes and squares. . . .''[3]

We may experience our formation again as dreams of enormous masses of matter wedged right up against us, passages which go out interminably and change the size and scale of our organs as we follow them internally. We feel ourselves being twisted and rotated in sleep, but we cannot see the background of the motion. It is as though we are lying in a wall and the wall is moving. In other dreams we may burrow through long passageways and suddenly emerge in a hollow field of stars, or find ourselves climbing through tiers of broken planks beneath a house, unable to reach the bottom or make our way out. In one archetypal dream we stand beneath the entire starry heavens, and the sky slips—not by much, just by a fraction of an inch. The lynchpin is surely close at hand.

The darkness within emerges physiologically in gastrulation, for clearly the unconscious of thought and dreams is also the unconscious of digestion and respiration. The rolling folds of tissue create all the organs from the same cloth and in one synchrony. Margins plunge into deep tissue, free edges seal, and a lining forms over them, spheres twist out and over spheres as they cover them: these are the rudiments and the boundaries of mind—not of any one thought but of the continuum of mental activity. The mind reflects the gastrulated field; the canyons and gullies of thought arise in a whirlpool of heart, intestines, lungs, and brain, even before baby has a self or identity, or brain has a central nervous system. Thought goes out from there to encompass the universe, or at least that is what we think it does. Somewhere in the bloody tissue of the womb lie the raw id, the unconnected asterisms of personality, the sparks of sexuality and ego awareness, the moist body of psyche, and even the opaquity of the collective unconsciousness. They incarnate through the organs.

Gastrulation shows us the way in which we put ourselves into our bodies. They are not rigid things with fixed structures; they are active reorganizing layers. If mind is there, love must be there too, and anger likewise must begin at the beginning. If not there, then where? These folds and introversions of tissue must be the grounds of our experience, our girth and pulsation. Our excitement and our desire form in the same way as our gut cavity formed. Sadness and joy, as well as rage, have an original shape, divorced from any object or goal.

Layers of cells are absorbed and translated into new structures, regions of tissue fuse into other tissue: so images arise, change, and disappear, and living animal flows into living animal. And this is where "all of the days of our lives" have gone, as so many folk songs still tell us. We can't lose any of it, but we can't save any of it either.

11. Morphogenesis

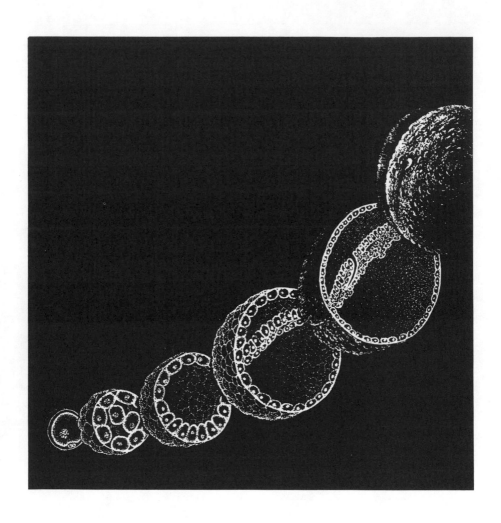

Development remains a mystery. Experiments have shown us that cells are individual creatures compelled to specialization by nuclei; tissues are aggregates of these cells influencing one another to become organs. Embryogenesis happens spontaneously trillions of times a day throughout the biosphere. The seemingly innate intelligence of cells and tissues is the collective effect of simple chemical and mechanical events regulating and regulated by each other in hierarchies established in the evolution of membrane-enclosed systems of energy. Intricate three-dimensional matrices (plants and animals) arise not because protoplasm knows how to behave but because it cannot escape the homeostasis of the self-made fields of membranes.

When we stand in a rainstorm the droplets are driven against us by the wind. If we take a plane above the storm we realize that it is actually a coherent layer of clouds. Astronauts see the bank of clouds as part of a larger system in which clouds arise in continuous cyclonic patterns from shifting centers. Scale is crucial in a way that defies simple explanation. When we look at cells through a microscope we understand that they are discrete animalcules working like ants to build hereditary structures. But we do not live among cells; we cannot communicate with them, we cannot relate our existence to theirs on a day-to-day basis, we cannot even see them without elaborate artifice. We have established a kind of diplomatic relationship instead, a scientific neutral zone in which we pretend that they and we are entities in the same phenomenological realm.

We see wholeness because we stand far enough away; the resonant waves of material are invisible. The pink and purple plumes of the orchid and the speckled skin of the butterfly and fish are the same to us as the banded Cyclops face of Jupiter—patterned chaos.

The lines on the tiger's face are etched there beyond correction and beyond error. Every mark is ingrained by billions of cells sustained in a symmetry within another symmetry. Every thorn and hair on the beetle's leg as well as the flecks in its eyes, the signals from its ganglia, the junctions of its shell requires the same unmonitored care.

135

They are hewn, with a rugged specificity that reveals intelligence even as it denies it.

To William Blake too, in the "fearful symmetry" of "The Tyger":

"What the hammer? what the chain?/In what furnace was thy brain?"

And then the question that will plague us till the end of time:

"Did he who made the Lamb make thee?"[1]

"What is development?" write the authors in the opening of a contemporary textbook on embryology.

"No one has completely defined it, any more than the organism itself has been fully defined. In all cases a cell or a group of cells becomes separate from the organism as a whole, either physically or physiologically, and progressively becomes a new complete organism or a new part thereof. A fern spore settles and develops into a fern gametophyte. An insect egg may become a caterpillar which transforms into a pupa and emerges as a butterfly. The stump of a salamander leg regenerates a new limb. A microscopic cell, the human egg, proliferates and develops into a giant creature able to contemplate its own nature and origin."[2]

Another embryologist waited patiently for his child to become old enough to understand how animals are formed from tiny eggs. He had taught this scientifically rigorous sequence to graduate students for years, and no one had ever challenged its essential adequacy. When his son was six years old the father decided to see if he was ready. First he showed him pictures of cells gathered in different planes as tissues, then cross-sections of the tissues assembled in organs. But the perplexed child asked: "Daddy, how do the cells know which of them are going to form those organs?"[3]

What could he say? Initial questions have not been answered without evasion while we have moved on to incidental matters. It takes a child to see that the "emperor" is naked.

The same physical laws that mold stars and planets and shift the tides are the basis of all cell movements. Gravity has its finger in every chink and cranny. Originally, cells and groups of cells conserved their free energy by presenting the least possible surface for their volumes—a sphere. This is how the egg first manifests embryogenically, and, despite its cleavage divisions, the blastula remains a sphere, a sphere made up of spheres. Even cells artificially separated from distant regions of a gastrula will come together in a ball. If there were no other internal or external influences, life forms would remain as round as most stars and planets. But spiral energies and longitudinalizing forces have been present from the beginning. Even the tilt of the Earth on its axis and the irregular phases of the Moon provided a subtle series of oblique axes for living tissue to orient around. Waves of energy were displaced and fragmented in all directions, and, right from the beginning, cells struggled to keep in contact with the environment from which they were metabolized and fed, to prevent themselves from getting trapped in solid cores.

D'Arcy Thompson derived many of the fundamental shapes of living things simply by applying the rules of Cartesian transformation. In his book *On Growth and Form,* he showed that both animate and inanimate configurations arise from the relationships between surfaces, contours, and volumes. Living fields begin in the electromagnetic and gravitational fields that generate atoms and stars. As fields of different size interact with and distort each other, the forces behind symmetry, polarity, and gradient rearrange and order them.

The internal dynamics of cells and tissues are modified by their simultaneous resistance to and movement through currents, so ancient polyps and crustacea, and even the prototype fishes, are compromises between the organization of protoplasm and the tension in which their bodies form. The gullet of the minute ciliate *Stentor* arises, according to Thompson, from a combination of membrane tension and its own cilia beating currents of water around it, so its "curved contour seems to enter, re-enter, and disappear within the substance of the body . . . bounding a deep and twisted shape or passage which merges with the fluid contents and vanishes within the cell."[4] The ripples, density, and gravity of water and the simultaneous resistance to them shape the umbrellas of jellyfish, the arms of starfish, and the fins of fishes. Insofar as our long sinewy bodies are derived through the phylogeny of aquatic creatures, the impact of ancient waves is retained in our shape (though we have tilted the lines upward against gravity and now must resist it along the deformations of a spine of vertebrae).

At the center of large creatures lie the nuclei of small creatures. Each time we swallow we feel an ancient worm without knowing it.

The same mathematical properties are embodied in phenomena of very different origin and duration. The gullet of the *Stentor* could also be a formation of spits and tombolos in the crosscurrents around an island, or hydrogen swirling into a star. A flower is a slow splash sup-

ported in membranes. Falling droplets from spouts recur in hydro-
zoan polyps in lakes and seas. Similarly, tatters of gas coalesce as
planets even as bits of protoplasm are drawn into the nucleus of the
cell.

Muscular forces may sustain the kink in the chameleon's tail and
generate the intrinsic spiral paths of insects proceeding toward a
light, but the helical configurations of separate florets within a sunflower and the winding course of a snail's shell are maintained by
intricate composite forces working over time. "The growth of organisms is more complex," wrote the early twentieth-century
embryologist Ross Harrison, "since they grow internally and are
made up of many different components with different rates of increase."[5] That a spiral geometry is an attractive resolution for a
variety of interrelating currents is shown by its ubiquity in the
"transitory spirals in a lock of hair, in a staple of wool, in the coil
of an elephant's trunk, in the 'circling spires' of a snake, in the
coils of a cuttle-fish's arm, or of a monkey's or chameleon's
tail."[6]

Layers of tissue achieve their complexity and functional integration through continuous interpenetration and coalescence. When cells
recognize their kinship they unite in epithelia, integrated layers of tissue. At the same time, they move away from other epithelia, and
crevices form between them. Such polarized motion lies at the heart
of gastrulation: the neural plate and mesoderm sink inward together,
away from epidermal and endodermal cells with which they were previously fused.

Although the morphogenesis of epithelia is extremely complicated, there are only a few basic mechanical possibilities and these
are propagated in endlessly novel contexts within living organisms.
Tissue essentially moves where there is least resistance or where it is
drawn by other tissue. Layers of cells can thicken, thin, separate, fold
up or down, or break apart. As these movements follow one another
in intricate juxtaposition, organs are formed.

Sections of tissue become folded and refolded in the construction
of quite separate organs. In the cerebral cortex of the brain and the
duodenal and intestinal loops of the alimentary canal, the pure folds
take on distinct meanings and functions in terms of their relationships
to adjacent tissues and the unity of the organism. The ears and pharynx are both twisted labyrinths. The lungs and kidneys emanate from
continuous branching. Heart, ovary, and tear ducts all originate as
hollowed-out regions. There is no biomechanical rule for the forma-

tion of homologous organs. Within the same phylogenetic lineage an ear may either be a pocket in an evagination or a solid mass of cells subsequently hollowed. The human archenteron is a cavitated morula; the salamander archenteron is an invaginated cavity. The embryological translation of the genetic message can change radically without loss of function.

It would seem that something so fragile, at the beginning of such a long and delicate assembly, could be easily disrupted, but the zygote is remarkaby adaptable. Embryologists have centrifuged fertilized eggs, removed sections, pricked them with needles to make multiple activation points, and poisoned them with extraneous chemicals. In a large number of cases the embryos reestablish their symmetries and polarities and develop normally. It might take four waves of tissue movement to accomplish what one fashioned initially, each wave "correcting" a single factor, but the primary theme prevails, riding out, as it were, the distortions without losing the melody. In some cases the foetus absorbs the disruption, becoming an anomalous creature—a freak—until the effects are lethal. A tadpole embryo centrifuged perpendicular and then parallel to the animal-vegetal axis forms as Siamese twins joined ventrally at the gut. They have two mouths and two throats but only one midgut and anus.[7]

A developmental process which has undergone so much violence in the currents and storms of the planet has obviously survived by incorporating multiple pathways to the same configurations. Embryogenesis is a cumulative interdependent progression, so a chain of amino acids may translate to the phenotype in a variety of different ways. A mutation affecting one gene can generate a ripple of changes throughout the organism as proteins from other genes will behave differently in order to maintain the functional unity of the biological field. Mutations thus combine existing elements in unforeseeable ways. The webbed feet of the duck, the talons of the eagle, and the pig's hooves are subtle variations of similar fixed pathways that come to express utterly different personalities and life-styles. There is much latent potential for new species in any organism.

Mutations may be random, but the proteins are highly structured. Because the genes continue to recode only in terms of a closed amino-acid field, old patterns inevitably return at new levels of structure. The ciliated tentacles of the simple entoprocts so suggest the bryozoan lophophores that they have been jointly named "moss animals," despite the fact that the former is more allied to rotifers and the latter to clams and snails. But disintegration and recurrence are the morphological repertoire of phylogeny. The worm is deformed into a mayfly, its sister into a lobster. Fish become frogs and turtles, and these new creatures can breathe the straight atmosphere of the planet with their protein deformities. Even the duck-billed platypus is an organized whole, not a collection of separate intentions (a reality sometimes missed by the nineteenth-century naturalists who discovered it); its unity is expressed in its single personality and life plan.

Intelligence arises because the protein field has the capacity for sorting and storing information and, under conditions of bulbous inductions of the ectoderm, can integrate the excess cells into a new harmonious activity. Not all creatures survive such deformation; in fact, few do. Not only must the new morphology be able to inhabit its tissue in the world, but the creature must be able to grasp its wholeness and translate it into meaningful actions. If a being can use the increased internal light of its cells to find a niche on the planet, it will make it possible for myriad of its kind to arise, and for other diverse branches to sprout from them.

The thickening of epithelia

A regional surge in mitosis (or a migration of cells to a single site) provides material for the formation of new organs. The neural plate appears first as a thickening of ectodermal tissue toward the mid-dorsal region of the gastrula. The brain begins as a series of thickenings at the front of the neural plate. The gonads of the vertebrates arise as carbuncles on the mesodermal tissue lining the back of the body cavity. As this germinal ridge extends into the coelom it is squeezed laterally so as to become suspended and brought into contact with the excretory organs.

The lungs and pancreas originate in regional thickenings of endodermal epithelia which continue to branch out and fuse in individual patterns, in some places forming secondary ridges. The heart rudiment forms when free edges of the mesodermal mantle come together in an area outside the mesoderm and consolidate, creating an endocardial tube with the heart cavity inside. Hair, feathers, scales, and glands all start as little round pegs, raised placodes. If they continue to buckle out, scales and feathers emerge, but they can also invaginate to produce glands and teeth. Thickening of the epidermal cells generates hair. In fishes longitudinal thickening of the epidermis becomes fins.

The thinning of epithelia

Cells can also simply degenerate. Regional cell death causes the separation of bones at the joints (such as between the radius and ulna) and apparently occurs between digits of feet; among ducks and other water birds a webbing is left behind. Interestingly, once cells have received the signal to die they will deteriorate even if the tissue they are part of is transplanted to an expanding region.

Conventionally, epithelia thin out when the number of layers of cells in them is reduced; for instance, during epiboly the epidermis expands and covers the outside of the body; cell fission simply does not keep pace with the extension of the layer's territory. As a vertebrate elongates from head to tail, its epidermis continues to stretch, though it will eventually fold and thicken in spots as it forms organ rudiments in combination with underlying tissues. In fishes and amphibians the endoderm at the front of the alimentary canal is dilated. This region later becomes the epithelial lining of the pharynx and propagates a series of pouches just behind the mandibular arches, the rudiments of gills, which are converted to glands in the land vertebrates.

Epithelial layers become separated

As groups of cells lose contact with one another (no longer recognizing each other as part of the same layer), gaps develop between their tissue; for instance, the coelomic cavity between the parietal and visceral layers of mesoderm. Many of these crevices are subsequently filled with secretion.

The somites of the mesoderm are also formed by loss of contact between groups of cells. As a series of gaps develops perpendicular to the surface of the epithelium any individual cell must gravitate to one clump or another.

Mesenchyme

Epithelia may also become uncoordinated and break apart; this process simultaneously opens cavities and provides critical raw material—free cells for a new generation of tissue. For instance, the neural crest cells individually break from the edge of the neural folds and migrate throughout the body participating in a wide variety of structures; at the same time they leave a "necessary" gap between the epidermis and the neural tube which is in the process of fusing.

Cells arising from the break-up of epithelial layers are called mesenchyme and, unlike epithelial cells, travel on their own outside any homogeneous layer. The tissue from which they arise is usually left intact but is somewhat depleted of cells. In many species invagination of the endoderm and mesoderm occurs through the systematic disassembly of layers of tissue: the cells of the primitive streak of the chick slip away from the surface one by one. Because they do not adhere to one another they cannot form a smooth-wall cavity, and that is why the archenteron must be hollowed out secondarily.

Mesenchyme cells often accumulate in a colloidal substratum, as they are unable to move through fluid. If the fibers of the substratum are disperse, the cells are likewise scattered, but if the fibers are organized, the loose cells are led to a specific site, oriented by epithelia, and held in place. Cell movement is induced by solid surfaces. In the early part of the twentieth century, Ross Harrison learned this while studying protoplasmic activity at the ends of nerve fibers grown on spiderweb frames. These fibers required surfaces, so migrated along paths created by solids (even along the glass covers of laboratory dishes).

Although mesenchyme movements are amoeboid in appearance they do not use the internal streaming of protoplasm to fill a pseudopod. The ruffled cell edge makes contact with a surface and pulls the body along, then relaxes. In orienting toward the newly formed interior of the body, cells of the primitive streak send out superficial pseudopods which are withdrawn if they contact blastocoele fluid. New edges are sent until they locate the cavity.

Mesenchyme is a sort of cosmological substance, like the dispersed gas of first- and second-generation sun-stars; it is continuously supplying raw tissue for critical vessels, bones, and nerves throughout the body. Cartilage begins as mesenchyme on the inner wall of somites and later captures some of the free-moving neural crest cells. The rudiments of blood appear first as angioblasts, mesenchymal aggregations in the mesoderm. The limb-buds of the vertebrates originate in dense packs of mesenchyme; others of these renegade cells participate in the sense organs, the sex cords, the neural arches, and the heart. Not only does the primordial embryo provide the shape for the organism, its debris is the unhewn material from which continuously more elaborate bodies are integrated.

As we have discussed earlier, many eighteenth-century biologists assumed that the characteristics of body parts arose from the elemental constituents of the different tissues. The advent of genetics drew the attention of mechanists from abstract elemental determinism to individual cellular potentials. However, when severed blastomeres developed as whole organisms and dorsal skin cells became brain cells after being transplanted from one frog embryo to the neural plate of another, hopes for a simple linear geometry of development were disappointed. The determinant was located neither in the body parts (as eighteenth-century mechanists had thought) nor in the blueprint from which they were assembled (as late nineteenth-century geneticists had intended to demonstrate).

A living creature is coordinated in a single cell of its species. Initially, in the blastula, each cell is identical and only loosely joined to its neighbors. Each blastomere is a prism reflecting the whole organism.

The discrete characteristics of differentiating cells can originate only in their nuclei, transmitted to the cytoplasm by RNA molecules. The genes are the "script" for morphogenesis. However, the regional differences in the cytoplasm of the fertilized egg alone "tell" the nuclei which genes to activate. Thus we have a paradox of a classical embryological persuasion: the cells are genetically under the control of their nuclei (and have no other source of specialized information), but the nuclei are regulated by substances in the cytoplasm (which should not become regionalized without the nuclei). Apparently, differentiation is determined by both the genes and the cytoplasm in such a manner that one requires the other to express itself in the overall design.

Embryologists now speak of selective gene activation and inhibition as the mechanism behind all cell differentiation. Specialization occurs because either only some genes in each differentiating cell are copied by the RNA, or, after transcription, some of the information is destroyed in the nucleoplasm and thus does not affect protein synthesis. There is differential transcription of genes or differential passage of RNA into the cytoplasm. As small variations continue along gradients groups of cells deviate from one another and begin to behave idiosyncratically; the regional heterogeneity then initiates a new cycle of differential gene activity.

The internal chemistry (thus the behavior) of cells changes as sequences of proteins, including enzymes, hormones, and structural materials, are synthesized. For instance, the pancreas becomes a discrete organ and participates in carbohydrate metabolism as its cells differentiate to assemble insulin molecules. Some cells become cartilaginous as their nuclei impel them to manufacture collagen. Muscles form when the large protein myosin (requiring fifty to sixty unit

polyribosomes) is produced by an alliance of cells; these polyribosomes gather in filaments approximately 200 molecules long to knit
striated tissue. When the genetic message calls for red blood some
cells must suppress their own nuclei, attract iron, and assemble
oxygen-bonding hemoglobin molecules. When the code requires
tracks of neurons, large numbers of cells develop the electrochemical
properties of their membranes and extend them grotesquely in fibers.
In the region of the retina some cells fold their plasma membranes
and become rods. Where pigment is specified, cells specialize in the
production of appropriate enzymes like melanin. Other cells must secrete large amounts of fluid, and their Golgi complexes swell and
multiply while their endoplasm is roughened.

Cells also change their shapes through the production or rearrangement of organelles (such as cilia, microtubule, microfilaments), or
through the accentuation of some other aspect of anatomy, for instance the expansion of the vacuoles in cells clustering together to
make the notochord. In cells not involved in morphogenesis, there
are very few microfilaments; the microtubules crisscross at random.
However, the microtubules can become ionized into parallel sheaths;
then the affected cell must elongate. This is a crucial factor in the separation of epithelial layers, for when clusters of cells form bottle-
heads they must contract inward homogeneously and the outer edge
of their layer collapses. Since the extenuated ends of the cells adhere
to one another, the contraction is then transmitted across the layer.
This happens dramatically during gastrulation when the epithelium of
the marginal zone invaginates along the blastopore. The surface is
pulled deep into the cytoplasm in a smooth concave pocket which becomes the archenteron.

The sheer unremitting power of the microfilaments is realized
in the high-speed beats of the wings of insects as these organelles expand and contract in alternation. In morphogenesis they
commonly tighten like purse strings and contract or narrow the
cells of which they are part. The microfilaments are a major element in neurulation, squeezing out the necks of prospective neurons while intervening cells flatten into epidermis. In the formation
of glands microfilaments pull the inner open surfaces of cells together so that a cone shape imbeds in a cavity, for instance, the
pituitary in the stomodeum.

Cells are aware of one another's existences and define their own
natures by the properties and locations of the cells around them. As
they mingle in their closely packed world they read their neighbors'
shapes, charges, and the chemistry of their surfaces. Molecules pro-

truding from the surfaces of adjacent cells provide sites (antigens) for bonding. Each family of cells apparently produces one kind of surface protein and its antibody, so cells bond at two points (each cell being "bipolar") and then deepen their connections with gap junctions. The electrical aspects of cells also enable them to recognize one another and stick together after incidental contact (the divalent cations calcium and magnesium establish the underlying charge). But cells are not planets; they do not attract one another from a distance; they respond only to direct touch (and if their coats are removed experimentally they will not aggregate).

Tissue structure is always regulated by surrounding hierarchies—the patterns of other tissue and various extracellular secretions, including mineralized three-dimensional matrices in feathers and bone. When cells from different layers of tissue are combined artificially in a laboratory, those from the same layer find each other and adhere. Ectodermal cells gather on the surface of the culture, and endodermal cells retreat to the interior. Mesodermal cells meanwhile form a layer between them and begin organizing themselves around coelomic cavities. Whatever disorder they find themselves in (and experimenters make their plights as exotic as possible), they will seek their counterparts and attempt a gastrulation. If true introversion is impossible they will form a series of layers resembling the rudiments of organs. Biologists sometimes speak of "the order of Nature," as if there were a persona seeking the most elegant arrangement of living substance and embodying creatures with the raw tools and instincts for survival. But, in the realm of pure science, no such intelligence exists. Cells do not have motives; they are not part of the mind of "Mother Nature." They react only to different degrees of free energy and adhesiveness as their surfaces collide, and they are coordinated by hormones released from other cells.

The affinity between the cells of the layers can transcend even species. If the cells of a mouse and a chicken at the same stages of development are centrifuged together, they will fuse according to their origin in ectoderm, endoderm, and mesoderm rather than according to parent. Chicken and mouse disappear altogether, and there arise strange chimeric organs that look like parts of bones and dissociated tubules and abortive kidneys and hearts. In a breakdown of the whole natural order cells return to the ancient cosmogonic layers in which they began.

The known properties of molecules and cells never seem quite sufficient to explain the elaborate synchronization of the organism, so scientists have (at different times) suggested morphogenetic molecules, biological clocks, and electrical depolarization waves, but there is no evidence of another hereditary system or a timing device in the cell, and electrical waves would accomplish in seconds what takes cells hours and even days.

George Oster, an entomologist at the University of California in Berkeley, offered a curiously nonchemical explanation for morphogenesis during the early 1980s. Combining elements of nineteenth-century mechanics with computer technology he derived the effects of fertilization, gastrulation, and other embryogenic motions from models of cell interaction within the high-viscosity cytoplasm. Tension, compression, and sheer force, transmitted gradually from cell to cell in the form of waves of motion and globally coordinated over the embryo, generated the same structures in computer projections as fertilization and gastrulation in actual embryos. In the thick turgid environment of cytoplasm, motion follows geometrically determined paths, stopping only for natural boundaries or from the redistribution of force. The laws of low Reynolds numbers apply, leading to a kind of Archimedean physics in which floating bodies stop instantly when impetus is suspended. The archenteron, mouth, and teeth (in their downward undulations) and the rudiments of the liver, pharyngeal pouches, and pancreas (in their folding upward) are all generated by the "same" wave constricted in different directions. The sperm becomes a winch polarizing the egg's contents back through the cell, the acrosome a moving hydrodynamic boundary encountering cytoplasmic drag.[8]

The objection to this model is that it is a computer version of sterile geometric cells in a "mathematical" rather than a protoplasmic medium. Unconsciously, the computer could have been programmed to our own image of a two-dimensional cartoon of gastrulation, with little relevence to an unknown three-dimensional event in living tissue. However, it is still striking that Oster was able to generate artificial embryos with his video caricatures. In an era of psychotropic drugs, tranquilizers, synthetic vitamins, and varieties of street heroin, a mechanical model is an unexpected relief. Waves of tension (not enzymes) set herds of antelopes galloping and fish torsos squiggling. We can imagine all motion as the contacts of ancient creatures echoing and attenuated over time and space—an Archimedean hologram in which touch alone imparts kinesis. The motions of ancient extinct worms are thus captured in the beating hearts of gulls over the city, hearts that beat only as the blood islands disperse and hemoglobin contacts their cells, motion captured by membranes in the baseball pitcher's arm in the stadium below. If invagination can be propagated by a compression wave across a high-viscosity fluid, then no wonder break dancers and musicians have the rhythm, and capoeristas tumble and spin to the sound of the berimbau.

12. Biological Fields

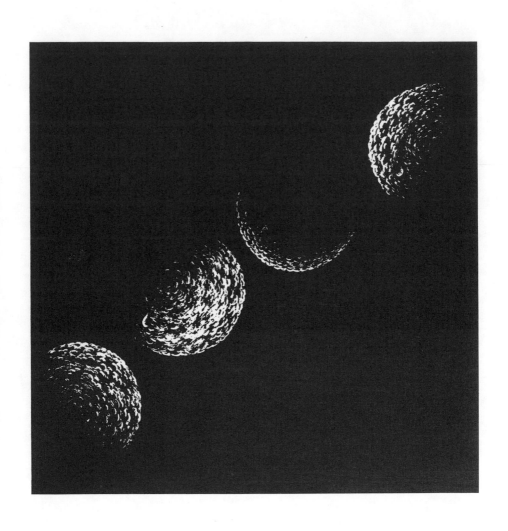

Outside of the exquisitely precise orderings of tissue patterns the genes, and even the cells, mean nothing. Eye cells do not (by themselves) see. Brain cells do not think. These highly specialized entities live typical protozoan lives—they metabolize, they synthesize protein, they divide or die. Neither genes nor cells are farseeing sentinels who make sure that creatures are formed. They produce building block materials, and something else assembles them, or they assemble each other.

Even educated people misunderstand the complexity of development. A popular illusion is that geneticists have mapped the internal structure of the gene and translated the hereditary code so that embryologists merely have to trace the effects of the genes through tissue formation. The double helix and accompanying chromosome maps do not, however, tell us how genetic molecules express themselves in tissue structure, and they certainly do not explain how mutations lead to evolutionary transformations. Genetic space does not translate in any simple or consistent fashion into organismic space.

We may learn by experimental deduction that a specific gene is responsible for blue eyes in a cat. But this does not mean that the gene *colors* the eyes blue. Its only link with the world outside the nucleus is its transcription of itself onto RNA, which is the template for the assemblage of amino acids. Neither nucleic nor amino acids color the eyes. By the time that a chemical substance reflecting blue light has located in the iris the direct expression of the gene has been ''lost'' in the complexity of the system. The genes for all the intricate aspects of the eyes like the genes for the blood and cells of the heart are scattered through every embryonic organ, and their expressions in tissue come together haphazardly and unpredictably.

The sequences of nucleotides contain single bits of genetic information. As these stream into the organism, they are read in terms of existing meanings. Thus, the same piece of genetic information may be interpreted uniquely in different regions and at different times; it may have one expression in the developing neural tissue and another in cartilage. Invaginating mesoderm is similar to invaginating neural plate, but they have different organismic meanings. Arms and legs

151

are regional interpretations of the same basic instructions. If the cross-eyedness, eye color, and shades of fur pigmentation of the cat are linked in one gene, this would be a lucid case of the capability of multiple expression that exists in all genes.

The experimental embryologist Paul Weiss felt that many of his colleagues overused terms like "information," "control," and "regulation" when describing living systems. He preferred paradigms based on "networks," "fabrics," and "fields." At the age of seventy-one in 1969, he wrote: "We encounter here the phenomenon of emergence of singularities in a dynamic system—unique points or planes—comparable, for instance, to nodal points in a vibrating string."[1]

We might compare a gene with a flashlight beam which is most discretely identifiable closest to its source and disperses from there over an increasing area through time and space. As new genes are activated, areas of tissue expressing the interactions of previously synthesized proteins are changing their relative positions, and groups of cells once remote from each other are brought into proximity. The meanings of cells within the embryo are constantly changing. Although morphogenesis is defined by precedent, neither the genes nor the cells act from habit; they carry out each embryogenesis as if it were the first, and the so-called programmed molecules discover one another anew each time. Sherrington describes the formation of the eye:

". . . the whole structure, with its prescience and all its efficiency, is produced by and out of specks of granular slime arranging themselves as of their own accord in sheets and layers, and acting seemingly on an agreed plan . . . two eyeballs built and finished to one standard so that the mind can read their two pictures together as one. . . . That done, and their organ complete, they abide by what they have accomplished. They lapse into relative quietude and change no more."[2]

Collectively, the genes are like musicians in an orchestra. Each of them can play only one instrument, and it is the collection of instruments that defines the "meaning" of a particular clarinet or piano note at any given time, a meaning which changes as the symphony progresses. The musicians in the human half of the metaphor can at least hear and appreciate their playing, but the "genetic" musicians are deaf and blind and do not even know thay are playing. Some genes play only once, some twice, some many times, some on and off, and some continuously. This is how the fly and the albatross are composed. From the first instant following fertilization until the death of the organism, messenger RNA is selectively led into the polyribosomes for translation, but there is no conductor.

Since the 1920s, embryological experiments have most significantly revealed the role of prior tissue configurations in the formation of subsequent ones. The embryo "creates itself" by moving from one field state to another in precise developmental order. As early as 1921, Hans Spemann grafted small sections of epidermis from a newly formed amphibian gastrula into the neural region of one more advanced and found that they differentiated according to their surroundings—as neural plate. Other sections of transplanted ectoderm developed into normal parts of the notochord, the kidney tubules, and the lining of the gut. Subsequently, sections grafted from other tissue layers differentiated according to their new places in the organism.

The ability of cells to change their commitment if their position in the field changes (known as prospective potency) gradually narrows as the organism develops. Neural material will not behave as ectoderm if transplanted at the end of gastrulation; it will sink from the surface and develop a vesicle with thickened walls, a functionless brain and spinal cord adjacent to the integrated one. Competence is the pliancy of early embryogenesis; as the body forms, its underlying field rigidifies. Medical surgery can never replace the smooth original molding of a figure, which, by law, happens only once—and then more like thought than like substance.

In 1924, Hilde Mangold, a student of Spemann's, transplanted the dorsal lip of a blastopore of an early newt gastrula onto the lateral lip of a blastopore of an equivalent newt gastrula of a different species. The graft invaginated and ultimately developed a whole second set of organs (notochord, ear rudiments, kidney tubules, gut lumen, etc., missing only the anterior section of the head). Some of the organs were purely from the host newt, others from the grafted newt, and still others had mixed cells. Although development proceeded no further, the central role of the dorsal lip of the blastopore was demonstrated. Its ability to spawn a virtually complete second organism when transplanted led Spemann to call it "the primary organizer." Its forerunner was eventually identified as the gray crescent of the oocyte.

Experiments with newt eggs allowed embryologists to trace the sequence of primary organization. During oogenesis differential chemical properties at the animal and vegetal poles are expressed in regional separations of charged molecules, a gradient of cytoplasm and yolk made visible in a gray crescent. During gastrulation this gradient becomes the dorsal lip of the blastopore and sinks to form the roof of the archenteron; there it establishes the primary axis of the embryo, inducing a series of organs from the head to the tailbuds. In birds, the anterior zone of the primitive streak is the primary organizer, and it is induced from the underlying hypoblast (a pattern likely repeated in mammals).

Spemann and his successors observed that the center of organization originates as a singular global feature which is then distributed with gastrulation until it spawns a series of regional centers, each one inducing a hierarchy of major systems with their organs and subtle functions. In gastrulation, the first material to invaginate induces head organs which in turn induce brain ventricles, eye and nose rudiments, and ear vesicles. The last material to enter becomes the trunk and induces posterior organs. Spemann tried various combinations of transplants in an unsuccessful attempt to pinpoint precise causative factors. He learned that head organizer grafted to the head region of another embryo forms a second head with eyes and ears, but the same material grafted onto the caudal region of another embryo generates a whole second embryo (as does the cranial grafting of tail organizer).

Essentially, the embryological process works by inducing other tissues out of ectoderm and then specific organs from those tissues, often in the context of ectoderm. Uninduced ectoderm will simply produce mucus-secreting epithelial cells. This sequence of course has phylogenetic significance, for animal life began as pure ectoderm (even the vegetation of plants is a kind of skin), and then induced deeper structures from this original epidermis. The equivalent to a primary organizer in a protozoan is the cell cortex. When the whole cortex is taken off, the cell dies, but if just a tiny bit is left, it is able to grow around the edge and restore itself. Thus, it is quite likely that organelle replication in mitosis is induced by the cell's surface.

There was initial enthusiasm for the primary organizer as the solution to the developmental riddle, but as early as 1932, experiments showed that even a dead organizer, one that had been boiled or treated with alcohol, could induce organ rudiments. In addition, when the dorsal lip was replaced with other tissues, induction still occurred and the same structures formed. Subsequent experiments showed that reptilian, insect, and human cells (including human cancer cells) induced organs in newts. Guinea pig bone marrow is an excellent inductor of mesodermal organs (including the spinal cord) in many species. The liver of the guinea pig, on the other hand, induces brain vesicles and eyes. The kidney tissue of an adder will induce hindbrain followed by ear vesicles.

The variety of potential inducing substances belies the early hope of embryologists that simple proteins guide tissue along particular paths. If inductors do not create tissue patterns, then they must trigger preprogrammed sequences of cell activity. Regions of tissue arise in juxtapostion to each other by providing the framework and timing for development. The entire system evolved in such a way that contextualizing tissues created fields to which they and their neighbors responded. The fact that other substances can trigger the same responses does not make the system hopelessly nondiscriminatory; after all, evolving organisms are finalized in terms of the events which generate them. Those landmarks that were used historically remain landmarks, embryologically. Artificial interference in the laboratory creates false histories and thus triggers established sequences out of order, but this pattern shows a powerful embryogenic resonance toward development more than nondiscrimination.

Induction is not a specific force or event. It is a property of the relationship between fields of proteins and the changing potentials of cells as embryogenesis continues. The capacity of tissue to respond to induction in a morphogenetically specific way (its competence) is determined by its location and degree of specialization. Neural competence becomes spinocaudal competence in a mesodermal context. Ectoderm which can no longer join the neural plate has secondary competence as ear vesicles, if induced by the hindbrain, and nasal pits, if induced by the forebrain.

Organs emerge along gradients of induction. In insect embryos, for instance, the neuralizing gradient forms brain and sense organs, then mouth, appendages, and subesophageal ganglia; the mesodermalizing gradient originates caudally and induces legs, wings, claspers, and copulation parts. Where neuralizing and mesodermalizing influences meet, trunk structures and spinal cord develop. In areas of strong neural influence, medulla forms in mesoderm; and in areas of prevailing mesodermalizing influence, notochord and muscles occur in neural tissue. The spinal cord itself is the thinning of the neuralizing gradient through the mesoderm.

What appears to be nondiscrimination may actually be the inherent complexity of overlapping fields of influence. Even when set in motion chaotically the fields resonate in a reintegrative series of waves. Ross Harrison reminded us long ago that the limb rudiment "may not be regarded as a definitely circumscribed area, like a stone in a mosaic, but as a center of differentiation in which the intensity of the process gradually diminishes."[3] The embryo is made up of such overlapping centers, not concrete organ rudiments but moirés and waves. The more delicate branching organs are radiated from the vortices of central organs. Limbs are induced by a thickening of the ectoderm; fingers and toes by limbs; blood vessels are induced in the mesoderm by endoderm; blood cells by other blood cells; the lens of the eye is induced by the optic vesicle; and secondary nerve branches are induced by primary nerve branches. It is through these gradated hierarchies that sections of the embryo are brought into unity as they are formed; the continuity of induction and biological fields ensures functional wholeness at every stage.

Competence still has a genetic basis; it cannot transcend its lineage to supply parts to strange animals. Tissues grafted onto a foreign species will not develop organs idiosyncratic to the host. For instance, if salamander ectoderm is transplanted onto the site of the ventral suckers of a frog embryo, it will develop the balancer organs it would have developed in the equivalent site on a salamander. It matters little if the ectoderm is removed from a region distant from the prospective balancers; it will form these organs if it has not lost its competence through maturation. Likewise, a frog sucker will develop epidermally on a salamander if competent frog ectoderm is transplanted to the balancer site.

One embryologist writes: "It is as though the transplant says, 'I recognize my new position but I must respond in my own way.' "[4]

For decades biologists felt that they had to decide whether cells and tissues found their locations by differentiating, or differentiated according to their locations. By the 1930s, physics provided a new paradigm—the particle (or position) within the field. Position of any one entity in a biological field affects the positions of all other entities just as in a gravitational field. Paul Weiss wrote: "Fields divided into smaller fields during development until the embryo was virtually a system of equilibrated spheres of coordinated action."[5] Each field was simultaneously whole and partible, and each subfield retained those characteristics. Hence, brain tissue could develop within the general neural field and give rise to the subfields of the individual sense organs. Limb regeneration and wound-healing could be explained as the field compensating for a deficiency, just as the split-off blastomere recreated the entire field of the blastula. Ross Harrison's statements during the 1930s portray the emerging model:

"The egg and early embryo consist of fields—gradients or differentiation centers in which the specific properties drop off in intensity as the distance from the field center increases, but in which any part within limits may represent any other."[6]

He cautioned against putting too much weight on something as unilinear as the primary organizer:

"The organizer, itself a complex system with different regional capabilities, merely activates or releases certain possible qualities which the material acted upon already possesses. The orderly arrangement which results depends not only upon the topography of the organizer but also upon that of the system with which it reacts."[7]

". . . the emphasis upon 'determiner' and 'determined' leads to a very lopsided and often erroneous view of the process, for it is questionable whether one factor can influence another without itself being changed."[8]

Cells are differentiated by tissues which arise from differentiating cells. We can no more assign priority here than to call light a wave or a particle, or define matter separate of energy. There can be no physiological activity without structure and no structure without physiological activity.

Joseph Needham wrote in the same spirit as Harrison and Weiss in 1931: "Determination or chemodifferentiation takes place with reference to the whole organism; what any given part will develop into depends upon its position with reference to the whole."[9]

To mechanists interested in living systems only as the ultimate machine, this all seemed a disappointing regression to vitalism. But the paradigm that was to emerge in place of either pure mechanism or vitalism was a structural organicism first proposed in the 1920s as "general systems theory" by Ludwig von Bertalanffy. Organicists defined regulation in terms of wholeness and organization. The basic laws of matter must, of course, operate at each level of organization, but the levels are hierarchically ordered so that each also has its own discrete set of laws. The overall system is stabilized by feedback between the levels (but also destabilized so that new forms arise from the dynamic equilibrium). If "general systems theory" sounds like embryogenesis, it is no accident. Bertalanffy was in contact with Weiss in the early 1920s, and he no doubt realized that the embryo (i.e., the developing organism) is the link between complex social and symbolic systems and the basic physicochemical processes of the planet.

Paul Weiss's description of the living organism is "the exact antithesis of a classical machine"[10]: a biological field is "a rather circumscribed complex of relatively bounded phenomena, which, within those bounds, retains a relatively stationary pattern of structure in space or of sequential configuration in time despite a high degree of variability in the details of distribution and interrelations among its constituent units of lower order."[11] Driesch's sea-urchin eggs had restored whole functioning systems out of parts; they had regulated themselves. Organisms also "have the power to take inactive materials such as carbon dioxide and water and to produce preferentially one optically active substance to the exclusion of all other possible forms."[12]

Joseph Needham adopted this organicism in the context of his own biological Marxism:

"From ultimate physical particle to atom, from atom to molecule, from molecule to colloidal aggregate, from aggregate to living cell, from cell to organ, from organ to body, from animal body to social organization, the series of organizational levels is complete. Nothing but energy (as we now call matter and motion) and the levels of organization (or the stabilized dialectical syntheses) at different levels have been required for the building of our world.

". . . every level of organization has its own regularities and principles, not reducible to those appropriate to lower levels of organization, nor applicable to higher levels, but at the same time in no way inscrutable or immune from scientific analysis and comprehension."[13]

Organicism is simply a modern cybernetic version of mechanism. To Paul Weiss's question, "Coordination by what?,"[14] general systems theory answers implicitly: "Coordination by the whole." Cause and effect cannot be reduced to the properties of elements, DNA molecules, or inducing proteins. It is all of these, but none of them individually.

In 1960, Weiss added: "The patterning is inherent and primary; . . . what is more, human mind can perceive it only because it is itself part and parcel of that order." [15]

Despite the sophistication of these models we cannot solve the mystery of coordination. We do little more than invoke the implied inherent complexity of trillions of orderless entities in stacks of organismic fields. Declarations of "holism" do not magically convert physical systems into biological systems, and "general systems theory" itself is merely a sister to "probability theory" and cannot explain the translation of blithering randomness into law and order. Something is missing between the inanimate drawing of the cards and the emergence of the game complete with players and rule books. Inevitably (though unconsciously), some scientists return to vitalism and neo-Platonism in a vain attempt to escape the labyrinth. Within the underlying polarity of our Western belief system, there is no other place for them to go.

In the 1960s, the biologist C. W. Waddington proposed an individuation field in multidimensional space. In its transformation into an organism, according to Waddington, the fertilized egg crosses an "epigenetic landscape." This landscape, invisible to us in the usual sense, is an unfolding space-time wave. The organs are its ripples. Their fields are self-organizing and mutually integrative; they emerge as though time were running backward and their wholeness assured in advance. In fact, time *is* running backward: phylogenetic time is running backward into ontogenetic time which is running forward. This is how coordinated tissue emerges so mysteriously in our space-time and seemingly, without mechanical cause. The physicochemical identity of the embryo is simply a manifestation of a functional whole.

According to Waddington, causal events are integrated by evolutionary paths or process chains called chreods (based on the Greek for "necessary" and "path"). Chreods are not visible layers of tissue, or cells, or even molecules; they are shapes of space and time with complex attractor surfaces that bring displaced points back into line so that biological stability is maintained. Thus, molecules and cells are compelled into patterns by multidimensional topologies. The chreods can be represented as valleys canalizing traits by their naturally sloping sides. For "events" to escape they would have to get over the ridge into the next chreod, but the tendency would be for them to roll back toward their original field. This tug then appears as a resonance, or moiré, of tissue patternings.[16]

The individuating power of these hypothetical paths is illustrated by the butterfly/caterpillar system. It contains two different chreods which are activated at different points in space-time. From caterpillar to butterfly (in our linear time frame) the brain, hindguts, and heart alone survive; the rest of the morphology is broken down into molecules and transformed from one chreod to another.

In 1981, Rupert Sheldrake, a British biologist, translated the chreod into a general theory of mind and matter.[17] Calling his version a "chreode," Sheldrake postulated a kind of morphogenetic field able to operate from the past and influence events by a "morphic resonance," an asympathetic coinciding of vibrating systems. Since morphic resonance does not involve mass or energy in the usual sense it need not act according to the laws of thermodynamics or even quantum physics. Influences can be transmitted from any point in time; in this fashion the instinct of a bee travels across generations to reach newborn bees, and a turtle "chreode" impels recently hatched babies to rush into the sea. These "chreodes" would explain the existence of highly refined social patterns in insects at birth as well as the migratory routes of birds inherited from generation to generation. They would not be chemical substances or molecules but fields of resonance associated with shapes in multidimensional frameworks.

Distances would be covered instantaneously *in* space as well as *across* time.

For instance, Sheldrake claims that things learned by one population of monkeys are simultaneously known by other monkeys elsewhere in the world, clams and humans likewise. This "telepathy" is not even a property only of living things; when a previously unknown crystal pattern occurs in a factory it may then begin turning up at other factories. The present "illusion" is that seed particles are blown globally through the atmosphere from the original crystals, but, according to Sheldrake, it is the crystals' "chreode" resonating throughout the world.

Sheldrake uses this type of resonance to explain simultaneous inventions and discoveries. Just as ancient ants transmit to modern ants through "chreodes," human society is in touch with its prior stages of development, and we ourselves are generating chreodes across time to our unknown descendants.

Nature, a British scientific journal, called Sheldrake's book, *A New Science of Life,* "the best candidate for burning there has been for many years."[18] Another British magazine, *New Scientist,* is offering a prize for anyone who can devise an experiment that will suitably test Sheldrake's hypothesis.

But the irony is that Sheldrake's notion is not new. All that is novel is his attempt to present it as an experiementally testable hypothesis. We have known his "chreodes" by other names for millennia. They resemble the talismans and star images of sixteenth-century natural magic, and they appear unnamed in the third century A.D. writings of Plotinus:

"Particular entities thus attain their Magnitude through being drawn out by the power of the Existents which mirror themselves and make space for themselves in them. And no violence is required to draw them into all the diversity of Shapes and Kinds because the phenomenal All exists by Matter's essential all-receptivity and because each several Idea, moreover, draws Matter its own way by the power stored within itself, the power it holds from the Intellectual Realm."[19]

When science pretends to concretize archetypes and spirits, it loses its value as science. Once we are discussing chreodes we might as well be involved in the primal forces of Taoism or the angels of Meister Eckhart; these at least have the power of tradition; they come from the masters. The chreodes sound too much like the latest computer game or "new age" cover story, i.e., more marketing and repackaging than true vision. The poet Michael McClure offers a contemporary scientific version of the old wisdom:

"A ribosome in a liver cell in a salmon might relate to a field of energies or a 'position' within a quasar or in a distant sun. There might be interlocked and predisposed relationships of these, and other constructs. . . . If the universe is a single flow, vibration, or aura, it seems highly likely that such interrelationships exist. And the Universe IS indeed an aura of trillionically multiplex interrelations. And it is primarily comprised of natures of matter that we do not, certainly not consciously, contact . . .

"All life is a single unitary surge, a single giant organism—even a single spectacular protein molecule. In the four billion years life has grown on this planet it is not possible to imagine (in view of its whirlwind energy and delicate complexity) that there have not developed interacting fields, forces, auras, within the behemoth topology of it."[20]

The way our body takes shape is a summation of all universes, known and unknown, imaginable and unimaginable, till now. Our deepest feelings and images are strung out and illuminated in individual cells on embryogenic paths, cells which are objects in fields, and fields which are ruled by larger invisible fields, to the end of nature as we know it in particles and suns.

13. The Origin of the Nervous System

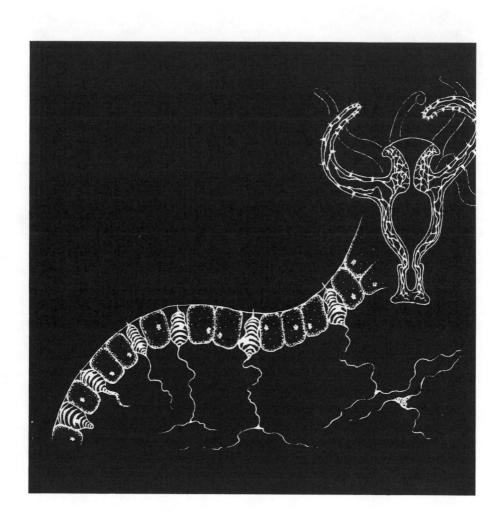

The world manifests only through properties of a nervous system which is itself an obscure genealogy of cells inherited from more than a billion different types of creatures, most of them extinct. We are the continuous firing of neurons into neurons throughout the dimensions of our substance. Most people equate this electrochemical massage with reality, and, for this reason, some may well expect to awaken again after death in the same kind of world as this one, despite their lack of sense organs and a body. They "forget" that, even apart from ontological issues of reality and illusion, the receptors which make us conscious are activated by only selective aspects of the physics of the events around them. The rest of the universe of light, color, sound, and even psyche rushes by, beyond our receptors. The transcendent universe, the "true reality" of Buddha, would lie outside even this universe. It could not be penetrated by our equipment the way X-ray stellar sources and chromosomes are.

There is also no extraneous biological computer. Neural systems evolve from within only. Living tissue came into being with neural properties and, through the aeons of its differentiation and divergence, has compounded and interpolated those properties with every layer and branch of itself. The impervious shields of skin, muscles, and viscera are imbedded with sensory endings whose charges pass into tributaries and collect in different hierarchies of plexuses and ganglia.

Nothing in nature suggests the possibility of intelligence outside this network. From primitive ancestors whose closest contemporary branch is likely the jellyfish to the music of Bach and the formulae of Einstein (or, for that matter, the totemic pantheons of prehistoric man and woman) there is only a homeostasis of neurons interrupted by quantum leaps in their numbers and transformations of their arrangements. In each individual embryogenesis the neurons likewise multiply and gather in nodes, condensing the hierarchies of evolution if not the actual phylogenesis of their lineage. Our sense of reality must encompass the history of our species and the mode of our reconstruction (and these are the same in the phenomenological moment of our lives).

165

Our nervous system is neither new nor superficial. Its branches penetrate every fiber of flesh and every lacuna as well, the substance and its shadow, the residue and the obliteration. The somatic therapist Stanley Keleman defines this condition in therapeutic terms:

"We are not flesh with a spirit or genetic code dwelling in us. We are an event that sustains a particular life style. We are not a machine with a mind or with a spirit. We are a complex biological process that has many realms of living and experiencing. The body is a layered, ecological environment of ancient and modern lives, just like our planet's strata of life. The Human Body has many bodies: a water body, blood, lymph; an action body, bones, muscles, ligaments; a reflective body of brain and nerve. . . .

"Life is old, very old, and we are old in it. The continuum of existence we experience has no discrete beginning.

"A living process is an eternal process.

"A living event is committed by a continuum to billions of years of existence, an infinite chain of living events."[1]

There are bird calls and the distant hammer, the honeysuckle bush and the vastness of space stretching out into fields, the first drops of rain in the warm breeze. We are immersed in breath and hunger and tingling, the ever-present hint of nausea, the warm blanket of memory and the gnawing cracks of amnesia. We are also permeated by an unconscious prehistoric residue that forms the background of our existence. If all but the conscious aspects of the organs were extinguished, everything we consider commonplace would vanish. We subliminally experience the hormones that drip from glands and the skeleton ossified within our viscera. The production of proteins in our cells and even the birth and death of cells themselves must contribute to the overall proprioception of being.

Amazonian Indians in boats and British scientists in laboratories contain the single sensorium of our species. Vietnamese soldiers carrying guns and teenagers dancing to a jukebox in Tulsa may be distracted in different ways, but the experience of being alive and human is fundamental and complete, and takes priority over their activities. Life is religious without prayer or yoga and pleasurable without pleasure. The nervous system is itself a visionary condition, a state of meditation and self-inquiry. Keleman recognizes this universal stasis in his clients:

"The act of living is a reward itself.

"Does that sound strange? It is just a natural by-product. Ask people why they like being alive. They just like it.

"If not, people die like prisoners of war die; they just lay down and die. Young and strong, yet you couldn't wake them up.

"It's an important thing I am saying—most people who enter your office do not recognize *that the fact of their existence is satisfying; there is nothing to look for, even though they may be in an existential pain situation.*"[2]

Nervous systems have two polar but balanced aspects: they are diffuse collectors of information about the environment, and they are compact storage nodes and integrators of sensory information. As collectors neurons tend to spread through the body. As integrators (and ultimately interpreters) neurons entwine and tangle in ganglia. In the former instance we can anthropomorphize them as expeditions to the frontiers, gathering data about light, smell, taste, gravity, movement, etc., with no goal, always instantaneous (within the limits of the physics of transmission). In collaboration with skin, muscles, blood, and with the whole network of which they are part, neurons create the tissues through which they explore. In the latter instance, as centralizing clumps, they translate their own waves of pure sensate information into knowledge and being.

Time, as such, is an abstract and inherently biological requirement. If evolving creatures could not simplify the cacophony of their senses, interpret it, and respond to it selectively, they would perish instantly amidst the perils of the natural environment, and of one another. The entire process of nervous integration produces the illusion of a present moment and an imaginary flowing time-line, but physicists have discovered that our experience of instantaneous reality is idiosyncratic and relative. Within our "present moment," billions of atomic particles are born and die and, in general, carry out complex activities the mere shadow of which we see through our finest instruments. Meanwhile, stars and galaxies live and become extinct on a time scale many times the length of the whole history of life on our planet. Even our experience of these events (as well as the daily ones of our lives) must occur within the internal frame of reference of our nervous system, which is our bias of reality. Ordinary sensations of everyday events mesmerize us into thinking that time flows evenly and is irreversible. That impression is intitially confirmed by Newtonian thermodynamics. However, these are all biological provincialities, built into living tissue by selective evolutionary rhythms during the neuralization of tissue. Although we cannot escape this trap we have begun to perceive that it exists and to imagine what the universe actually looks like on its own terms.

The neural quality of life emerges in free-living cells even without nerves. Paramecia respond to light, find food, reverse their directions after collisions, and, in general, know in one part what the other parts of themselves are doing. The coordination of their cilia, even over intervening spaces, suggests neuromuscular apprehension of an environment; they adapt their motions to events just like metazoa. Amoebas have been observed changing direction to track prey and maneuvering to escape when engulfed by other amoebas.

No one is sure how protozoans receive sensory information and respond to it. The most likely "neural" organelles are various subcellular fibers; yet these cannot be the precursors of nerve fibers even if they conduct excitation because the transmission of information in metazoa involves only the electrochemical properties of the individual cell membranes.

Such a system of conduction likely emerged with multicellularity itself as one of the aspects connecting cells and maintaining wholeness. As living creatures began to fill three dimensions, cells became dependent on the coordination of tissue, and some regional membranes began to express their intrinsic conductive capacities. Although all cells maintain a standing charge, a mutation in one lineage allowed those cells to specialize in the transfer of current along uniquely distended membranes. This current became the basis of intercellular awareness, and it has been passed down through the generations as the impulse of animal awakening. The specialized membranes, the primitive axons, came together in linked pathways of synapses, points at which the excitation passed from one cell to the next through the close contact of their membranes. The depolarization of one membrane would be transferred to its neighbor, and so on in a chain. Thus, does *being* emerge from *nothingness*.

The electrical potential of the cell is the effect of the differing environments within its membrane and outside of it. There may be thirty times as much potassium in the guarded internal medium as in the fluid surrounding it, but the amount of sodium in the cytoplasm may at the same time dwindle to one-tenth of its concentration in the external pool. Although the actual chemistry is more complex than this one disproportion, the sodium-potassium equation is at the heart of the standing electrical charge of the cell. In the unrestrained circumstances of free-flowing material, sodium and potassium will tend to equilibrate by draining away from their respective areas of concentration. It is not enough to say that living tissue *prevents* this equilibrium, for disproportion of internal and external milieus is the very basis of the cell; without imbalance there is no life. Electrical energy is an inseparable aspect of life energy.

It is unclear what role the cell membrane had in the cohesion of the primordial cells, but its descendants are vigilant in their defense of electrochemical identity. The complex lattice of molecules in the membrane is able to distinguish between the very similar positively charged ions of sodium and potassium and able to bind potassium selectively. The negatively charged proteins, which are too large to get out through the pores in the membrane, hold the potassium ions in the cytoplasm while negative chloride ions slip back into the cellular medium in their place. Of course this kind of vigilance is a great deal of work, and the cell must use the energy from the conversion of glucose and other foodstuffs to keep pumping out the sodium and protecting its boundary differential. It is an instant-to-instant priority in the continuum of life.

Each cell is negative inside and positive on the outside, the potential difference across the cell membrane being in the range of seventy to one hundred millivolts, a very small amount of potential but conducted across a space only a millionth of a centimeter in thickness. The cell charge is thus a hundred thousand volts per centimeter—potential energy used to embody the texture of the world, in worms, crabs, and fish, as well as mammals.

The depolarization wave of the nervous system is literally a breakdown of the electrical potential of the membrane of the cell, set in motion by a sensory stimulus. As the wave passes along the axon, sodium and potassium suddenly change their positions, the sodium flooding through the membrane and the potassium vacating. The cell then must oxidize glucose immediately to restore the imbalance, which is depolarized by the next wave. As exhausting as it sounds, every nervous impulse passes only by means of this change in the resting potential of the membrane. Waves, lasting from 0.3–10 milliseconds each, may occur from dozens to hundreds of times per second in a given neuron. These result in quantal pulses that conduct excitation from its origin over immense distances (at least by the scale of the system). There is no way to comprehend such activity on the scale of our own lives in time and space (even though it is precisely incremental to us). Every level of matter has its own vibration, from the electrons that make up the molecules in cells through the waves of neurons, to the sudden pirouette of a ballet dancer coordinated in her cerebral cortex which alone contains ten billion of these neurons.

The axon is the extension of the membrane that transmits the depolarization currents—the impulses—away from the cell body to another neuron, or to a gland or muscle. This long-branched process is basic and universal from jellyfish to mammals. Although simpler invertebrates have naked axons, these membranes are usually surrounded by nutritive cells without nervous properties (tissues called glia in the central nervous system and Schwann's cells elsewhere).

Axons transmit their impulses through synapses, links between neurons (the name, bestowed by Sherrington in 1897 from the Greek for "clasp," expresses a dynamic interchange). The synapse is not a concrete object like a wire or buckle; it is a statement, from our vantage, of the quantal nature of the axonal firing. The internal cell cytoplasm does not simply flow from neuron to neuron (as was believed early in this century); instead there is a potential-measuring valve between them. If the firing at a particular synapse crosses the threshold of excitability there, the message passes through it and the next cell fires. It is known famously as the all-or-none response and has been built into all artificial binary systems and computers (and was used even before we fully appreciated its biological origin). What other model did we have, either consciously or archetypally?

Images and ideas are built primordially from sequences of neurons firing and not firing. There are no prepackaged events in the biophysical realm: All forms must be reduced first to single all-or-none responses within the neuron network and then reassembled in ganglia (if at all) as interpretations of increments. Likewise, imagination, dreams, philosophy, music, and other modes of thought must be assembled from the firings and cessations of neurons within the brain.

To the degree that each of us contains an imaginal realm of memories, feelings, and forms, we must in some way steal this from the unrelenting passage of stimuli to responses. Our first gods were perhaps our images, stretching by fragmentations back into our animal heritage. These images once occurred as numinous objects within the solitude of the prehuman psyche, and their unconscious propagation encompasses phylogenesis itself. The irony is that cerebralization *limits* sensory activity, or at least must inhibit the bulk of it in order to translate the rest into shapes and events. It is no wonder that images are beginningless and nowhere. They are the means by which we have escaped the ocean of not-having-been-born, and they are the present buoys that keep us from drowning in the vast chemical transaction of life.

It is hard to believe we must go through so many neuron firings, use so much glucose, and transfer so much sodium and potassium even to think the smallest part of one thought; it is beyond imagination how these neurons can pass on such complete information, such sustained images and emotions simply by saying yes and no. But this is apparently how things are. The cybernetic revolution confirms the power of the binary system; even the Mona Lisa looked at under a microscope reveals only the presence and absence of dots and the presence and absence of degrees of color and texture.

There is another factor often overlooked, that without the synapses we would have a dull, nonprobabilistic nervous system in which all streams of information would come together equally and all data would converge. Evolutionarily this did not happen. Neurons fire; they do not stream; and, when they fire, the next neuron has a choice (through the synapse) of whether to pass on the impulse or not. Thus the axon may embody raw sensory information, but the synapse is the unit of consciousness, regulating ceaseless unrefined data.

The passage of information across the synapses actually involves the secretion of a complex protein. Although ancient nerve cells may have dispersed hormone-like secretions throughout the organism, their descendants eventually formed paths of secretion in the context of depolarization waves and shot their substances across the minute gap junctions between cells—the nascent synapses.

The oldest synapses occur directly on the bodies of adjacent cells, but as creatures became more complex another cell process developed for integrating the synaptic input—the dendrite. Dendrites were apparently an evolutionary response to the complexity of data flowing into single neurons. They are branched thorny extensions of the cell body (the soma), often indistinguishable. The axon of a cell is a long smooth branch usually insulated in a fat called myelin and dividing infrequently. The dendritic spines are short, irregular, and occur at each synapse upon the individual neuron. Among the invertebrates and in the vertebrate spinal ganglia, dendrites are interwoven with axonal terminations, but otherwise vertebrate dendrites are fully segregated structures developing on the soma.

The dendrite is subsequent to the axon and soma, not only phylogenetically but ontogenetically insofar as it becomes histologically distinct rather late in embryonic life or after birth among mammals. Since synapses in many creatures occur without dendrites these briery processes are assumed to be later refinements of the primordial nervous system; they subtilize the passage of information. Neurological recording devices seem to measure the axons as sharp spikes descriptive of bursts of excitation; the synaptic regulation of the dendrites would then generate the slow, smooth brain waves characteristic of normal mammalian metabolism. In fact, it is not known exactly what electroencephalograms measure, but the proposed dichotomy is suggestive: Mind is first a scatter of distinct firings, a pure excitability, and then it is a modulated flow of bits.

The vegetal quality of the nervous system is apparent. Neural tissue forms like weeds in a field, as if seeking a sunlight within. Even well before their integration into formal ganglia, neurons sprout jungles of tufts, tassels, and tap roots, bloated cells climbing one another like ivy on a grapevine, or spreading through each other like overgrown shrubs and panicles. Other neurons interdigitate like rosette fingers and dig into the soma with claws or cups like petals. Varieties of impulses flow into the dendrites and soma, often many thousands on a single process. The collective foliage can either excite or inhibit the neuron. The axon may already have a presynaptic potential accumulated from earlier events; it considers all the synapses impinging on it at a given moment and then fires, or not. The simplicity of the system vanishes into its secondary complexities. The brain itself is a vast arborescence almost beyond terror, for if we think it we must somehow contain its wildness.

Although there are many creatures with few and simple sense organs, the coelenterates are considered the animals closest to the ancestral line through which multicellular systems first developed. In jellyfish and their allies, excitation pulses in all directions through a nerve net; in other metazoans the pulses themselves are coordinated in ganglia and hierarchicalized. Excitation is then returned selectively through effector fibers to muscle tissue that responds to the interpretation of the data.

The nervous systems of jellyfish are simple not because they have been reduced but because they were inherited from ancient primitive creatures who also gave rise to the other metazoan phyla. The jellyfish have retained the primeval life-style and its body. Only the medusoid forms have true sense organs, notably photoreceptor cups with lens-like cuticular masses and equilibrium-measuring organs called statocysts. These sensory pits and vesicles make contact with the animals' nerve nets, allowing generalized responses to light and currents. In the related phylum of comb-jellies specialized sensory structures of apparently mixed modalities form under the comb-plate rows; upon stimulation they change the rhythm of the ciliary beat and spread luminescence through the combs.

In most simple metazoans awareness is an incipient property of the ganglia, the aggregations of nerve-cell bodies, but the jellyfish nerve net is disperse enough that the number of neurons per zone is for all intents and purposes unvarying. Where there is no "difference" there is no awareness. The coelenterate mazes form a series of crisscrossing paths duplicating one another's links many times over. Cuts in the net do not disturb the animal; the same information continues to flow through remaining pathways. Impulses originate in sensory cells, even in polyps, but there is barely any distinction between the electrical flow of information and the neuromuscular response. The life beat of the animal is the throbbing of its whole bell, modulated by mechanical contact, vibration, and hydrostatic pressure. Located in the same epithelia, the food sensing cells make up a separate nerve net.

Even through such disperseness, there must be an emergent center. Jellyfish creatures feed, defecate, and contort and sway with seeming intention, separate of an external stimulus. Where such awareness is located is a difficult question to be asked even of human beings; at the jellyfish stage its physical basis is generalized throughout the entire epithelium. Perhaps primitive ganglionic forerunners exist in the local nerve rings close to the margin of the bells and in touch with plexuses in the sub- and ex-umbrellas.

The jellyfish-anemone is almost like one continuous axon driven by a wave of excitation. The global throbbing, punctuated by sudden bursts of electrical activity during spontaneous contraction or when exposed to light, suggests the universal autonomic consciousness at the base of all living systems, i.e., the universal unconsciousness.

From the neural pathways of the jellyfish to the complex thoughts of mammals, there is no opening in the circuit, no return to original cause and effect. A newly conceived creature is already wrapped in a fully operating nervous system that has been molded and defined by very ancient events. Experience can never be original or unique. Instinct and responsiveness are inherited from creature to creature and from species to species; the neurons of anemones are the neurons of chimpanzees and hawks. Even as nervous systems change in capacity and configuration, they incorporate prior habits and modes of behavior within their new networks.

The coelenterate nervous system is evolutionarily fundamental; the elaborations that occur (not only along the progressive line to the vertebrates but throughout the invertebrate phyla) are always toward a central nervous system with an anterior brain. Once the deep syntax of the computer is somaticized it tends toward circumscription and discretion. Complexities converge in the capacity for spontaneous action, either because hierarchical control is such a successful mode of animal survival (and thus differentially favored in all lines) or because mind itself is an archetype pressing upon the solidity of matter. The cerebral ganglion may have initially been no more than a specialized statocyst or an aggregation of receptors, but it became the organ of focus for animal identity.

Despite the tendency toward centralization and cephalization in most phyla of worms, the arthropods, the mollusks, and of course the vertebrates, reduced systems with a paucity of sense organs, and a distinct tendency toward decentralization also exist throughout the invertebrates, even in some of the most advanced phyla. The echinoderms, close relatives of the chordate phylum which includes the vertebrates, have no brain to speak of and their radial pattern of nerve cords resembles in many ways the coelenterates. Likewise, they have few sense organs, and those they have are of simple construction. But echinoderms are not "jellyfish." Note the coordination when a starfish chases and captures prey or rights itself. In light of their primitive receptors, echinoderms have an advanced central nervous system with a circumoral neuromotor ring and linearly arranged bundles conducting excitation.

Obviously, relative complexity of the nervous system is only partially a consequence of phyletic position; the other dominant factor is habit of life. Reduced systems exist in sessile forms, parasites, and to some degree, sedentary animals in all phyla. Sense organs are often lost when they are not used, and the seemingly hard-won centralization slides back to regional ganglia. Clams, fleas, chitons, bryozoans, and sea squirts all show what is apparently secondary reduction of their nervous systems as consequences of different types of natural selection (their more neural larval forms expose the collateral nature of the loss).

Nervous systems likely have developed independently from nerve nets countless times. The path to neuralization and ganglion formation appears to be highly versatile, perhaps because of the intrinsic neural potential of universal ectoderm. It is almost as if neurons invade the expanding layers of the body and penetrate tissue at every embryogenic opportunity. This stretching out of the nervous fabric gives creatures a physical opportunity for new behavior, an instrument on which to invent themes. Although the same laws of natural selection must apply to sensory and ganglionic structures, these tissues seem more volatile; they leap into being like synapses themselves, often overshooting the requirements of an ecological niche, a paradox recognized by Alfred Russel Wallace at the dawn of the Darwinian era.

If diverse blueprints have the potential for neuralization, then styles of sentience may occur genus by genus, ecologically, according to each life plan. The nerves themselves may be universal and preexistent, but the mode of being is unique to the animal. Creatures are not just mechanically circumscribed programs or the quantitative outcome of synaptic decisions. The experience of the world transcends, phenomenologically, the circuits of the nervous system. Classes and orders of animals have their own identities; even species have distinct personalities. Among humans individuation has proceeded to a degree that each organism has its own personality, its "I," or ego.

Creatures are little more than water, coagulated, compressed, pulsating, electrified. Through the jelly of cells, currents of protoplasm flow outward into organs and limbs and are realized as sensations and motion. The billionfold contacts of layers of tissue with one another are transmitted collectively by neurons in an inexplicable streaming that turns matter into mind.

Just as nerves rush out with blood vessels to organs as they are formed and areas of brain awaken to nerves that penetrate them, so do lives themselves seem to bring their own minds into being with the nerve designs to feed them. An ant is what it is, a snail too, each of them idiosyncratically and as witness to the mystery; for where the very tangible nerves contact the abyss, the realm of nothingness, fully equipped creatures of knowledge shuffle into the world.

14. The Evolution of Intelligence

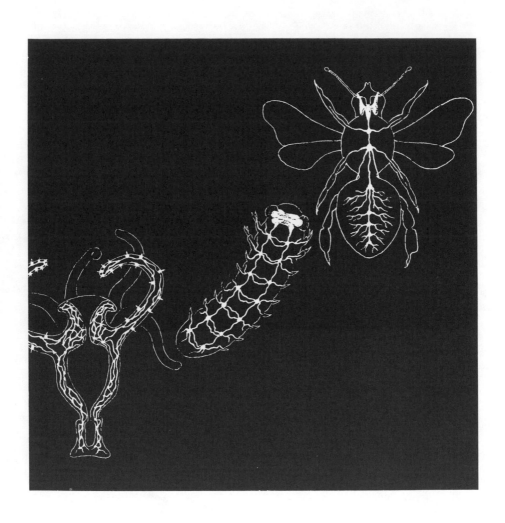

Our experience of ourselves contains at its core a ganglionic integration known as the brain. The word itself gives us trouble, for we intend it to denote the center of consciousness and identity, but we see that the brain is a lumpy mass of cerebral guts having little to do with our imagination of who we are, finally. There is nothing about the brain viewed from outside that suggests our experience of the mind, although in this computer age we are lulled into believing that microcircuits can be packed into any shape, size, or space and can generate the full repertoire of possible images and information. Mind is a subtle and multilayered experience which we cannot assign to any other concrete entity in the world, so we conclude (by default) that we exist as an epiphenomenon of a meaty organ. That is, the billions of cells which flow toward (and away from) the brain generate their collective mental apparition within its lobes.

Although there is no clear point in the history of creatures when number and concentration of neurons cross the threshold and ignite mind, we presume that it is the great number of sense cells and their ramifications in grids and hierarchies that generate thoughts and awareness. What could never be explained qualitatively is now given a solely quantitative justification. The machine synthesis of intelligence is reapplied to nature; the more circuits, the more complete the "mind." The chimpanzee remains forever subhuman because it does not have enough circuits in its brain; the wolf has even fewer, the frog fewer than the wolf, and the worm is presumed not even to know it exists. The jellyfish and the sponge, for all intents and purposes, *do not exist*. This is the current official status of our troubled relationship with the animal kingdom.

Computers challenge our other flank. It has become difficult for us to say what distinguishes our seemingly natural intelligence from the artificial intelligence of electrode brains. How can we know if machines have minds without knowing what gives us *our* minds? The existential crisis in our political and economic institutions mirrors and is mirrored by the crisis of our identity. The number and relationship of neurons becomes not just the threshold of mind, it becomes a statement of being and not-being. It is because it is such a precarious

179

statement of such a crucial matter that we cling rather desperately to the remaining shards of humanity. If we are a mere quantitative degree removed from bestiality, then we are not removed from it at all. The more conscious we become, the more we perceive the imminent deterioration of consciousness.

Although mind occurs in a variety of creatures, there appear to be two major lineages of intelligent life forms representing distinctly different orders of mind. One path leads to the mollusks, echinoderms, and chordates, hence to the land vertebrates—to sentient animals as different from one another as are octopi, birds, whales, and humans. The other leads to the annelids and arthropods, hence to the great insect and spider societies of this world.

Species along both paths have migrated from the seas to the land masses and colonized every microenvironment of the globe. Both lineages appear to have emerged from a primitive ancestor whose closest present day counterparts exist among the Platyhelminthes (flatworms). The chordate line, however, postponed complex social orders and intelligence until it had accumulated enough neurons to escape the trap of pure genetic behavior (at least so we believe, though sociobiologists tell us otherwise).

In many ways a flatworm is like an anemone that has been distended into a thin tube. An underlying nerve net remains, but it is penetrated by two long cords leading into a cerebral knot. The flatworm has degrees of peripheral apprehension in a number of submuscular nerve plexuses, including pharyngeal, genital, and visceral ones, all connected to the rudimentary brain. On either side of this brain, taste and light receptors sprout, while additional neurons recording light, "odors," and tactile sensations differentiate ectodermally across the scant body.

Within these neural concentrations the nerve processes have begun to separate from their nerve cell bodies forming a dense tissue known as neuropile, with glial cells filling the spaces between neurons. This is the ground state of all central nervous systems and brains.

The diffuse nerve nets and simple plexuses of the polyps are superseded elsewhere by clusters of neurons forming hierarchies and actual ganglia. Ganglia are historically the control centers for sets of organs; in more sentient creatures they themselves are overridden by a cerebral ganglion, but, as we shall see, such centralization is relatively weak throughout the invertebrates, and in trauma the local ganglia take over even in animals with a central network connected to a "brain." So the fragments of the worm go on crawling and the penis of the mantis keeps copulating after decapitation. The arm of the octopus swims about like an independent organism after amputation, for it has its own "brain." This regional independence continues in the peripheral and autonomic regions of our own nervous system where the stomach and intestines digest food like independent self-sufficient creatures, and the lungs and heart breathe and beat within their own frame of intelligence which is far older than our cerebral stratum.

Diploblastic creatures like jellyfish, flatworms, ribbon worms, and nematodes are plant-like in their immobility. Neuralization is first a means of coordinating movement through space, so the more elaborate nervous systems develop in concert with mesodermal tissue. Nerve-packed muscles give animals greater range and independence. There is no duality in evolution—no "first, flesh; then neural control." The nervous system is a means of translating the new mesodermal tissue and mesodermally induced organs into proficient activities and life-styles, thus avoiding the undifferentiated bulk of the jellyfish and the seamless length of the ribbon worm. Intelligence is a by-product of the deepening of tissue.

Annelids and arthropods share a body plan based on the formation of their successive sections in segments. The individual segments, called metameres, arise from a kind of embryogenic reduplication of mesodermal somites. Subsequent segments bud off existing ones to create a reinforced length of neuromusculature. Metamerism was likely generated by a series of mutations in a flatworm-like ancestor. Integrated segmentation produced stronger faster animals with central cavities. Some of their survivors then developed new specializations to become insects, crustaceans, and spiders.

Metamerism is genetically repetitive but not superficial. The divisions of the earthworms go through the grooves in the flesh and the mesoderm into their internal organs. Each metamere contains all three layers of body tissue and the three pairs of nerves that occur in the larval segment. The coelom is molded when cells on either side of the worm pull apart; the resulting cavities merge until the wall of mesoderm is hollowed out (except for the mesentery of the gut and thin sections of tissue between segments).

Primitive ocean-dwelling annelids discharge their sperms and eggs into the sea, and the fertilized zygotes develop into ciliated gastrulas with apical sense organs (called trochospheres). The metameres bud from the anal region, pushing the formed segments forward. This bilateral symmetry overrides the sphere and establishes the grid on which all segmented worms will emerge (including Frank Herbert's huge imaginary sandworms of the planet Dune).[1]

The segments of these worms are integrated through the muscles of their mesoderm; as suckers (called setae) make points of contact with external objects, the muscles contract in response to stimuli; they then must be restored to their precontraction lengths by antagonistic muscles. The hydraulic torso bulges backward in forward movement; pulling is turned into pushing as one set of muscles uses its tension to operate another.

An exoskeleton of chitin distinguishes arthropods from all worms. The chitin arises from the epidermis and lies within it as an intercellular matrix of nonliving cells—long unbranched proteins with a lipoprotein waterproof layer. The retention of moisture within this shell provides an internal milieu in which develops a pumping organ, a heart with ostia. The bearers of this shell, the insects and their allies, were thus the first major group of animals to escape the sea.

The outer skeleton is not part of the living body; therefore as the animal grows, it must excrete a new cuticle and molt the old hardened one, removing calcium and returning it to the blood. The entire ectodermal epithelium is then renewed, hardening around the changed dimensions of its bearer. These metamorphoses are fired by hormones similar throughout the phylum, from crustaceans to insects, and occur several times during a creature's life.

Crustaceans begin as nauplius larvae with three pairs of jointed limbs. As they molt, appendages are added at the rear (similar to the annelids). The nauplius becomes the head, and the rest of the crustacean forms from budded sections. Metameric development can also lead to unrefined bulk. In the Decapod order (which includes shrimp and crabs) the nauplius may be suppressed, causing a more developed and mature zoea larva to emerge from gastrulation, a creature with legs, chelae, and abdominal segments that may grow into a fifty-pound lobster.

In the annelids the serial organs repeat from segment to segment with some regional differentiation for respiration, excretion, and copulation. In the arthropods, however, the different metameres specialize and take over the incipient functions of organs. The three major regions (tagmata) are the head, the thorax, and the abdomen. Embryonically, the insect head consists of six segments, the thorax three, and the abdomen seven. In spiders the prosoma (corresponding to the insect head) begins in the embryo as separate segments but then becomes fused in adult forms so that the metameres seem to vanish. The head, appendages, and eight legs are all borne on the prosoma; the twelve segments of the rear opisthosoma also become a single tagma. The fossils of the ancestral trilobites show all three tagmata, and the compacting is a particularly modern feature, only partially developed in scorpions and other primitive arachnids.

The limbs of land arthropods are efficiently arranged in a series of levers around a fulcrum. Although the exoskeleton is hardened throughout, it is flexible in the joints. The trunk moves forward without swaying, and the angles between the parts of the limbs change continuously, keeping the same distance between the midline of the animal and the point of force, the backstroke thus propelling the body. A single mechanical pattern is used in crawling, running, digging, jumping, climbing threads, and running upside down. Multilegged, wingless insects like millipedes and centipedes likely evolved from walking worms with paired limbs. Other creatures evolved from these, i.e., ancient insects and spiders with long limbs and articulations closer to the center of gravity in their bodies. The "psychology" of the aboriginal worm is preserved in many insect larvae, but it is transformed in the adult creatures, often in a bellicose and inflexible direction. When living parts (like the walking legs of a spider) link together rigid plates of armor animal and machine seem almost interchangeable.

Flapping of insect wings comes from opposed contractions of muscles deforming the thoracic segments that bear them. The two sets of muscles work in rhythmic alteration, changing their angles during both upstroke and downstroke and twisting the wing so that its leading edge moves down during downstroke to give thrust as well as lift. The indirect flight muscles do not contract always in synchrony with their motor axons. Instead, impulses may set the muscles in excited states from which thirty or more contractions occur. Dragonflies have only one stroke per nerve impulse, but hive bees can beat their wings 250 times per second, and some midges can flap a thousand times in a second.

The nearest contemporary expression of an ancient worm brain occurs in the trochosphere larva of the aquatic annelid. The larva becomes a peristomial segment around a mouth from which palps, antennae, proboscis, jaws, and eyes emerge in the adult. However, the concentration of nerve rings under its apical sense organ develops into a rudimentary brain (while the organ becomes the *pro*stomial segment).

A different primitively bilobed "brain" arises in the miracidium larva of the flatworm fluke. Two medullary cords run out of a ganglion to the rear of the animal, each radiating side branches to the margins of the body and periodically crisscrossing nerves like the rungs on a ladder. This ladder will recur in all instances of higher development, though it will be differently constructed and oriented. The medullary cords in flatworms and ribbon worms develop as direct outgrowths of the ''brain,'' but the equivalent cords in annelid larvae originate quite separately, from a pair of ectodermal ridges. Only later do these unite with the ''brain'' through circumesophageal fibers. This would appear to be the more primitive situation; yet, as seems endemic in the invertebrates, it occurs in the more advanced nervous system.

In the arthropods the cerebral ganglion originates as an anterior continuation of the neural ridges of ectodermal material at either side of the stomodeum. Other ectodermal cells migrating inward form peripheral ganglia. The central ganglion is divided into three parts: the protocerebrum, which is made up of several neuropile masses and includes the optic lobes and the association neurons for self-initated action (the optic lobes emerge as a series of neuropile masses joined to each other by tracts between the retina and the protocerebrum); the deuterocerebrum, which forms the neuropile for the first antennae; and the tritocerebrum, which includes the nerves to the anterior alimentary canal, the upper lips, and, in some crustaceans, the association center for the second antennae. A subesophageal ganglion controls chewing and maintains the tonic excitation of the other ganglia.

Insects have developed their own compound eyes for recording light and their own wing-beating synapses and jumping twitches (in locusts and fleas). But they attained ''intelligence'' with very few neurons and virtually no associational ones, so they must follow ancestral paths. Their intricate wiring is inherited and inflexible. They have no memories or intentions; they cannot learn anything. The functions of the intelligent insects have been imprinted into their tissues for generations. They act automatically, without mind, or at least without a mind by human standards.

Insect bodies move by generalized kineses causing patterns of flying, feeding, mating, attacking, and caring for the young. These kineses are genetic pathways requiring motion in random directions based on the intensity of a stimulus. As a rock is turned over, exposing insects to light, each one of them rushes about until it finds itself in the dark, at which point it is quiet again. Taxes are another kind of genetic pathway, prescribing movement directly toward or away from some stimulus. The blowfly larva turns its head alternately right and left while swinging its body away from the side of stronger light. Ants maintain a regular gradient in relation to the Sun, integrating successively experienced angles, like surveyors carrying out mathematical operations.

Even mating languages are fixed. Crickets are compelled to make and respond to a repeated sound. Male phosphorescent beetles emit light patterns, and females must respond after the flash. Moths find themselves emitting and chasing aromas. They are "compulsion energized," their neurons inseparable from the things about them. They cannot distinguish the exogenous map. A bee trapped in a room and shown the way out, even a hundred times, cannot learn it, and will die there eventually, batting itself against the photo-attractive window. Beyond human parameters, though, there may well be an unknown phenomenology of insect experience.

Bees can tell one another the whereabouts of nectar and run "nurseries" and "factories." The "queen" bee (if that is who she is) arises amidst thousands of drones in her mating "dance." A single "husband" from among them fertilizes her; then she kills him, tearing his genital out of its abdomen and carrying it off with the sperm. Ants and termites "fight wars," "build cities," and "manage selective breeding programs." Periodically, termites swarm, fall to the ground, chew off one another's wings, and then form couples. After a period of "chastity," these mates begin an interminable intercourse. The male is attracted to a secretion at the female's rear; and as she lays a fertile egg every two seconds they breed thousands of workers who care for them and for the young and collect food. The workers also dispense hormones that produce additional workers and "soldiers" who repair the nest, scout for danger, and drive off enemies.

When we look at bees and termites we are still looking into light and not into darkness. It takes mud a long time to become as fiery and sweet as a hive of honey-bees. But Maurice Maeterlinck describes the nightmare that must be replayed whenever the hive is cleaned of drone overpopulation: "Each [drone] is assailed by three or four envoys of justice; and these vigorously proceed to cut off his wings, saw through the petiole that connects the abdomen with the thorax, amputate the feverish antennae, and seek an opening between the rings of his cuirass through which to pass their sword."[2]

Is this the war of bees among themselves or some ancient Eurasian war Maeterlinck is restaging with bee actors?

"No defense is attempted by the enormous, but unarmed, creatures; they try to escape, or oppose their mere bulk to the blows that rain down upon then. Forced on to their back, with their relentless enemies clinging doggedly to them, they will use their powerful claws to shift them from side to side; or, turning on themselves, they will drag the whole group round and round in wild circles, which exhaustion soon brings to an end. And, in a very brief space, their appearance becomes so deplorable that pity, never far from justice in the depths of our heart, quickly returns, and would seek forgiveness, though vainly, of the stern workers who recognize only nature's harsh and profound laws. The wings of the wretched creatures are torn, their antennae bitten, the segments of their legs wrenched off, and their magnificent eyes, mirrors once of the exuberant flowers, flashing back the blue light and the innocent pride of summer, now, softened by suffering, reflect only the anguish and distress of their end. Some succumb to their wounds, and are at once borne away to distant cemeteries by two or three of their executioners. Others, whose injuries are less, succeed in sheltering themselves in some corner where they lie, all huddled together, surrounded by an inexorable guard, until they perish of want. Many will reach the door, and escape into space, dragging their adversaries with them; but, toward evening, impelled by hunger and cold, they return in crowds to the entrance of the hive to beg for shelter. But then they encounter another pitiless guard. The next morning, before setting forth on their journey, the workers will clear the threshold strewn with the corpses of the useless giants . . . ."[3]

Is this consciousness, or unconsciousness? Do the bees even dimly intuit that they embody a cruel and implacable design? Do they feel fierce and bellicose, or does it all merely go by as a flapping of wings and a siren of neurons?

Our language repeatedly betrays us. The insect mind is totally alien, but in the active expression of its ganglia we see ourselves as in a warped mirror. Through postwar and prewar crises we may wonder if we are not more like them than even like ourselves. Our wars are as grim and pointless as two attacking columns of ants. Our genocides resemble the mayhem of the bee hive. Our cities and factories seem to be great termitaries and anthills.

The romantic dismissal of any kinship with the insects during the last century now becomes our satire of modern society. Less and less can we escape our own prophecy that the animals embody raw nature and we, for all our fineries and exotic evasions, must be reclaimed by the "call of the wild." Dachau was more ant than human, perhaps Detroit and Tokyo too.

But just as we cannot overlook the mirror the insects hold to us across the aeons of phylogenesis, we cannot become hypnotized by it. There are no insects in our lineage, no insect stages in our ontogenesis. The king and the queen sit together in the termitary, she swollen and fat and laying her eggs continuously; every now and then he squeezes under her body and mates. A stream of workers licks up their secretions and carries off their eggs. It is intelligent not on the basis of single entities but as a whole termitary. It is a superorganism, sharing hormones, breeding specialized cells, controlling its societal metabolism, arising in a cloud like the cloud from which it proceeded. It is no wonder that Eugène Marais spoke of "the soul of the white ant,"[4] and described an invisible influence emanating from the queen into the "minds" of the other termites.

The individual bees in a hive or ants in a colony may not have enough neurons each to achieve consciousness, but they themselves are like single neurons, or nerve plexuses. The worker bees resemble cells moving in embryogenesis, creating the "cells" of their hive, each symmetrical polygon the same as the previous one, as in tissue. The swarming of bees resembles the primitive streak, the clustering of chromosomes on the mitotic spindle: half of them follow the old queen, half of them are drawn to the new regent. Perhaps they actually think of themselves as one creature rather than many. Everything about their world suggests not only the unimportance of the individual but the nonexistence of the individual. The larger Bee experiences the pleasure of the single male and female in intercourse and communicates it to the hive; the deaths of bees at the hands of their kin do not diminish hive consciousness, which is sustained telepathically. The individual bees suffer pain, but they are no more individuated than neurons, so their death is never experienced in a personality.

Though they participate in the same general consciousness as we do, sharing the same axons and dendrites, insects are not part of our psyche, only of consciousness and psychic life in general. They have a mode of consciousness related to earthworms, crabs, and spiders. They are the highest form on a whole separate branch of the history of consciousness.

The entire intermediate range of the evolution of consciousness occurs within the Mollusk phylum alone. The chitons are little more advanced than flatworms whereas octopi are as sentient as bony fish. This diversity of intelligence is accomplished with the same underlying morphology—six ganglia usually around the gut, each one paired and crossconnected by long commissures. Constellated as rinds of cells with central cores of fibrous nerve processes, these ganglia respectively form the neural centers for the head organs, the visceral mass, the pedal musculature, the ctenidia (gills), the mantle, and the radula (toothed tongue). The cerebral ganglia send connectives to and receive them from the other neuropile centers, innervating the sense organs, muscles, gills, genitals, and, in the cephalopods, the tentacles.

In the chitons, small intertidal-zone mollusks in colored shells, the cerebral ganglia are little more than medullary strands; the animals are sensual leathery heaps of viscera with tactile, tasting, and balancing receptors over their ectoderm. Gastropods (snails) show increased cephalization through the subesophageal fusion of their ganglia, but they are still, basically, sluggish "worms" without quick reflexes or apparent learning capacity. Their hydrostatic skeletons transmit force only as retractor muscles contact remote antagonists, so their functional level is predominantly ciliate and annelid. Their intelligence manifests in their awareness of locales and their classically molluscan ability to sort through particles and separate them from one another, the edible from the inedible. Tactile sorting surfaces occur not only on the feeding organs but in the guts of many mollusks, so that the interior and exterior world are profoundly merged.

The submerged intelligence of the mollusks is dramatically realized in the most advanced of all invertebrates, the cephalopods. The octopi and their kin have many well-developed sense organs: statocysts that measure direction, speed, position, pressure, angle of acceleration, and sound; olfactory pits; chemoreceptors and mechanotactile sensillae on their arms and suckers; and eyes with corneas, iris diaphragms, lenses, and retinas that are not inverted. Backed by massive optic lobes, these eyes come together, in some classes, to form a full third of the body surface.

The cephalopod nervous system is the result of an ancient and deep-seated phylogenetic pattern. The embryogenic twisting of internal organs into loops is an inherited characteristic of all mollusks (concretized in the familiar spiral shells of the gastropods). Although the ecological benefits of torsion are obscure, it may be of survival value to the larvae—the reorientation of organs allowing the head and velum to be withdrawn first in life-threatening situations. Twisting profoundly rearranges nerves and viscera. In the many dextrally coiling gastropods the right-hand member of paired organs in the mantle-cavity degenerates as the foetus rebalances body with shell.

Torsion is likely the result of series of related mutations in different classes. The primitive limpets are conical, but most gastropods are coiled and helicoid as well. Although this extreme torsion does not persist in other classes of mollusks there is still a proclivity toward reorienting internal organs and realigning the neural commissures. Through cephalopod evolution the head-foot region and the mantle cavity have come into contact in the anterior portion of the animal as the two main molluscan axes of symmetry have converged. The central nervous ganglia, distinct in the other classes, have been transposed into one another in a mass around the esophagus forming the one true invertebrate brain. In addition, various new ganglia have appeared in the cephalopods, structures with no homologs found elsewhere among mollusks; these have merged with the esophageal brain and subesophageal neural centers to form a fat fleshy organ. In effect, the decentralized molluscan ganglia have become lobes in the larger octopus brain.

This brain germinates the talents and gestures implicit in the animal, i.e., coordination of head tentacles and suckers, integration of the mouth organs, and storage of memories and knowledge. The subesophageal region, subject to some hierarchical control from superior lobes to which it is linked, activates motor controls for the chromatophores, mantle, ink sac, gut organs, arms, the general movement of the mantle, and jet propulsion. Stellate ganglia on the inside of the mantle contain hierarchies of giant neurons which run to effector muscles and synchronize subtle movements of the mantle wall associated with the jet, including contraction of both circular and radial muscle fibers. With these hierarchies of neurons octopi and cuttlefish are able to propel themselves at extremely high speeds in all four directions, spin horizontally, hover, and perform a repertoire of subtle water maneuvers.

With up to four million classifying cells in their optic lobes, octopi can distinguish and remember shapes. In experiments in which crabs are fed to captive octopi in association with different shapes (one shape leading to shock and the other innocuous) the animals discriminate and quickly learn the meanings of subtly varying figures. The eight tentacles for which the animal is named are in continuous motion, their own axial cords provocatively reminiscent of the vertebrate spinal column. Rich supplies of nerves run from these medullary cords to the skin and muscles of the arms and to the suckers. The integration of movement and sensory data among the tentacles represents the pinnacle of this skill in the animal kingdom. One author describes these organs as "rather like the fingers of a blind person sorting through the contents of a jumbled drawer."[5]

At least three cranial regions have an enormous density of cells but no sensory or effector connections; these are, likely, areas of association and learning, the silent realms corresponding to the cerebral cortex of human beings. We do not know what images and feelings pass through the octopus brain and how they are stored as memories; in the eyes of the octopus we see no mammalian intelligence but rather the collective being of billions of generations of jellyfish, worms, and clams coming to an apperception of their single lonely existence in the world ocean. Long before the vertebrates appeared, the ancestors of the octopi and squids were the mute seers of this planet.

We tend to think of the brain as the center of identity. During the executions of the French Revolution a morbid curiosity led some officials to pick up newly guillotined heads and address them; it was claimed that many of them knew their situation and tried to answer. But this is a vertebrate obsession. In just about all invertebrate animals the cerebral ganglia can be cut or ablated with relatively minimal effect on the creature's activities.

Removal of the entire brain of the flatworm slows locomotion and makes it harder for the animal to find food; instead of turning around an obstacle after a number of unsuccessful encounters, it will persist. Most insects continue to feed and move about after loss of their brains, but because their eyes are coordinated with each other and because of the proprioception of their legs, the loss of visual contact leads to hyperactivity. Even in cephalopods, where there is clear centralization, the highest vertical lobe—the most convoluted region— can be removed without any noticeable effect on the behavior of the animal. It can even continue to learn most mazes and tasks. However, it learns much more slowly and cannot do anything which goes against well-established behavior.

The secondary ganglia and motor centers are very strong in octopi, and the central brain, in fact, never becomes aware of the different weights or relative positions of objects held in various arms. To this degree the tentacles are more "intelligent" than the octopus itself. Higher cerebral lobes must always steal their consciousness from existing lobes or ganglia. If they grow large enough and incorporate enough functions, they can take over the body, but they can never assume all autonomic functions, and among the invertebrates they never grow large enough to become fully conscious by usual vertebrate standards.

Basically, the body plans of all invertebrates—annelids and arthropods as well as mollusks and echinoderms—necessitate esophageal brains; this is the context in which cerebral ganglia evolved in these phyla. Brains cannot grow indefinitely at the expense of guts or they would so condense the animal's esophagus as to make digestion impossible. The spiders are almost such "mistakes": in order to suck food through their guts in a thin stream they must reduce their prey to a fine liquid. Additional external limitations are imposed by the shells and hard chitinous exoskeletons of crustaceans, insects, and snails. When the shells are secondarily reduced, as among the cephalopods where they lose skeletal function and become mere bits of smooth in-

ternal cartilage, the brain may expand and centralize circumesopha-
geally, but it still cannot intrude upon the gut. Cerebralization appears to want to happen in many species in different phyla, and its
inklings are embodied in exotic varieties of ganglia and neuropile, but
it is finally realized only in an obscure chordate group of bottom-
crawlers. It is almost as if immanent consciousness struggled through
generations of frustration among worms, insects, and mollusks before it turned to a rather dumb side branch of all these creatures and
began a new design which would require aeons more for its realization. Of course, this is our anthropomorphization of a trend, but it
dramatizes how the development of consciousness first proceeded to
dead ends along separate paths. While vast arthropod societies covered the land and flew meadow by meadow across the planet, and
giant squids and prehistoric octopi ruled the oceans, the human brain
was dormant in a minor unsegmented worm.

The lineage of the vertebrates diverged from the Protostomate
phyla at a coelenterate stage. Thus, all of their internal organs developed according to unique Deuterostomate patterns. In Proto-
stomia the blastopore of the archenteron usually becomes the
mouth or is divided into mouth and anus, whereas in Deutero-
stomia the blastopore becomes the anal opening and the mouth
forms secondarily from a new perforation of the body wall. The annelids, arthropods, and even the mollusks are Protostomate; the
echinoderms and chordates branched off from the primordial
ancestors of both groups through a creature dimly hinted at by
starfish and sea squirt larvae.

The brain of the vertebrates is Deuterostomate and arises differently from any other cerebral ganglion. Only in the vertebrates does
the nervous system develop from an introversion of ectodermal tissue
to form a hollow tube. The vertebrate central nervous system is neither frontal nor oriented around the gut cavity; it is phylogenetically
(and thus ontogenetically) dorsal. This peculiar orientation arose in
simple worms that did not exploit it significantly or have neurally advanced descendants for thousands of years, so it may have had little
or nothing to do with intelligence initially.

The present-day chordate acorn worms are very similar in appearance to the more common Protostomate worms but have reduced
nervous systems even by annelid standards. They themselves are not
ancestral to the vertebrates but retain features which probably are ancestral including a hollow dorsal nerve cord and a notochord-like
structure in their proboscis. Caudal to these divergences, the acorn
worm is pure ribbon worm. However, some of its ancestors found

unique niches and were subject to slightly different selective pressures from the ancestors of the other worms; perhaps they were exotically transformed by a mutation. In any case, they developed a true notochord.

This supple rod of supportive tissue was the forerunner of the vertebrate spinal column. In the lower chordates it is a cylindrical sheath of tissue fibers enclosing a core of vacuolated cytoplasm and serving as a brace, a fulcrum for swimming movements. Tissues on either side of the notochord contract alternately as the tail wiggles from side to side. Such a notochord exists contemporarily in the lancelet, but also in the larval forms of the sea squirts (tunicates). As adults, these creatures are motionless lumps in "tunics" (hence, their name), with upper oral openings drawing water through sheets of glandular mucus into buccal cavities. Filter feeders like mollusks, the sea squirts become upright sacks with enormous pharynxes interlaced with elaborate basket-like gill slits, but their larvae indicate that this adaptation came secondarily from forms which were more annelid-like than molluscan.

Adult tunicates have a single small cerebral ganglion with simple sense receptors and plexuses of nerves on their body walls. There is even some evidence of a jellyfish-like nerve net between their siphons. Yet these primitive features occur in the context of very advanced structures like gills, a frontal nerve collar, and neuropile clusters. The true advanced feature, from our biased viewpoint, does not even occur in the adult. It is the hollow central nervous tube of the free-swimming tadpole and its propulsive tail which uses a classic notochord as its axial skeleton. Dorsal to the notochord is the nerve cord, enlarged anteriorly into a light receptor and an organ responsive to tilting. As the nerve cord extends into the tail it loses its neural character. Embryogenically, it is an inward continuation of the excretory canal formed by an invagination of the dorsal pharyngeal surface. A vestige of this structure remains in the adult form behind the neural gland, but it has lost its neural character.

We have seen in metameric phyla that, when larvally derived ganglia fuse, neural structures repeat along the segmented length of adults. In the echinoderms and tunicates this does not happen; the neural features of the larvae are isolated and ignored by the mature animal, and ultimately deactivated. Sluggish molluscan/coelenterate styles of creatures take the places of the promising starfish bipinnaria and the sea-squirt tadpole. A mobile and neural way of life is consequently abandoned—a collective act of evolution that may appear retrogressive to us but which produced a great variety of species that continue to participate in the rich autonomic currents of oceanic life. In a Ferenczian sense, the tunicates and brittle stars and sand dollars "threw off" the tyranny of the backbone and nerve cord—their constant neuromuscular tension and insistence upon awakening—and rediscovered the nurturing thalassan archetypes of the deep. They never had to enter the painful and imperiled worlds of the squid or the hive bee. But one of their lines chose (unconsciously) to continue splashing about and to brave such intimations of witchcraft; and we are, for better or worse, their last will and testament. Our shadows vanish yet beyond the waves.

During gastrulation, starfish are molded as advanced Deuterostomes. A blastocoele forms, and the blastopore becomes the anus; the stomodeum breaks through later. This is the heritage of the bilaterally symmetrical ancestor common to all echinoderms, but it is long extinct. Those starfish that have returned to bilateral symmetry have done so only secondarily through a radially symmetrical larva. Most mature starfish become radial animals with elaborate arms bearing tube feet, and, in some species, feeding tentacles. All these structures arise almost exclusively from the left anterior region of the brachiolaria stage of the larva and are nonchordate specializations. The rest of the originally bilateral and neural animal is absorbed into the "star."

A fully centralized brain evaginating at the anterior pole of the neural tube is a chordate and vertebrate hallmark. The central nervous system is part of the neural tube; it emerges embryonically with the notochord and is imbedded in it from end to end. It is not a ring or a hierarchy of ganglia. The tube merges with the spinal column when their epithelia come together after gastrulation; the muscles and nerves are then attached to the backbone at vertebrae. Afferent nerves shoot out to the sense organs, and efferent nerves impregnate the muscles and limbs.

When the head of the neural tube expands to become the brain it has no point of contention with the skeleton or gut, only with the structural requirements of the spinal column and the rest of the facial skeleton. The basic vertebrate brain has three frontal swellings, and the facial organs form in concert with them, emerging as subsidiary lobes. The forebrain develops with the olfactory lobe, the midbrain with the optic lobe, and the hindbrain as part of an organ of balance, vibration, and general equilibrium and kinesthesia.

The hindbrain is the ancient compass of the invertebrate realm and of primitive vertebrates long extinct. The forebrain evolved as a center for the strong sense of smell in the early mammals. However, it continued to expand beyond its olfactory base to become the cerebrum, and in the higher primates it has swelled out over all the other structures of the brain, engulfing prior strata of intelligence in its own. It is in the cerebral cortex that mind lays claim to a replica of the cosmos and establishes a human universe.

15. Neurulation and the Human Brain

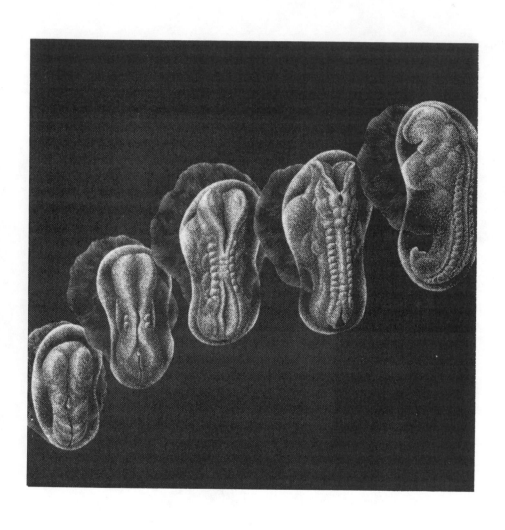

Through early embryogenesis cells are equipotential; the same cells that will become brain and eyes could also be hair or liver. Until they are differentiated from their neighbors they share identical genetic potential. The expression of the senses in flesh comes almost exclusively from the contexts in which cells differentiate. Without induction by nonsensory tissue, neither eyes nor tongue would form.

Mind is also potential in all. The blanket of skin through which we make contact with the outer world is simply unsensitized nerves, calloused brain, deadened tissue that will be impregnated with axons and synapses. The brain and sense organs are polarized epidermis, cells which would become fur, or scales, or teeth if they were not induced by the notochord and the surrounding mesodermal epithelia. There is no such thing as pure uncontaminated sentience or (within living beings) totally unconscious flesh or bone. All nonsentient tissue is potentially neural, and all neural flesh is partially epidermalized.

Our awakening is not simply a global wave of neural induction. If these sensualized fields were not organized precisely down to the smallest fiber, innumerable indiscriminate (and mostly blind) eyes and fingers would form all over the body. In a sense, viscera, spine, and nerves induce one another in exquisitely synchronized sequences throughout the body. The nerves hold the organism together as the spherical unity of cells and clusters of cells is absorbed in a systemic unity of tissue.

In the young tadpole gastrula, the prospective mesoderm, notochord, and somites are all arranged around the coelom; the germinal nervous system covers the dorsal hemisphere, lying atop the somites and notochord in the rear and above the endodermal lining of the foregut. As the blastopore closes, the neural plate forms by separating from general ectoderm. A dorsally moving epithelium thickens and the cells that are to become neuralized elongate within it while the cells of the future stratified skin remain flat. Then the edges of the neural layer thicken further and fold above the plate along its length. As they ascend, the region between them deepens into a groove and the entire plate contracts transversely. The folds touch at their midline and fuse to form a tube with a frontal opening, the neuropore.

Ontogeny *must* recapitulate phylogeny here. As in the origin of the vertebrates the neurulation of the embryo is concomitant with its elongation. The animal stretches out to fill the spine it suddenly embodies. The sphere becomes a tube, and its solidity unravels.

One can recreate these mechanics with a piece of clay. First we flatten it; then we squeeze all its free edges so that they become lips. At the same time we make the piece smaller by pushing its substance to the center. As the lips come into contact we round their surface over the gap to represent the neural tube.

Once formed, the tube proceeds to separate itself completely from the overlying epidermis. The outer layer is now free to become skin and joins at its edges to cover the entire back of the foetus.

In the human embryo, the ectodermal surface of the neural plate gives rise to the central nervous system—the spinal cord and the brain. The elongation of the notochordal process induces extension and thickening of the neural plate. On the eighteenth day after conception the plate invaginates and a groove pushes up along its main axis, secondary folds forming on either side of it like buttes adjoining a plain. A strange new topography is squeezed out of the depths of matter. The cells of the old hydrosphere form valleys along which rivers of neurons will run and dense foliage spring up. The folds of neural tissue reach over the plain and fuse to form a tube; its already fat anterior end will become brain and brain cavity (subterranean ventricles), and its narrower tail will develop as spinal cord.

The rapid formation of the mammalian nervous system in the embryo is a brief transposed memory trace of a discontinuous evolutionary history. The abrupt translation from a radial "jellyfish" to a "larval worm" does not mean that an equivalent change occurred suddenly and dramatically during phylogenesis; it does show that this specific divergence in body form, however gradual in the ancient oceans, eventually took on a new life meaning and distinct ecological consequence. The vertebrate foetus now embodies the retroactive profundity of these layers of tissue.

In the quantum leaps of intelligence that occurred in the phylogenesis of the vertebrate brain, functions were passed hierarchically upward, to superior lobes which swelled over and encapsulated ancient ones (even as foetal tissue reenacts it in condensed ontogenesis). Whatever their origin, mutations continued to provoke neuroepithelial expansion. Centers of identity in fish and newts became mere relay stations for subsequent cerebral hemispheres, their raw sensory and proprioceptive data translated upward into lobes that themselves became superseded by the cerebral cortex. Over millions of years the countenance of intelligence changed as much as the visage of a lamprey into the Renaissance face of Albrecht Dürer.

The human foetus relives this extraordinary hallucination in a matter of weeks. The epochal bursts of neurons are complete by the end of the first six months of foetal life. On the average, twenty thousand neurons are formed every second before birth, recapitulating whole geological eras. These cells cannot divide, so each of us is born with all the nerves we will have. Though 10^4 of them will die before the person does, that is still only 3% of the total.

The glial cells alone proliferate after birth, their lipid sheets lining the gray matter of the infant. The crucial postnatal event in the nervous system is the sudden florescence of dendrites and the spread of synapses through the expanding tissue of the cortex. Axons also crawl outward, their long neurofibrils responding to the transmitter proteins of the synapses. By three months after birth the cortex has begun to convolute, its cells interacting along their new pathways. Nerves from the lower chambers of the brain penetrate this lobe, and more and more regions are annexed under cortical control.

It is no wonder the infant is in a reverie. She is experiencing, almost hourly, a change in consciousness, a cinema of the fleeting minds of extinct creatures. Her identity is still animal and universal. She may sit there and coo in undisguised delight.

Histogenesis becomes noogenesis—in us as in the lineage of our species.

The small burrowing fish-like lancelets are generally regarded as the living chordates closest to the vertebrate line. Virtually headless filter feeders spread through the world's oceans, they have flexible notochords with muscle units on either side. Their major pathways and central ganglion are located in a neural tube which runs the length of the lancelet and is fed by afferent and efferent fibers.

This old spinal cord of primitive sea vertebrates continues underneath the brains of more advanced animals, its essential figure recreated ontogenetically each time a vertebrate is conceived. A core of neural fibers from a tail (or vestigial caudal zone) through the spine, the cord is latent to the brain. In the vertebrate line the spinal column gradually enclosed the neural tube, and the so-called gray matter (pink in living creatures) was covered with white myelinated fibers carrying sensory information about touch, temperature, and muscle kinesthesia, and transmitting instructions to coordinate the arms, legs, shoulders, and neck. The occasional butterfly-shaped sections of exposed gray matter along its length indicate where nerves flow off to the body's peripheries. The trunk of the quadrupedal and bipedal mammals is a primitive fish.

Where myelin covered the neural tube no further expansion of nervous processes was possible. However, a zone of gray matter swelled out over the head of the tube and formed the palaeocortex, the forerunner of the brain. This cerebral hemisphere probably began as fibers in the olfactory bulbs of fishes and was no more than an amplifier of smell. It was inherited by now long extinct mammals from ancestors they shared with amphibians and reptiles, and is recapitulated in the human foetus after gastrulation when the front of the neural tube—the brain stem—thickens into distinct sections. These will become the rhombencephalon, mesencephalon, and prosencephalon, i.e., the forebrain, midbrain, and hindbrain, respectively.

The simple neural tube and brain stem are the whole chordate brain. Its original central cavity has become the ventricles of the human brain, which are filled with cerebrospinal fluid similar to blood but without red and white cells.

The contemporary human brain stem begins as the medulla oblongata, a vertebrate expansion of the spinal cord as it enters the skull. The medulla coordinates our basic chordate existence: heartbeat, breath, digestion—the mainstays of ancient swimming. Bundles of nerve fibers that converge here are combined in systems and conveyed to the higher lobes.

Early in the development of the human foetus the neural tube forms as a straight line with a slight bow in it, but as the brain bulges anteriorly a number of deeper bends occur. The so-called cephalic flexure forms at the spot where our forebrain bends downward in front of its midbrain. This hump, which obtrudes at the end of the first month after conception, is so pronounced that the whole organ is almost bent in half. Its dramatic early appearance must reflect a monumental event in the vertebrate lineage.

In all the more cerebral animals, a cervical flexure develops between the medulla and spinal cord, and a pontine flexure twists the opposite way, thinning out the roof of the hindbrain. This bend becomes distinct in humans during the middle of the second month after conception, at which time the midbrain is most enlarged in relation to the historically subsequent lobes. The pons develops its own thick bands of connecting fibers between the cerebral and cerebellar cortices and the spinal cord. Some neurologists believe that a zone buried in the lower pons is the physical location of paradoxical sleep. In this state (by contrast with slow-wave sleep) the muscles relax, the closed eyes scan excitedly, heartbeat and blood pressure decrease, and the sleeper dreams deeply and often. Apparently two chemical transmitters, natural mammalian hallucinogens, are involved—serotonin during slow-wave sleep and noradrenalin during paradoxical sleep.

When the cortex goes to sleep, it goes to sleep; and something else, something very mysterious by waking standards, takes its place. The memories of the mind, the organs of the body and the glimmers of their ancient vestigial functions, the emotional traces and traumas of the tissues, and various environmental and cosmic rhythms all contribute to dream formation through the pons and brain stem. This timeless and uninterrupted stream of motion, of internal shape and form, molds the dreamwork. As Freud warned us: There is no time in the unconscious. A dream has the power to turn every mote of dust into a universe, not its own universe, but the universe it invokes in the neural space of the dreamer.

"The mind can make/Substance," wrote Lord Byron, "and people planets of its own/With beings brighter than have been, and give/A breath in forms which can outlive all flesh./. . . a thought,/A slumbering thought, is capable of years;/And curdles a long life into one hour."[1]

By now we can tell Byron that it is even less.

Without dreams we would hallucinate continuously and not ever be fully awake. Our ancestors could not have survived in the oceans and forests in this condition—unable to dream while asleep, so forced to dream while awake. Very early, the brain had to mediate conscious and unconscious phases of mind, and the pons may be the living relic of the onset of nocturnal appearances. Surviving creatures have all perfected some form of hibernation or rhythmic sleep; other animals have long since sleepwalked off the body of this world.

The other major organ of the hindbrain is the cerebellum. Buried in folds with only about a sixth of its surface lying open to the outside, it is one of the most convoluted regions of the brain. Formed as a series of bulges in the brain stem, the cerebellum receives connections from many points of the body, including muscles, joints, sense organs, and the cerebral cortex itself. Because it sends back only a third of the number of fibers it receives, we assume that the cerebellum forms interneuron links between afferent and efferent paths, condensing and coding multiple sets of impulses in reflex arcs, and returning them as single messages.

Although not particularly an organ of higher consciousness or creative action the cerebellum integrates the precise movements of the motor system, fine tuning muscular activity and adjusting equilibrium. Without this organ people would not be able to gauge distances and would undershoot or overshoot what they reached to; the trembling of older people is a breakdown of cerebellar regulation. The calligrapher inscribing each line of the character with his pen, the Eskimo carving scenes of minute animals and flowers in a pebble, the astronomer tuning a telescope to a remote galaxy, the baseball batter timing a ninety-mile-an-hour fastball—are all located preconsciously in the cerebellum.

This organ appears first full-blown at the back of the brain in animals travelling on legs. Birds have an especially enlarged cerebellum, no doubt for maintaining balance during flight and for alighting. The cerebellum in fact consists of lobes from three different phylogenetic eras. The small archicerebellum at the lower rear (just above the medulla) has fibers connecting to the chambers of the ears. The next oldest palaeocerebellum is an anterior lobe that deals with sensory data from the limbs. The intermediate neocerebellum, the actual posterior lobe, developed last and is the source of subtle limb movements and timing.

The midbrain evolved with the bony fishes, probably 450 million years ago. With their vertebrate motility and sharp eyes, these intelligent carnivores established themselves in the Ordovician and Silurian oceans of the planet. The midbrain was for them a forebrain, a ruling lobe, and it incorporated their newly acquired powers of sight. These eyes may have begun as simple pigmentation spots in chordates at the dawn of our lineage, but the neural tube of of the vertebrates grew out to create the retinal surface and the image-forming brain. The eye spots of the lancelets deepened into pits with light-recording cells at their bottoms. These gradually came to meet the stalk of the optic nerve in the brain stem and culminated in the two swollen lobes of the optic tectum which dominate the midbrain of the fishes.

Sight is an inherent capacity of the raw ectoderm. On a sunlit planet, tissue will "see," and "sight" will induce image-forming crystals and storage chambers for its remembrances. The sense pigments of protozoans and jellyfish recall the dawn of visual receptivity in roughened ectoderm. Actual eyes with lenses have evolved three times independently of one another: in the insects, the cephalopods, and the vertebrate line leading to the mammals.

Although these fishes have brains like ours, they retain an invertebrate separation of functions—distinct "brains" for olfactory, optic, auditory, and visceral functions, and for the surface of the skin. Their cerebellum is undeveloped, and a cerebral cortex is not present at all. In mammals the optic center has been shifted to the forebrain, and the tectum is reduced to four small swellings—the colliculi. These represent the brain of the fish ontogenetically and are formed from the same plate of neuroblasts as our afferent nerves; a superior pair relays visual impulses and an inferior pair relays auditory reflexes.

The new forebrain of the primitive amphibians separated almost at once into two hemispheres (and repeats this revolutionary act in each vertebrate embryo). The primacy of this event, ontogenetically in each foetus and historically in the fish-like ancestors of the newts, signifies a fundamental duality of consciousness originating somehow in the "schizophrenic" neural response of these creatures. Outside their watery ancestral home the echoes, aromas, and far-flung landscapes stimulated a new mode of intelligence. Amphibians and reptiles may have "invaded" the land, but the land also invaded and reshaped *them.* In primitive land-dwelling vertebrates the forebrain split into a central diencephalon with the vesicles of a telencephalon on either side of it. The diencephalon then took over as the brain.

In modern mammals the diencephalon is permeated with blood vessels and primarily involved in bringing food and oxygen to the nerve processes of the brain. Originally, though, it was the dominant hemisphere, and during amphibian and reptile ascension it governed the hormones and emotional pathways. This olden aeon is relived in the human embryo when neuroblasts accumulate in great numbers in the lateral walls of the emerging diencephalon and their swellings protrude into the underlying brain cavity as the thalamus, hypothalamus, and epithalamus. The thalamus continues to grow until the third ventricle of the brain is a mere slit. In mammals and reptiles this organ coordinates all incoming afferent pathways, a role preserved in the mammals where it serves as the last ascending relay center for messages below the cerebral cortex. It also harbors the initial optic synapses outside the retina. But the thalamus is not just a passive electrical line; it fuses and coordinates impulses. Sensory modalities are disperse prior to entering the thalamus but leave as complex gestalts travelling along fibers into the mind of the cortex.

The level of awareness that exists now in the primate cortex must already have had a simple existence in the thalamus of noncortical animals. Human babies born without a cortex function in a normal fashion at first. The thalamus "remembers" its culminative role in the evolution of amphibians and reptiles and can reenact it as the upper lobe of the malformed offspring of creatures in which its reigning function has been totally superseded. In the absence of a cortex it is still brain-like enough to organize rudimentary behavior.

From mid-dorsal diencephalon sites where neural folds come together in the foetal human brain, the pineal body and parietal organ later evaginate. Ocular in amphibians and reptiles and their ancestors, these have become glandular organs in mammals. Where the ectoderm of the primitive mouth cavity fuses with the diencephalon floor a pituitary gland is induced. This bilobed organ which hangs down from the base of the brain joins in the homeostatic regulation of the entire organism. Although a gland, it is anatomically a part of the brain, particularly of the hypothalamus which lies just above it. Axons from the hypothalamus flow into the pituitary and synapse. The pituitary participates in multivariate networks of chemical feedback, manufacturing not only basic life and growth hormones embryogenically but hormones which regulate the formation of other hormones. The endocrine and autonomic systems are reciprocal from their amphibian beginnings.

One convoluted cluster of cells sends axons from the spinal cord and the thalamus into the cerebellum; this is the inferior olive which synapses the proprioceptive data of the shoulder girdle and the neck. An accessory olive is singularly developed in vertebrates that swim, integrating the reflexes of their trunk and tail muscles. These olivary nuclei are ontogenetically the products of neuroblasts migrating ventrally from the same neural plate as the colliculi of the midbrain and the afferent nerve fibers.

The basal ganglia (including the globus pallidus) form to the front and sides of the thalamus; they are the processing center for the discharge of thalamic responses to the cortex. Fibers from this zone branch out to the thalamus, hypothalamus, and the cortex, as well as to the brain stem and reticular formation. Instinctual activity is probably coordinated in this center, including the mating, nesting, and territoriality of reptiles and birds, creatures which have highly developed basal ganglia.

This becomes the area of the human brain referred to as the R-complex by the biologist Paul MacLean; its highest functions are pure reptilian. Within our own higher lobes we inherit the brain of a crocodile almost intact—aggressive, ritual, ever prowling for food, guarding fiercely its stature in a dominance hierarchy. It has "its own intelligence," says MacLean, "its own sense of time and space, and its own memory."[2] Its world is one of pack hunting, hoarding, and boastful exhibitionism. In an experimental test of this supposition MacLean cut into the globus pallidus of an unfortunate squirrel monkey and ostensibly it stopped all ritual display.

When human beings perform reptilian-like functions, it is not (by MacLean's premises) that they are reliving actual reptilian lives but that the ancient brains are continuing to send their instinctual configurations to the higher lobes. The basal ganglia express themselves through our social classes, armies, ceremonies, and compulsive ordering. The Sioux Sun Dance, with prayersticks, feathers, beads, and chants; the Saint Patrick's Day Parade, with flags, floats, and marching bands, are (in MacLean's system) reptilian displays embroidered with the ornate symbols of the cortex. The basal ganglia can produce extravaganzas when the higher lobes are at their service. The crowd standing as one and roaring at the great catch and touchdown run during a crucial game is perhaps a corroboree of prehistoric animals on another continent. The sense of cruciality and triumph (or defeat) is truly millennial and cannot be explained by the rationales of the cortex.

"The repitilian brain is filled with ancestral law and ancestral memories," says MacLean, "and is faithful in doing what its ancestors say . . . it is not a very good brain for facing up to new situations. It is as though it were neurosis-bound to an ancestral super-ego."[3]

The next layer in this psychoanatomy is the limbic system, which is made up of a variety of structures, including the amygdala, the hippocampus, and part of the cortex and the olfactory bulb. The amygdala is a bulge imbedded in the temporal lobe of the cortex, a relay center for afferent messages from the motor cortex, olfactory lobes, reticular formation, and other proprioceptive areas. So many sensory signals converge on even single cells of the amygdaloid nuclei that it is impossible to guess what separate impulses are being fused there.

The hippocampus is a seahorse-shaped section of gray matter folded into the cortex and connected to the higher brain by small fiber bundles. Aboriginally, it is the limbic brain, with its origin in the dominant archicortex of the early mammals. In humans the hippocampus is a chamber of short-term memory, more concrete and linear than the cortical zones where selected long-term memories are stored. The hippocampus contains the enigmatic "counting cells" that regularly tap out rhythms of four or five numbers or are activated only when a discrete number of stimuli have occurred. It also includes novelty-recording cells, silent except when a new stimulus arouses them; they fire, but not again if the stimulus is repeated. Such "idea formation" in actual neurons suggests the archetypal basis of both morphology and temporality. It also appears that short-term and long-term memories overlapped in the first mammals and that these beasts experienced in their hippocampus brain an unbroken eternal time (in which most of their descendants still reside).

The limbic system is MacLean's old mammalian brain, not by itself but through its interactions with the hypothalamus and the autonomic nervous system. It is a coordinating center for the homeostases of viscera and glands; thus, it is an organ complex of gut emotions, ancient drives and feelings. Fear, anger, and even *eros* are partially embodied by their passage through the limbic nodes. We would like to believe that these emotions are sublime extracts from our highest consciousness; aspects of them are, but other aspects are unconscious and uncontrollable, feelings without thought or meaning, as sudden as the epileptic fits which arise with inexplicably intense moods on the borders of the limbic system. We sublimate, enrich, and transform these prehuman bursts, but we cannot abstract them from their bestial quintessence.

According to MacLean: "Affective feelings provide the connecting bridge between our internal and external worlds and, perhaps more than any other form of psychic information, assure us of the reality of ourselves and the world around us. The limbic system contributes to a sense of personal identity integrating internally and externally derived experiences."[4] In fact, he ascribes the Nazi fury of World War II Germany to a sudden irresistible appeal to the limbic system, all the more powerful because we do not experience its archaezoic roots, only its vivid cortex symbols.

The hypothalamus maintains the basic internal milieu—coordinating information about body temperature and blood pressure, instigating panting and shivering to alter the flow of blood to different regions, and measuring water balance and fullness by sending out signals we experience as thirst and hunger, or satiety. The hypothalamus also conducts upward states we interpret as erotic (stimulation of the hypothalamus causes monkeys to begin mating). In other experiments conducted in the mid-fifties a so-called pleasure center was discovered in the hypothalamus, a region in themselves which captive rats chose to stimulate by electrodes. Even when parched and starved the animals preferred the "pleasure center" to food, drink, or sex, and activated its button until they passed out from exhaustion. Electrical stimulation of other areas of the hypothalamus have produced, ambiguously, rage and docility, terror and total absence of fear; so it would appear that it is a mediator rather than an originator of emotions.

Experiments are misleading, for they destroy complexity in order to discover its componential aspects. As one scientist remarked of the cyclotron, we smash Swiss watches in order to find out how they work. We split atoms apart in great violence to understand matter, and we ablate and sever tissue in order to isolate the functions of the brain. In the end, though, we can assert little more than that all levels of the brain embody rudimentary feelings and phenomenologies, some of them ancient, some of them newly arising but in the context of the ancient ones. In the emotionalized centers of reptilian higher consciousness and cerebral hierarchies of ascending mammals, it is impossible to know when and how each organ arose, what phenomenological and neurological factors in the lives of these creatures served to induce new structures of brain in their descendants.

Meanwhile, within the foetal brain stem and spinal cord other clusters of neuroepithelial cells proliferate in reenactment of past events. Some differentiate as neuroblasts, some as glioblasts, and others line the central canal. The lateral walls of this canal thicken irregularly, producing a *sulcus limitans,* a long shallow groove between two lamina—the alar plate in which the afferent functions of the spinal cord originate, and the basal plate which is the source of efferent spinal functions. The cerebellum develops late in embryogenesis from symmetrical bulges of the alar plates which protrude into the ventricle of the forebrain. In general, wherever limbs form peripherally, there is corresponding development in the nerve centers on the spinal cord supplying them.

The peripheral nervous system consists of spinal and cranial nerves connecting the central nervous system with skin and muscles; it is the sum of nervous connections not located in the spinal cord or the brain. The formation of nerves joining the spinal cord to the organs and limbs of the body requires cells that originate outside the entire structure of the neural plate and subsequent neural tube. These are the neural-crest cells—original ectoderm that separated itself from both the epidermis and neural plate during the formation of the tube and then travelled through the organism as mesenchyme. These migrating "creatures" stream into spaces between the epidermis and mesoderm, between the neural tube and the somites, and among the rudiments of the organs. Some locate behind and within the eyes to become cartilage and ciliary muscle; some contribute to the skull and teeth; some become sheaths of nerves and membranes around the brain and spinal cord (meninges). Others migrate from above the neural tube and cluster segmentally in groups along the spinal cord; these become spinal ganglia and participate in the formation of the peripheral and autonomic nervous system. Still other neural-crest cells are induced into innumerable different states by regionally developing tissue.

Brain is spread throughout the body before it has become brain, so mind is present in all the organs not only as a potential of the individual cells but as a specific feature of the neural crest's dispersion. While the nervous system is still being formed, these partially neuralized cells are migrating throughout the body and viscera, lay-

ing the basis for connective and skeletal tissue associated with neuromuscular coordination. If this could have been done without such a
complex odyssey of mesenchyme, then surely the muscles and bones
would have developed in local layers without immigrant associates.
Obviously, complete coordination and comprehensive intelligence—
conscious and autonomic—require a series of polarities and disjunctions, a coming together of different layers and states of tissue. The
sensory organism embodies such subtle apprehension and kinesthesia
that it must be put down in layers of sequentially neuralized cells.
Thus structure becomes immersed in its own experience as it forms.
Mind emerges from matter, but mind also sinks down into matter and
reconstitutes itself as organs. Individual tissue responds to its own existence.

The connections between the spinal cord and the peripheral organs
are at least partially induced by the organs themselves. Nerve processes originate in the spinal ganglia and creep outward until they
contact a limb or a gland (at the same time they travel back to the spinal cord so that the connection between the central nervous system
and the organs is through the spinal ganglia). Some of these axons
grow incredibly long: nerve processes will detour around obstacles or
leave their customary paths to intercept a limb that has been transplanted. The attraction between organ and nerve is so generalized in
overlapping centers of differentiation that the processes may be
drawn into any proximal tissue mass, even an irrelevantly transplanted limb bud, an eye grafted onto a torso. Actual neuromotor
awakening follows only when there is a match between nerves and
their terminal organs.

The longer neural processes growing out to the skin are afferent.
They converge with efferent nerve fibers travelling out of the ventral
columns of the spinal cord. Together they form sensory-motor
nerves, and, as these mixed afferent-efferent processes, they branch
out to the different regions of the body. The afferent fibers bring sensory information from throughout the skin, viscera, and other organs,
including proprioceptive messages from muscles, tendons, and
joints. The motor fibers carry impulses back to limbs, muscles
viscera, and glands. The child will be born with all its segments coordinated, its organs interconnected and functioning, and a lucid transcendent mind.

A separate branch of the peripheral nervous system is involved in
regulating a series of subliminal activities including blood pressure,
body temperature, cardiac rate, dilation of the pupils of the eyes,
blood sugar amount, and erection of the penis. These autonomic

functions are coordinated in a homeostasis of nerves, glands, smooth muscles, and internal organs.

The autonomic nervous system is composed of two complementary branches: one called sympathetic, the other parasympathetic. In general, those activities stimulated by one are inhibited by the other. The sympathetic accelerates the heartbeat and invigorates the lungs by dilating their bronchi. The parasympathetic sedates these processes, but, on the other hand, stimulates peristalsis and gastric secretion in the digestive tract and arouses vegetative functions. The sympathetic retards these same functions. Their opposition is not happenstance; between them the systems maintain a complex internal balance regulated through such organs as the hypothalamus and medulla.

The sympathetic ganglia begin to form in the foetus even before the spinal nerves start their migration outward. They originate in neural-crest cells migrating in dorsal relationship to the neural tube. After connecting in pairs the ganglia spawn links between one another and construct a twin longitudinal nerve cord which synapses with the spinal column alongside it.

The parasympathetic system originates more obscurely in different clusters of neural-crest cells, some of which migrate as far as the mesencephalon. For a long time it was believed to be a separate autonomic system arising directly in the neural tube or local mesoderm. Its ganglia do not form chains but are located individually next to glands and muscles. Its paired longitudinal columns lie alongside the spinal cord, with one ganglion each in the visceral efferent branch of the cord and another per column extending to a muscle or gland. A sacral section regulates the lower colon, bladder, sphincter, and genitals, and a cranial one branches axons out to the head and face, even to the lens of the retina where the autonomic system works through ciliary muscles in focusing images.

The sympathetic and parasympathetic functions are not just brief nerve reflexes; they are sustained internal states initiated by hormones and ganglia; they embody the visceral component of the emotions and transmit it to the spinal cord. Although essentially unconscious, the autonomic nervous system is subtly responsive to tension, anxiety, desire, fear, and other emotional states (some of which it participates in generating). It is no wonder that psychosomatic links have been mysterious and imponderable from Ice Age shamans to physicians with ultramodern diagnostic equipment. Just in the brief history of Western psychiatry we have been informed (by successive generations of doctors) that the identical diseases originate in the body, then the mind, then in the body again. Neuroses once thought to be brain pathologies (and so treated by phrenologists and pathologists in the nineteenth century) were rediagnosed by the first twentieth-century psychologists as a mental etiology expressed through somatic symptoms. Modern psychotropic therapists and neurologists now explain these disorders as themselves the symptoms of hormonal (autonomic) imbalances.

In truth, we can no more discern the origin of so-called mental diseases than any other mental states, including dreams or thought itself. Mind, fantastically, seems capable of altering tissue directly, even consciously, so it is something more than an epiphenomenon of the proximate chemistry of the body. The secret of faith-healers, yogis, and psychic masters has ever been to penetrate the autonomic nervous system and transform the body by enlightened mind. The Taoist alchemist induces his consciousness down the brain stem into the mouth of the famous kundalini serpent coiled in the spine. This is what is meant by "true awakening." Nothing is considered permanently unconscious, nothing is forever lost, even deep cells of the bone marrow and gut, and the relics of childhood diseases.

The central and peripheral nervous systems are a single emanation, the medulla and nerve bundles inducing each other right up into the midbrain. The efferent cords of this supersystem wrap around and through the whole body like an aura, and its afferent fibers spiral into a knot at the center of the brain. It is, at the same time, a giant bolt of lightning in the skies of an unknown planet and a mammoth oak whose roots extend into the humus underneath the forest of light.

There is also another prehistoric unconscious system in the brain, even more disperse and shadowy than the peripheral systems. The reticular formation is an extremely diffuse core of gray matter, generalized and undifferentiated, running from the medulla and midbrain right up through the thalamus. Fed by fibers from the cerebellum, the colliculi of the midbrain, the hippocampus, and the other chambers through which it passes, the reticular formation sends inputs to the thalamus and the cortex itself.

The cells of the reticular formation make up as many as ninety-eight different clusters. Its ascending influence is described as a wakefulness, an alertness, a general "take note of." When the reticular formation is stimulated, animals respond more quickly. The descending influence of the reticular formation both facilitates and inhibits motor activity (with the lower medulla being the most inhibitory). Lesions in the reticular formation will cause a victimized cat to go to sleep, and it is impossible to arouse the animal no matter how great the disruption. At best, it will briefly awaken and then go back to sleep.

Apparently, we can be immersed in the world but remain unaware of its phenomena without this faint background stimulation. Our link to reality is a matter of tuning within the old chordate brain stem, an original message that tells us anything at all is worth our attention—not only heavy objects bumping into us and throwing us over but the mere end of a night's sleep from which we arouse ourselves in the morning. The change of internal rhythm alerts us that a dream reality created by brainwaves and internal stimuli has been succeeded by an outer sunlit world.

Through the evolution of the mammals the control center continued to be translated upward into the neocortex, its lobes apparently induced by olfactory rudiments so that its hemispheres, in gradual and coordinated fashion, "hemorrhaged" out over the rest of the brain. The ascension of the telencephalon is relived ontogenetically: Nerve cells migrate up from the other lobes—the more advanced the animal, the more abundant and encompassing the sheets of cells. By the third month they dwarf and almost cover the diencephalon in a separate layer. In another month this cerebral cortex spreads over the cerebellum which has also begun to expand. At this point it is smooth, but suddenly its texture changes. It develops folds and irregularities, increasing its effective neural surface by convolution. The fully developed human cortex is three to four millimeters thick and enfolds most of the external surface of the brain.

The new mammalian brain captures the highest functions of the other lobes—coordination from the cerebellum, visual integration from the midbrain, memory from the hippocampus, and creature identity from the limbic system. This is somewhat of a misstatement; the cortex does not take these functions away; it incorporates aspects of them, and in so doing, transforms them. The cortex redefines activity at a higher level. It also grows in another way—silent regions of gray matter, association areas that have no direct motor outputs, no historic functions in the old mammalian and reptilian lineages; they do not project outside and they primarily interact with each other, ostensibly processing information and creating imaginal reality. Each cortical zone associated with a sensory modality is also surrounded by an orbit of cells involved with the integration of symbols and ideas.

The cerebrum is divided into four highly convoluted lobes which arise (at least in part) from their internal hydrodynamics and the geography of the underlying brain. They are separated from each other by fissures. By cutting connections between regions, brain research scientists have deduced approximate functions for each of the zones. The parietal zone, the most developed lobe at birth, coordinates motor input and output and the senses. The occipital lobe is a region of visual reception. The frontal and temporal lobes are associated with speech, learning, memory, and symbol formation. Numerous sophisticated processes germinate here and interact with one another—representation of objects, naming, simple classifying, higher orders of classification, and philosophical abstraction from classes. Just for the basic evolution of speech, all of these discontinuous symbolic functions must be coordinated with sensory data and motor control of facial, thoracic, and other muscles.

But the cells of the cerebrum are neither as predetermined nor regionally distinct as this mapping seems to suggest. Experiments have shown something else; i.e., brain tissue is equipotential before it is regionalized by neural pathways, and after injuries it regains embryogenic flexibility and can take on new functions. Abilities learned in one region of the cortex can be transferred to another, intact, even as memories are passed from one molecular formation to another throughout a lifetime. There is enormous redundancy of tissue, and we hardly yet understand what it means.

From the first sensory modalities one of the roles of ganglia in achieving personality has been that of limiting biological consciousness. Although modes of survival dependent on intelligent behavior require a substantial reserve of tissue, apparently the mind cannot (or cannot yet) allow simultaneous expression of too many of these complexes of being. In inexplicable episodes a person may suddenly speak a language he was never taught or project images from his "brain" directly onto photographic film. Such paranormal phenomena are either doubted or denied by most scientists, but until their apparition is clarified, no laws of mind or matter can be finalized. More "normal" talents equally attest to the depth and multiplicity of the cortex: the concertos of Bach, the relativity theory of Einstein, the lines of perspective in Leonardo da Vinci's *Last Supper,* and the discursive myth cycles of South American Indians.

Unfortunately, this duplication of cerebral structures and skills has become an implicit justification for treatment of mental disorders by electric shock and lobotomy. If disturbing associations and emotions can be cut off without notable loss of function, then difficulties can, in a sense, be extracted from personalities. The only noticeable side effect from shock treatment is a partly subjective one—less profundity of character, more stereotyped and flatter emotions. It is as though the intention were to push a human being a notch down the ladder of phylogenesis, not far enough to become another animal but enough not to participate in the more painful paradoxes of the human condition.

The cortex is divided down its middle by a longitudinal sulcus; the medial surfaces of the two hemispheres come together embryogenically and flatten against each other, trapping mesenchyme in between. The hemispheres emerge as mirrors of one another, reminiscent of twins forming from the same blastula. We have two brains: a left one and a right one. The left hemisphere of the brain connects to the right side of the body, usually the dominant side; it is supposedly engaged in rational and analytical thought and is thus empirical and critical. The right hemisphere is said to be more intuitive and creative, identifying patterns and forming images; it connects to the left side of the body. These characterizations have been overdramatized in pop psychology, as if there were two selves waiting to be integrated through our personal development. The hemispheres of the brain are not exclusive in what they do; they have homologous topographies, and, to a large degree, duplicate each other's functions. There are eyes and ears in both, albeit connected to the opposite halves of the body.

Andrew Weil calls these "symbolic designations of the two phases of mind."[5] He laments that certain schools of creative training now prescribe binding the right arm so that the intuitive hemisphere is forced to develop. This kind of "new age" scientism is an almost ludicrous literalization of the dichotomy.

The two hemispheres are connected by the corpus callosum, a long band of fibers like that joining the hippocampus to the neocortex. Made entirely of white matter, this commissure is bent double in front and curved ventrally. Fibers continue to be added to it as the cortex expands. It is the basic line of communication between the two hemispheres and fuses their dual realities. When the corpus callosum is cut, one side of the brain does not see objects presented to the other side. This isolation is a natural occurrence in animals that do not have a band of connecting fibers.

Not all regions of the two hemispheres are homologous. The speech center appears in the dominant hemisphere of the brain only, which is, in most people, the left region (coordinating the right side of the body). Thus, when the corpus callosum is cut, a person may not be able to name an object presented to the left eye (for instance, a spoon) but still may use it properly. In the portion of the nondominant hemisphere corresponding to the speech center no function has yet been discovered.

A blood arterial system runs over the surface of the brain and through its interior, diffusing oxygen and glucose, and carrying away waste. Although the brain is only 2% of the body it consumes 20% of its oxygen in adults (50% in infants). The moment this torrent of blood from the heart ceases or is poisoned, consciousness ends. Our mental existence is as physical as a lake and involves a massive expenditure of the Earth's resources.

At an early stage of embryogenesis, the forebrain develops evaginations of its own: two lateral sacs, optic vesicles from which eyes will form. These islands are induced from the forebrain by adjoining islands of mesenchyme. They swell laterally while their connections to the forebrain shrink—mushroom heads on stalks. The vesicles now induce the adjacent ectoderm of the head into lens placodes. A deep telescoping begins—ancient eye within ancient eye. From the center outward each placode collapses, creating a lens pit; the boundaries of the placodes surge forward into each other and fuse as lens vesicles. The optic vesicles invaginate also and become the double-layered optic cups. Their outermost layer bears the pigment epithelium, which forms continuous with the pigment epithelium of the ciliary body. The inner layer differentiates as a thick neural zone, and, induced by the lens, it teems with neuroblasts. This becomes the embryonic image-forming surface, the retina.

The cells of the retina further particularize into rods and cones, and bipolar and ganglion cells. The axons of the ganglion cells travel into the inner wall of the optic stalk, making it a long nerve. The cones form images in daylight and are receptors of colors; the rods are receptive to shapes in dim light. The ciliary body is mostly a forward projection of the non-neural part of the retina, whereas its muscle is derived from mesenchyme beside the optic cup.

The lens vesicles meanwhile separate completely from the ectoderm and sink into the optic cups. The contractile membrane, the iris, is induced from the lens-covering rim of each optic cup: Its connective tissue is mesenchymal. The dilator and sphincter muscles of the iris differentiate out of neuroectoderm on the optic cup.

The long cell columns of the lens are induced by vascular mesenchyme, and the lens itself induces the ectoderm over it to become the cornea, which refracts light through it onto the retina. The vascular layer of the eye, the choroid, and the delicate membrane beneath the cornea, the sclera, both form from mesenchyme surrounding the optic cup.

The lens-ciliary muscle system controls the focus of an image on the retina, while the iris expands and contracts the lens at the pupil to accommodate decreased and increased amounts of light. Lying directly in front of the lens, the circular fibers of the iris can reduce the pupillary membrane to a minute pinhole when external brightness would otherwise make image formation impossible. This reflex is already developed in darkness before birth.

We have constructed all our cameras and light-magnifying instruments by imitation of these living principles, an event Sherrington reverses to make us realize how remarkable is the formation of the eyes:

"If a craftsman sought to construct an optical camera, let us say for photography, he would turn for his materials to wood and metal and glass. He would not expect to have to provide the acutal motor adjusting the focal length or the size of the aperture admitting light. He would leave the motor power out. If told to relinquish wood and metal and glass and to use instead some albumen, salt and water, he certainly would not proceed even to begin. Yet this is what that little pin's-head bud of multiplying cells, the starting embryo, proceeds to do. And in a number of weeks it will have all ready. I call it a bud, but it is a system separate from that of its parent, although feeding itself on juices from its mother. And the eye it is going to make will be made out of those juices. Its whole self is at its setting out not one ten-thousandth part the size of the eyeball it sets about to produce. Indeed it will make two eyeballs built and finished to one standard so that the mind can read their two pictures together as one. The magic in those juices goes by the chemical names, protein, sugar, fat, salts, water. Of them 80% is water."[6]

In an age of biomechanical determinism we have made the cerebral cortex the explanation for ourselves. An abundance of neurons alone seems to have made us more complex, more reflective, and more compassionate than animals, though we are still quite lacking in these attributes even by our own standards. So vivid is the image of higher intelligence in the convoluted cortex that we imagine higher beings, science-fiction creatures from our future or from other worlds, who have swollen crania, all sorts of new lobes bulging out of the neocortex, and thus are more intelligent and civilized than we.

We have translated symbols into artifacts in a mode unlike any previously on this world. We have made over the planet from the inside of our brain, but we have not made over the laws of nature, so we remain epiphenomena, tied to a physical evolution we cannot transcend.

We have enough wisdom that our situation is as exhilarating as it is desperate. There is no way we can truly solve the problems we bring into being, as people in relationships, as civilizations, or even within our own consciousness. Whether these could be solved with more neurons is impossible to know. I doubt it somehow. We have developed our culture and technology from the Stone Age without additional neurons (and perhaps with somewhat fewer neurons than Neanderthal and Cro-Magnon). We have no reason to believe that the physiological gradient of the cortex leads to intelligence beyond ours. Much of our brain is unused; a great portion of cerebral energy goes to inhibit consciousness, not to increase it. Perhaps becoming conscious in the way we are conscious is a plateau we cannot transcend along the same lines. More neurons would probably not be adaptive in the ecological niche we inhabit (we must not fantasize that mutations giving us powers to prevent self-destruction could arise from conventional evolution; the genes are not issue oriented). We have more than enough difficulty integrating what we already know. More signals and levels of abstraction might have driven us mad.

16. Mind

The cosmos has evolved through mind. Stars and worlds come into being through the minds of their creatures. Sun and sand and seas are the source of consciousness but have their only real occurrence within it. It is needless to belabor the tree that falls unattended in the forest. Whether it is real or not, whether it falls or not, signifies a paradox deeply rooted in our condition. Being-and-nothingness is the quintessential problem for the brain as it develops; and certainly enough such "trees" fall on other worlds, eternally unconscious, to warn us against any modern presumption that we have become cosmic witnesses.

At the moment life begins, mind begins too. Even in the primitive sponges where cells are barely associated, matter begins to astonish itself by its own existence. "How?" it asks. "How am I here?" We can see the question forming in crabs and snails, in the eyes of the sullen fishes. Predators may prowl, kill, and feed unconsciously, but there is a hint of consciousness, a doubt, a brief dissociation between the self and the act. "It is ludicrous that I should be this thing," all things think. The bear whines restlessly, the lynx looks curiously to the clouds and blue sky—whatever it calls them to itself; even the walrus wonders why it has to cart this immense blubber across the rocks. Each of the animals is a combination of something and nothing, a question that refuses to ask itself, an obliteration that floats on the margins of waves of air before crumbling in a heap of feathers and protoplasm. The frog sits on the shore and chants. Lion and zebra face and become each other at the kill. Cats and dogs occasionally look to us for the answer because we act as though we know what is happening. Monkeys in the Amazonian foliage look down at the Indians painting themselves for the ceremony, not aware that they are equally not aware.

At first it must have seemed like voices of gods. The first humans could not have known what it was. They awoke into an empty space, perhaps filled with the chattering of the animals. We, on the other hand, awaken into a room filled with language and symbols, and we identify our own spirits with those things that have already been named around us. The first humans had to break through the silence of the Earth.

Many of our deepest interior senses are so ancient and primitive that we have little sense of where they are located and if they even exist as discrete receptors at all. We feel, as watery tissue, the gravity of the Earth; the equilibrium of the Earth, Moon, and Sun; and, to some degree, the minute but profound gravity of the universe itself through the weft of the Solar System. We experience the subtle pressure of underground water, the vibration of the rotating planet, the diurnal-nocturnal polarity expressed in the circadian rhythms of our bodies. Time itself permeates our cells and organs as octaves of a biological clock. We may also inherit instincts from various layers of the brain—senses of number, courtship, mating, child-rearing, fears of dangers long past, love for something no longer incarnate. And then there are possible ''extrasensory'' senses, still evolving or vestigial, through which people ''read'' the minds of others (or the mind of matter itself). It is almost impossible to distinguish our own discrete being from all the ancestral layers inside us. They overwhelm children with consciousness like an immortal intimation, and then our lives swallow the mystery leaving only a faint nostalgia.

Men and women come into consciousness slowly and also all at once, awaking after a long dream in the animal kingdom, and before that, a dream in dark matter. For all the millennia and lifetimes afterwards, they will try to go back into that passage to discover where they came from. It is closed, in some ultimate sense, to philosophers, psychologists, and even priests and artists, except for the glimmering or flash, of what we do not know.

Psychologists do not refute the long biological and anthropological prehistory of consciousness, but they are more concerned with its ostensible recapitulation in psyche—for each individual seems to undergo a birth of mind something like the original dawning of consciousness. Freud formalized what mankind had historically feared: Most energy is still unconscious, in nature and in self. He named the reservoir of this energy in us the *Id,* i.e., *it*—because it is beyond knowing or control.

The human being forms as a fragile membrane upon a fathomless dark stone. The unconscious is anything which had energy but is no longer experienced. Thus, most of the universe of stars and light and matter is permanently unconscious by our standards. The primary instincts of the animals in our lineage (and of life itself) are summarized ontogenetically and press upon our preconscious. Even if we do not recognize them, we experience and enact them.

Carl Jung came to a similar conviction for quite different reasons. He proposed that the unconscious universe and the unconscious organism coincide, and that forms composed within their synchronicity transcend mind and matter, i.e., are archetypes. The spiritualists call them gods. They may even be universal nonhuman waves of vital energy, embodied by plants and animals, and by crystals, minerals, and stars in other ways.

These intuited forces clearly transcend biological life, but they are dependent on the nervous system for their manifestation in personality. The id, the unconscious mind, and the archetypes *are* the embryo. Finally, what becomes conscious is what makes it through all this resistance, of matter, of protoplasm, and of our own individuation. Everything else floats bodiless through cosmos (i.e., through "mind").

For the last century, since Darwin's *Origin of Species* convinced us that all living creatures must have arisen from chance fortuitous mutations and blind selection of the fittest, we have assumed that mind on Earth evolved only as an efficient means of adapting to new environments. We have become conscious because lineages of creatures with rudimentary brains were uniquely successful at finding food and eluding hazards. There is no intrinsic justification for "mind" in biological science. If mindless moss animals, sponges, and jellyfish could have seized enough of the potential biomass on the planet, the waves could have washed to the shore and the trees fallen in the forest unheard for eternity. But neurons made more canny cell clusters, animals with novel skills, and these survived, mutated, and gave birth to even more intelligent creatures. Brains were the outcome of an accidental strategy pursued to its inevitable conclusion. This does not explain the existential and metaphysical aspects of mind, but biologists and anthropologists assume that these came later, or were incidental to the feedback of successful evolution.

The lineage of primates was emerging from the old mammals almost seventy million years ago (at the end of the Cretaceous and the beginning of the Palaeocene era). The order was derived from a simple reptile-like mammal much in the way marsupials, carnivores, and ungulates were; at this stage of the divergence of histories the insectivores and rodent-like animals were our closest relatives. Moles then dug their way into the mantle of the Earth and became dim-sighted underground predators. Otter shrews and marsh shrews took to the water. Anteaters, aardvarks, and other insectivores coevolved with groups of insects, their anatomy shaped by specialized diets. But some groups of shrews migrated into the trees, and from these evolved the flying chiroptera (the bats), the modern tree shrews, and also the extinct ancestors of the primate line.

The tree shrews are the living primates closest to the hypothetical ancestor of the order. These tiny foragers have few specializations, but the adaptation of their forerunners to a life in the trees, though hardly original for small carnivores, had epochal ramifications for the Earth. Through arboreal evolution the generations of primates slowly synthesized an array of disparate mutations into an efficient mode of survival and a radically new form of intelligence.

If one anatomical change preceded and influenced the others it may well have been the differentiation of the limbs into two distinct pairs. The fore and hind limbs (and their joints and digits) were modified very early in the history of the primates, most likely in response to their differing roles in locomotion through the trees. In jumping, an animal is propelled by its hindlimbs working as a lever, and in climbing, it pulls itself through the branches with its forelimbs. The primate limbs are more than a differentiated series of props and levers for support and motion; they are attached mobilely to the spine, its girdles, and each other and are almost acrobatically loose. The distal joints of the fibula and radius rotate, allowing orbits for hands and feet to explore at a variety of angles. Even the bones of the forearm (the radius and ulna) are separate structures. Comparatively, the skeletons of horses and leopards are fixed machine-like chassis.

Primate mobility is most restricted (at least among contemporary members) in the tree shrews, for they have continued to specialize in a direction counter to the generalized arboreality of their order. The fore- and hindlimbs of these animals are only moderately differentiated from each other, and their shoulder girdle is consequently somewhat more flexible than that of terrestrially oriented mice and anteaters. The shrews are alert and agile climbers with sharp vision and a cerebral cortex considerably more developed than the cortex of the insectivores.

By pure somatic standards, primates are strikingly ill-equipped —an order founded by the poor cousins of mice and moles. Their crooked limbs deny them the natural swiftness of their mammalian heritage; the diminishment of their teeth and claws robs them of natural weapons; and the reduction of their snout and concomitant atrophy of the olfactory lobes of the brain in the higher primates has cost them the exquisite mammalian tracking ability. The most notable thing about the direct human lineage is that it represents, first, the least specialized vertebrates, secondly, the least specialized mammals, and finally, the least specialized primates and anthropoids (at least in terms of athletic talents and inherent tools). If we imagine the various ecological niches being given out as kingdoms, the creature who waited the longest was ceded the strangest domain of all—one that did not even entail an explicit somatic resource but instead had to be imagined.

Sometime back in the Eocene some fifty to sixty million years ago, a second major line diverged from ancient primate populations; the present-day genera most closely resembling them are the lemurs and lorises; the tarsiers are close relatives. These animals are classically adapted to the trees and retain few insectivore features by comparison with the shrews. Their fore- and hindlimbs are strongly differentiated for climbing, thrust, and support, and their clavicle and neck-jaw muscles are oriented so that they can turn their heads and look about. At the same time, their center of gravity has shifted backward juxtapositional to their hind musculature and the tail has emerged as a rudder: lemurs are able to sit up. Without this rearward shift the bulky animals could not distribute their force properly and their leaps would displace into spins.

From fishes to hominids virtually no new bones are added, even in the head, so braincase must be molded from muzzle. Whereas the jaw of the average mammal is little more than a grasping, snapping instrument, the lemur jawbone is significantly atrophied and deflected toward a position below the braincase and at an angle to it. With quick snapping now less important, the powerful forces of compression in the skull can be used for chewing vegetation. But the changes in the skeletal proportions of the head also create space for the gradual expansion of the cranium and its cerebral contents. As the primate life-style becomes more coherent, one transition initiates another, disparate trends are fused, and anatomy is restructured allometrically by subtle degrees. Mind is the ultimate beneficiary.

Tree navigators must be able to see and interpret three-dimensionality; thus, for generations, random mutations expanding visual receptors spread through the order, and abundant nerve fibers from the corresponding optic areas of the brain were induced into the new tissue. Even in the lower primates, the nuclear elements of the thalamus have begun differentiating, particularly those involving sight. The cerebellum is becoming fissured, i.e., more complex, able to coordinate a greater variety of activities. But the biggest change of all occurs in the cerebral cortex and its sensory-motor projections to the limbs. This corticospinal web expands until the primates are quick, delicate acrobats, semiconscious tree-elves.

The topology of consciousness shows its distinctive folding in the lemur cerebral cortex. The separation of lobes, which sustains complexity of behavior, has begun in the temporal and occipital areas. But the frontal and parietal areas lack limiting sulci, and, overall, the prosimians are not as cerebral as monkeys and apes. Many are adapted to night foraging with special rod retinas, and some species hibernate through dry seasons.

The tarsiers are small, extremely wide-eyed prosimians with enormously enlarged tarsal bones in their feet (for leaping through trees). Their big eyes feed a huge visual cortex and correspondingly developed occipital lobe, and this gives them a humanoid appearance. Despite the fact their brain is otherwise primitive and lacks the convolutions of the lemurs, at least one zoologist has derived the human line indirectly from the tarsiers (rather than from the monkeys or apes) on the basis of their upright posture and peculiar brains.

The first anthropoids, the ancestors of monkeys, apes, and humans, diverged from all other primates some thirty to forty million years ago during the Oligocene era. The cerebral cortices of these animals were already deeply convoluted; in living ape and human brains, secondary sulci have overgrown and obscured the simian and prosimian layers underneath them. Even the most primitive modern monkey brain weighs three times more than a comparable prosimian brain.

The anthropoid brain has also changed qualitatively. A whole series of new nerve tracts run directly from the cortex to the spine. These fibers (called pyramidal) were apparently induced in the simultaneous evolution of the spine and the cortex, and they supersede the slower more diffuse extrapyramidal relays between the brainstem and the motor neurons. Many cortical fibers also skip their older internuncial relays and run directly to motor neurons— the greatest percentage of these to the distal muscles of hands and feet but some also to proximal trunk muscles. There is a concurrent development of the sensory fibers which record and adjust muscle contractions. Information from these proprioceptors and from the tactile neurons of the skin are transmitted in quick discrete quanta, at least by comparison with the disperse peripheral pathways in the lower mammals. The primates are strung very tightly and are on the verge of waking up.

On the faces of these animals the two eyes have moved toward one another creating an overlapping stereoscopic field. A whole new set of vistas rushes through them as these tree-dwellers leap and swing from point to point of proprioception. It is in the tree tops, perhaps, where space and time take on animal meaning. Large wingless mammals navigating perilously above the ground must interpret interstices of branches along the delicate slide of gravity; they must coordinate zigzag images. Birds do not require abstract algebra, for their wings lift them into infinite space; but the flight of arboreal primates changes unpredictably with the topology of the trees. Our own internal organs may be aquatic in origin, but the highest lobes of our brain are rough holograms of receptors hurtling through obstacles. We still use these synapses to find our way through abstract thought and to measure the distances between stars and electrons. It is no wonder that the "jumps" of subatomic particles confound us.

The limbs of the anthropoids are prehensile and used for clinging, gathering food, examining objects, and hanging from branches. The sensor pads of the manus and pes (hand and foot) become tenderly tipped receptors reminiscent of the octopus's tentacles, and some lines of apes have the ability to oppose their thumbs to their other fingers in preadaptation to tool making and counting. The forelimbs of the apes have been modified for brachiation, that is, for suspending the weight of their body from their arms and swinging through the trees. This requires muscular support in the thorax and freedom at the shoulder joint. The frontal clavicle guides the limb in a wide arc, and the dorsal bone of the shoulder girdle, the scapula, is able to move back and forth across the thorax—forward for pushing, and backward for pulling and climbing. The scapula has rotated vertebrally to permit the fullest elevation of the arms. Meanwhile the arms have developed the freedom to rotate at the joints of the radius and the ulna, and the hands have become flexible. Lengthening of the phalanges and reduction of the thumb has gone much further in apes than in hominids, as is appropriate for brachiation. The thorax of apes is broadened by comparison with that of monkeys and is flattened from front to back rather than laterally, which gives needed support to the trunk.

Although the line of extinct hominids leading to our species seems more similar to apes than monkeys (at least in fossils), we did not, in truth, evolve from either of them. Our ancestor most likely arose from the line leading to the apes sometime between ten and twenty million years ago at the end of the Miocene or during the early Pleistocene era (though we cannot exclude the possibility that hominids diverged earlier and that we share our last ancestor with other anthropoids at a point before the discursion of apes and monkeys). The apes today are represented by four genera: gorillas, chimpanzees, orangutans, and gibbons. We resemble these more than the monkeys in many characteristics: the dimensions and configuration of our brain, our general skeletal construction, our internal organs, the development of square cusped teeth at the back of the jaw (the molars), and our atrophied tail.

The brain of the ape lies, anatomically, somewhere between the monkey brain and the human brain, in size and in number and configuration of folds. The first ostensible humans, the australopithecines who appeared in Southern Africa some two to four million years ago, had brains only slightly larger than those of chimpanzees and gorillas (at least so far as archaeologists can tell from measurements of their fossil skulls). This means that individual chimpanzees and gorillas may be brainier than some of the first men; yet they must lack some subtleties of neural configuration not evident in mere skulls. Or perhaps culture is a gradual collective insight that modern apes might yet possess in rudimentary form if there were not other intelligent hominids in the way.

Man's ancestors were semibrachiators, their arms already freed for tool use before they left the forest. Their stout humerus bones were accustomed to weight, and their spine had stiffened as a vertical rod (a fossil australopithecine scapula falls, morphologically, somewhere between orangutan and human). Our ancestors could not have been full brachiators or their bone structure would have become too specialized for the hominid range of limb movements.

The early ''apes'' were unknowingly preadapted to a bipedal existence on the plains, but in no way would they have been tempted to leave their protected arbor for such a barren ecosphere. The actual human ancestors were probably marginal tree apes, outcasts. A number of factors might have forced them onto the savannah, for example, overpopulation, natural disasters, contraction of the forests (for which there is tentative evidence in Africa at the time hominids emerged). The earliest fossils that remain classified as hominid are those of the Indian and Kenyan creature Ramapithecus, known only from jawbones and teeth. He is not an ape, and he is not a human, so he is considered an extinct animal in the line leading to *Homo sapiens.*

From the time hominids entered the open plains their skeletal structure, head and limb mobility, sharp vision, and incipient intelligence took on radically new ecological meanings, contributing to survival in an environment in which fast-moving beasts both stalked and hid from them. But they were also handicapped, for agile tree-dwellers are not swift gallopers, and sensitive phalanges are ineffective weapons. Outside the forest the array of selective factors working against hominization (as we retroactively define our special preserve) would have been neutralized and, in some cases, immediately counteracted, but the positive effects of a new genotype would have taken generations to manifest in a life-style. At first his fate must have seemed like exile to the preconscious psyche of Ramapithecus and his forerunners. He had lost the bounty and protection of the trees and was now threatened by swift and brutal shadows. He was handicapped by the very features of incipient bipedalism that preadapted him to symbols and tools.

The ability to grasp objects—stones, sticks, bones—was the likely turning point in hominid survival. Even in the Pliocene forests anthropoids must have found and abandoned spontaneous weapons and tools. Modern apes, in experimental situations, use props to reach food placed out of their reach, and, in the wild, have been observed manipulating sticks to dig termites out of a mound. But the full potential of these objects eludes them, so they do not create appropriate symbols for imagined future uses.

Armed with sticks and bones, hominids were a fair match for the ancestors of dogs, lions, and pigs. The hurling of spears and stones—reenacted in Olympic competition and ball games—is one of the first sentient acts of our lineage. The hard-throwing baseball pitcher and accurate quarterback are still warrior-heroes. There is even archaeological evidence of the hunt at the early australopithecine sites—the bashed-in skulls of antelopes and other game animals (which may have yielded fresh material for future weapons). These hominids were barely, if at all, more intelligent than apes, and they limped and crawled by the standards of modern bipedalism. But they had entered the realm of imagination.

Tool-making required intelligence, and, at the same time, apparently contributed to the enlargement of the braincase. As stones were substituted for teeth, the human incisors and canines became smaller and the jawbone diminished, perhaps because random mutations reducing their size were no longer maladaptive. Meanwhile, mutations expanding the cranium were clearly adaptive. With their arms freed to bear spears and knives, bipedal creatures were more than just limping apes, especially as they began to put their new mindfulness to strategic use.

The first tools were probably stripped branches and rocks; eventually, creatures began to craft their own artifacts. The oldest such remnants of aboriginal man are the pebbles and small bits of river gravel chipped by australopithecines, likely used as hand knives, chisels, and scrapers. Tentative tool-making may have arisen thousands of times independently among different groups. Our own devotion to creating technologies might fool us. There must have been an implicit psychological resistance to an act even as innocent as tool-making, for to imagine one's self cutting a preconceived shape in stone is to become conscious, and to project, as well, all the borderline spirits that attend consciousness. The first tools may have been only crude unidirectional flakes cut onto the borders of small rocks, but they were images, even more powerful than those found millennia later on the walls of the Lascaux caves, for they were prior and seminal to them. If a creature could make an image at all, he could mirror his own identity, and then there would be no way he could avoid staring into that mirror and seeing himself. It was the beginning of *I*, the remote apperception that *I* and *It* were two different waves intersecting at a point of identity. The mirror was perhaps a carefully considered stone grasped between the opposable digits of an animal paw.

Australopithecus was eventually succeeded by the more cerebral fire-using creature Pithecanthropus. This far-flung race included Heidelberg man in Europe, Sinanthropus in ancient Cathay, and Java man in Southeast Asia. Fully bipedal, Pithecanthropus is considered the first member of our genus: *Homo erectus.*

In the perfection of bipedal locomotion, the human pelvis expanded in three dimensions and in breadth so that the flexor and extensor muscles were attached further out from their points of pivot. In general the blade of the ilium shortened, widened, and became curved; it now lies behind the point of articulation of the femur bone instead of to its side. The bending outward of the ilium created a new point of attachment for the muscles of the floor of the pelvic basin, but this curve has not proceeded as far in women, who require a pelvic opening for the birth canal. The sacrum also shifted its orientation to receive more of the weight distributed through the iliac spine. Muscles from the thorax to the ilium now lift the trunk over the spine at each step and transmit the center of gravity over the foot—bipedal locomotion with all the weight on one leg at a time. Meanwhile the human foot has narrowed, distributing its muscles along new load lines. Grounded in a tripod of the big toe, small toe, and heel, the appendage has lost its grasping tactile phalanges (which are shortened and weakened), but it has become an innervated volar pad with sensitive terminal receptors on its digits—the base for bipedal movement.

Driven from winter to winter, firepit to cave, Pithecanthropus enjoyed brief cloudberry summers in the North and long rainy autumns in the South. These ancient men and women are our grandfathers and grandmothers. They inhabited the epochs we call the Pleistocene, the Ice Age—glaciations hundreds of thousands of years long interrupted by brief interglacial aeons (of which the Holocene era of our recorded history may be only the most recent). The harsh climate no doubt speeded their dependence on fire and weapons and gods.

Pithecanthropus produced thousands of generations of hand-axes of lava, quartz, quartzite, and flint. His early Chellean tools were flaked around the edges with strokes in alternate directions so that two faces crossed in a staggered margin. These scarred bifaces spread with their makers across Eurasia, and by the second interglacial of the Pleistocene (some three hundred thousand years ago) they gave birth to more sophisticated styles and artisans. The later Acheulian hand-axes were roughened initially by flaking, and then finished with a second sharper edge by percussion from a bone or wooden baton. In the later Pleistocene, only one hundred thousand years ago, Neanderthal appeared, a fully cerebral omnivore with state-of-the-art Stone Age skills. He chipped fine edges on his side-scrapers and delicately retouched his chert blades. He also learned how to prepare large cores from which dozens of separate tools could be manufactured, each with the stroke of a hammerstone or baton.

During the last fifty thousand years, leading directly to the first written tablets and farming villages, tools became as labyrinthine and intricate as our image of the cortex itself. Cro-Magnon's "kit" included awls of antler, points, burins, concave saws, needles, fishhooks, willow-leaf blades, arrowheads, lunar counting devices in bone, small statues of either mythical beings or actual people, and pigments of crushed oxides and plants. Caves from Spain to Queensland and Argentina announce (at different points in global history) the birth of the gods. The hunt was now a ceremony, the game animals numinous beings. We had entered the imaginal world with no way out.

Although writing of a later phase of development, the Jungian psychologist Erich Neumann captures a certain aspect of what the twilight world of these creatures must have been:

"In the early phase of consciousness, the numinosity of the archetype . . . exceeds man's power of representation, so much so that at first no form can be given to it. And when later the primordial archetype takes form in the imagination of man, its representations are often monstrous and inhuman. This is the phase of chimerical creatures composed of different animals or of animal and man—the griffins, sphinxes, harpies, for example—and also of such monstrosities as phallic and bearded mothers. It is only when consciousness learns to look at phenomena from a certain distance, to react more subtly, to differentiate and distinguish, that the mixture of symbols prevailing in the primordial archetype separates into the groups of symbols characteristic of a single archetype or of a group of related archetypes."[1]

Very early in their lineage men and women must have formed communities, perhaps small bands of interrelated families or tiny migratory villages. This was an inevitable result of increased intelligence and strategic cooperation, but it was also a biological necessity. One of the hallmarks of primate evolution was the extension of the prenatal life of the embryo (a refinement of a mammalian trend). A long gestation permits delicate cerebralization and formation of neural structures in a protective uterus. The pelvic girdle, however, can accommodate only so large a cranium, and cerebralization beyond that point is lethal for both mother and child. In human development the cortex must continue its expansion outside the womb, so infants are born unfinished and must be cared for in their early years. Whether it truly happened that way, we imagine that the guardianship and education of the young led inevitably to families, sexual division of labor, and the tribe. As new animals reached manhood and womanhood they were initiated into the institutions and rites of their group. So powerful has this symbolic realm become today that we have almost forgotten our somatic reality in our attachment to it.

Incest taboo was probably one of the first belief systems—a collection of inexplicable customs with a precedential injunction. "Marry out, or die out,"[2] said the anthropologist Claude Lévi-Strauss in explaining an irrational belief so strong it might almost be an instinct. Perhaps this was how the long truces between Pleistocene hunting bands came about—by kinship (we assume such truces by our very survival). Neighboring communities consisted of sisters, brothers-in-laws, and nephews. They could not be enemies. *Eros* held back the wild beast. Beneficial genes flowed from society to society, from Africa to Europe, to Asia and Southeast Asia and back, later to the Americas and Australia. Through intermarriage humanity became a single family; we are all related, though we may have to go back to the Stone Age itself to find our common ancestors. No genealogies (but the genes themselves) bear this testament; the myths of the North

American Indians trace the same primal transformations as the myths of the Norse and the Celts.

The genes that spread during the Pleistocene reflected the "humanizing" and erotic influence of culture. With the reduction of the muzzle and corresponding enlargement of the cranium, the center of gravity of the human head retreated almost to its point of pivot upon the spine. This meant that the nuchal muscles, which attach at the back of the skull and support the head, gradually atrophied, and by the late Pleistocene the bony crest at which they attached on either side of the back of the skull had been almost eliminated. Men and women rose up out of the ground to stand erect like beanstalks in the Sun.

The human neck became longer, the face flatter. Bone and musculature receded from the sensitive regions of eye-to-eye contact and speech, and the face became a zone of subtle communication and contact. Its subtler etchings may be manifestations of profound neuroglandular changes that leave no archaeological trace. Community life required some cortical control of the limbic system and basal ganglia. Boastfulness, passion, anger, territoriality, jealousy, and other instinctual drives were masked in styles of personal ritual and play. They could still occur (in fact, they could not be prevented), but they had to be acknowledged, become partially conscious; otherwise, society would have been sociopathically torn apart at its inception. As adrenal and sexual functions were sublimated cortically, the facial features which supported their emotional outbursts subtilized. Through generations the ferocious or enraged look of the prognathous beast was replaced by the more dreamy gaze of a being impregnated with neurons and images.

The human body also shed its claws and heavy coat of hair. At the same time, abundant sweat glands developed in the layers of skin, their apparent function to moisten the epiderm and aid the tactile receptors. Our nakedness may be an indirect result of diurnal hunting which required long-distance marches in the heat, but it had psychological consequences, as did the sweat glands. It is no surprise that hair loss and fatty glands coincide in a creature who is becoming intimate both to himself and others. The glandular scents were erotic, and the whole epidermal complex expressed sexual preference as well as the hunt—how the awakening males and females wanted their partners, i.e., themselves, to manifest. Men and women were able to touch directly and to experience their sameness and difference as naked bodies. Ares and Aphrodite later emerge from the shadows of Olympus.

There was a place in culture, i.e., in nature, for one animal who named the stars and mapped their cycles; and there was a place for an animal who collected and classified esoteric signs; there was a place for the story-teller, the healing shaman, and the herbalist. The intelligence of these first doctors and seers transformed the tribe and began history.

The moment when man and woman became conscious is as ineffable as the twilight between being and not-being in ourselves. It is objectified as a ray beamed by an extraterrestrial intelligence in Arthur Clarke's *2001: A Space Odyssey* and Robert Anton Wilson's *Cosmic Trigger*. But until proven otherwise, that ray also emanates within. Wise ones oversaw our evolution through their proxy in the cortex. Whether this was a novel wisdom thrust into the world by bestially sired hominids themselves, the collective unconscious intelligence of the whole prior bestial world finally manifested somatically, or some other supernatural complex honored by vitalists and spiritualists, little matters at this point. It occurred on the Earth materially and physiologically out of its own necessity, and it shaped its nature as it evolved.

The awakening of culture is like a dream (perhaps it even proceeded in a dream). As man and woman begin to conceive of the clouds and the stars, the animals and the land itself, as things, things separate in their own nature; then natural objects reenter society as symbols and totems. Later, Cro-Magnon is to "invent" agriculture by the fact that everywhere he disturbs the climax forest, weeds and herbs grow; he is pursued by grains, amaranths and sunflowers, and his garbage heaps are bait for the animals he will later domesticate, animals which call attention to the possibility by beginning to domesticate themselves.

The Australian Aborigines consider the landscape itself a dream of their ancestors, a dream which gradually put the Sun, Moon, and stars in their paths and every stone and hill in place, and brought each creature and waterhole into being. Is this the dream from which all of the other animals flee, an interminable labyrinth in which we are doomed to encircle ourselves forever while they remain free?

Remember the cry of history as a nightmare from which we cannot awake.[3] For fifty million years primates existed in the forests of New World and Old without the hint of man. Millennia earlier dolphins and whales returned to the sea, and more generations of them have lived and died there than the whole of mankind back to the aboriginal tribes of Africa.

The disjunction between society and nature is never complete. This far into history, nature has been unable to reestablish pure unconsciousness, unbroken jungles and algae seas. Nor have we been able to extinguish, by cities and machines and industrial fires, the dormancy that constitutes most of this planet still. Unless we fail, both of us have a long way to go, hand in hand, approaching the light as we approach the final darkness.

17. Birth

Sperm and egg are separate germ creatures. After their merger a new organism exists but with no awareness of itself in the usual sense; it has no nervous system with which to apprehend its own being. Whether consciousness dawns on some transcendental level before the protoplasm knits nerves is unanswerable, but to modern science the blastula is simply a very complicated chemical reaction, a densely coordinated field of carbon rings.

By the third week of life the neural groove has driven the ectoderm down into the endoderm, and the heart is beginning to quiver. On the twenty-second day it begins to beat as the neural folds fuse. One day later, eyes and ears are forming. It is now a living animal turning in sleep. It knows nothing but the archetypal, the universal; but it is alive. In the fifth week it has hands and feet as well as a vestigial tail. Its lips are pursed. In the sixth week the neck contracts to withdraw the hand, a reflex arc. In the eighth week toe rays form in the webbed feet; the eyes open. In the eleventh week the embryo is swallowing. The head grimaces during the fourteenth week—the passage of a memory?, and if so, whose? At twenty-nine weeks the embryo begins sucking and making sounds.

When does matter awaken to itself? When does spirit merge with fluid and flesh? The question cannot be answered, either biologically or theologically. There is no first moment, either for spirit or for life, or the moment is outside of time and awakens as an eternal image without beginning. Consciousness arises from collective forces that seem to require it in order to exist themselves.

The lines we draw for the sake of argument will always be artificial. Does the being exist in a spirit world before it enters the embryo? Does it manifest instantly with fertilization? If it is not a human being at fertilization, how long does it take—seconds? days? weeks? When the nervous system develops? At what point in the development of the nervous system? If we have to answer these questions, we are in trouble, for they cannot be answered. And that is why our materialist society has produced the impossible debate between the right-to-life of the embryo and the right-to-abortion of the mother.

* * *

237

In conventional medicine and psychiatry the foetus is a subhuman animal, without a personality or an intelligence. It embodies only the most primitive instincts, the autonomic forerunners of emotions. Even when this creature emerges from the womb it is considered a virtual nonentity, the figment or raw material of a person. "Psychoanalysts have generally speculated that newborns are passive, unable to differentiate themselves from others, and behave as they do [only] to reduce 'drives.' "[1]

The newborn is not treated like an honored or even a welcome guest at most hospitals. It is "one more bean sandwich," slapped on the rump, separated from its mother, collected in a dehumanizing waiting zone with dozens of other newborns, and then put through a traumatic genital operation (circumcision) without anaesthesia. It is as though we want the child to be no one. We want it to begin as a pagan beast that we indoctrinate and educate, that we make human by our own humanity. We assume that until we teach it symbols it cannot know anything or be anyone. Some doctors even ignore its pain and suffering, treating it as a collection of unconscious semirandom nerve endings.

The child emerging from the uterus is no passive cipher, but an ancient being capable of directed movement, possessed of intrinsic power and strong will. In more spiritual tribes than ours this creature is the avatar of an alien order. It brings wisdom into the world.

Recent experiments (since the early 1970s) have shown traditional Western viewpoints to be simplistic and perhaps even self-serving. Newborn babies are not just packages of meat. When less than thirty-five hours old they start crying at the sound of other babies crying (but not at computer simulations of the sound), and *"stop* at the recorded sound of their *own* crying, indicating that they not only heard, but recognized their own voices."[2]

We routinely assume that a newborn has no language and that, until it can speak, it does not know what *it* is or who *we* are. Yet, in the hours after birth, children track speech variations in many languagaes (probably any language)—English, Cherokee, Chinese—but ignore nonsense syllables. In one 1980 experiment a just-born child regularly looked up at his father looking down, and, as the father began to move his right hand away, the baby reached up with his left hand and "grasped his father's finger."[3]

Although the newborn cannot speak, it apparently recognizes what is happening to it and internalizes the events of its birth in a coherent fashion. An adult returned to the moment of being born by either spontaneous dream or hypnosis can excavate the record of his or her primal self by bringing language to the memories. The resulting reenactments are often precisely true to the memories of the mother, the physician, or others attending. Years later, human beings remember being stuck in the birth canal, the pain of the forceps, the incidental comments by those present:

"She's holding me up, looking at me. . . . She's smelling me! And she asked the nurse why my toes were funny. . . . The nurse said that's just the way my toes are and that they weren't deformed."[4]

In various psychoanalytic systems of treatment individuals are put back through birth traumas so that they reexperience the terror of the bright lights, the cold, the rough handling and separation characteristic of birth in the civilized world. Both Primal Scream and Natal Therapy treat the neuroses that begin in our manner of entering the world. According to these systems, the baby does not forget; the forceps, the sudden premature delivery, the Breech, the Cesarean are all imprinted on the personality. A sterile anxiety arises at the moment of birth, and nothing resolves it, so it is simply transformed through the changing circumstances of a lifetime, to be revived anonymously in phobias, recurrent failures, obsessive compulsions, and sadomasochistic sexual activity.

"I didn't want to come out. Some kind of pulling and tightening. Movement, lots of movement; myself moving. . . . Safe inside; didn't want to come out. . . . Lots and lots of noises, and just confusion outside. . . . I'm in an operating room. . . . A lot of chrome instruments. A metal table. And my mother's on the table. And there's a lot of men and women—seven or eight—dressed up in gowns. They're all talking and rushing about. . . . And then there's light, lots of light. Really bright. Really bright. And it seems like something's on my head or by my head; seems like I'm just being pulled out by someone or something. There's one big bright light in particular. I'm being pulled out. And very scared; very scared! . . . I'm lying on my back, my legs moving, and my hands are scratching my eyes, scratching my eyes and nose. I'm crying, screaming, and nothing holding me. Too much open space! Too much freedom for my arms and legs. Air; too much *air,* too much freedom. . . . I'm not curled up safe."[5]

This awareness and coordinated intelligence could not originate *ex nihilo* at birth. It is explicable only as a system of mind assembled in the darkness of the uterus. From the condition of aborted foetuses scientists know that by thirty-two weeks in utero the nervous system is ready "to transmit signals back and forth through a complex mass of unnumbered cells, signals which miraculously arrive at all the right muscles, glands, and organs. How these electrochemical signals are ultimately transformed into meaningful messages, ideas, decisions or memories cannot be explained in physical terms alone.''[6]

From conception until about thirty weeks the embryo dreams almost continuously, its unopened eyes fluttering beneath their lids. At thirty-three weeks it remains awake about a third of the time (though the difference between waking and sleeping in the womb may not be great). At full-term the infant spends half its time in rapid-eye-movement sleep.

As early as the twenty-first week human embryos have been heard crying audibly in their wombs. They awake suddenly from dreams and begin to cry. What do they ''see''? What disturbs them?

It is almost as though the foetus is being fed the history of its race and the archetypal history of the cosmos, not in words but in biological concepts—realities indistinguishable from the dendrites and synapses in whose formation they come into being (much as baby Superman of the comic book was played the history and wisdom of Krypton as he flew through interstellar space in his capsule toward Earth).

From the third to the sixth month in the womb, the foetus ''now floats peacefully, now kicks vigorously, turns somersaults, hiccoughs, sighs, urinates, swallows, and breathes amniotic fluid and urine, sucks its thumb, fingers, and toes, grabs its umbilicus, gets excited at sudden noises, calms down when the mother talks quietly, and gets rocked back to sleep as she walks about.''[7]

The physical morphogenesis of the nervous system may be identical with the dreams of the embryo, requiring their imageless cinema for its own coordination and unity.

Diagrams of the day-by-day development of the human embryo are evocative. A little ball splits and unzips; a tail appears at one end, curling up; a head at the other, the face gradually impressing itself as the skull rises out of the gut. This sequence of pictures is also deceiving because it presents embryogenesis as a problem solved in advance, a *fait accompli*. Although we may know better, we tend to imagine that the reason the foetus individualizes is that a full-formed infant lies at the end of an irrevocable chain and there is no other way for ontogeny to unfold. The pregnant woman may imagine her own character imbuing and shaping her spawn; it will be biologically human finally because she is human. Anything else would be a violation of nature and the lineage.

Yet myths and fairy tales are filled with such horrors—a wolf crawls from the womb, a devil with pointed ears smiling malevolently, a changeling of one sort or another. Biology has its own specters of teratology: mongoloids, siamese twins, the "elephant man." The human guts and womb are also the territory of the shadow, and when it comes to the crucial moment of birth, one does fear that *anything* unknown could emerge from within, any one of all the nightmares and chimeras.

Some of these fantasies represent our modern pathology—the denial of our somatic basis, our fear of ourselves. Perhaps in ancient times (and no doubt in the Stone Ages) women honored the mystery of child-bearing and the changing Goddess in their own bodies. Giving birth then was more able to be (as the spiritual midwife Jeannine Parvati puts it) "a supreme passionate bliss and the major soul-making experience of their lives."[8]

There is also the industrial shadow cast over the act of birthing by the omnipresent patriarchal scientist of the modern hospital. The investigator (in the guise of the doctor) counting the heartbeats, operating from the various forms of electronic gadgetry, diminishes the woman's sense of her own mystery, hence her connection to the healing archetype. Without her aboriginal resources she encounters her pregnancy and forthcoming birth as an event outside herself, another object in male society. Her child-bearing becomes an act of science, a gift of Apollo rather than of his first-born twin Artemis (who delivered him). We are so socialized we often fail to realize our authoritarian benefactors gain power only by stealing it from us. The machines are not necessarily allies. "Fetal heart monitors violate the mystery," writes Parvati. "They are blasphemous."[9] "It is with the inner sight that we can see the mystery unfold, not with Superman's x-ray (and tetragenetic and carcinogenic) eyes."[10]

Our medical science, by mechanizing and displacing the experience of pregnancy and birth, has substituted a secular shadow for the holy one, and a sterile fruitless fear for the true awe of the unknown.

Birth is an act of memory beyond time, or of time beyond memory. It brings two creature cycles together in a cosmogonic moment.

"It is like women all wear masks—for *their* mothers and fathers, sisters and brothers—and in giving natural birth the masks come off. We are always surprised to see who She really is. These masks are personae, personalities; when in childbirth, the personality is stripped away. Off go the day-world clothes as we lie naked before God. Giving spontaneous birth shows the original face of both the baby *and the mother!*"[11]

But there is an archetypal shadow, one which haunted humankind at the beginning and will be with us in one form or another forever. Even a happy birth does not palliate *its* ghosts; it merely displaces them onto the experience of self, family, and mortality. This is not even a shadow we should try to escape, for it is the shadow cast into the cosmos by our existence, and it draws us toward an eventual midnight ceremony of our transmutation.

If they want to serve Apollo, the expectant parents could also watch an embryo that is in doubt, one whose metamorphosis is actual rather than imagined. Frog blastulas are available from many laboratories. Viewed through a microscope they are small puffy balls, like crystals more than animals. With each spasm their cells are more and tinier. They can be placed in a jar of dechlorinated water for viewing. By the first or second morning a dramatic change has taken place: the tadpoles have unravelled from their balls and lie as little lines on the bottom of the jar. They are neither alive nor not alive. Occasionally a nervous flutter activates one and it kicks up for an instant and settles back. Matter is awakening, but they are primarily asleep; such episodes barely penetrate the eternal bliss. As long as conditions are favorable they must be woken up; they will not be allowed to sleep much longer. Soon they will not only accept this inevitability, they will defend it fiercely and hunger for every drop of its metabolizing food.

By the second or third morning the sun shining through their jar will show their almost completely transparent bodies stretched up and down like worms hanging in space, at all different levels, quivering subtly. They are rootless plants dancing lightly in the sun. They are not awake, but they are acknowledging gravity and, through it, space itself, and, soon enough, time too. They will not choose life, but their nervous systems will summon them and how could they evade such a call? "Why do you want me so quickly?" they may ask in neither Hopi nor English like the primal creatures of a Pueblo creation myth. The stream of light through them is interrupted only by the little black dots of their eyes and the band of gut tissue from tail to cranium along their central axes. Long before they go after food their cells will filter-feed themselves, whetting their appetites, so we must add the smallest amount of plant sugar to the jar. It is difficult to awaken matter; some part of us wants to go back to sleep all our lives, some part of us kills us and numbs its self to its existence. We eventually experience only the hint of slowness, stupor, carelessness. The gradual collective suicide in which our culture and (in fact) our whole world engages is a relief to most—the poisoned air and water, the carcinogenic food, the weapons factories—for these make our trauma palpable and allow us to experience and even appear to oppose our mindless self-destruction. We can stand against death even as we contribute to it. But to that part of us that wishes to wake up, history is more and more discouraging.

It is no wonder that our civilization faces itself across the darkness— in each individual and through the imaginary tunnel of time, the aeons of linear progress. The trance within matter denies this all as hallucination. Nuclear suicide would put to an end something that was begun without our consent, amorally. Our awakening ego struggles to avoid such denouement, but our heritage betrays us, now as ever. In a week the tadpoles will swim around like little fish, a quiet phase in which they will remain for months. But once they are frogs they will be voracious, eating other animals and even each other. This is the dialectic from which consciousness is born.

Human life is in danger right from the beginning: It is promised neither on an archetypal nor a biological plane. It must come into being of its own material and make its psyche and spirit out of itself, its body. No matter how many transcendental planes there may (or may not) be, we cannot overlook or escape the sheer immensity of creation on a purely physical plane. The alchemists who sought to raise spirit and soul ever out of matter, uncommon substance out of common, and gods themselves out of ashes, understood implicitly that we are bound by our mortality. The decay and pain around us is not the wasting and grief of a mere residual world in which we chanced to occur; it is the decay and pain of us, the fact of our substance. We put on the mortal coil with great difficulty and minute precision; we cannot be abstracted from it, and we cannot idealize our wholeness and mentation as if it were solely angelic or metaphysical. The crisis of our becoming is real, not the image of a higher dimensional realm and not the symbol for another immortal mortality.

18. Soma

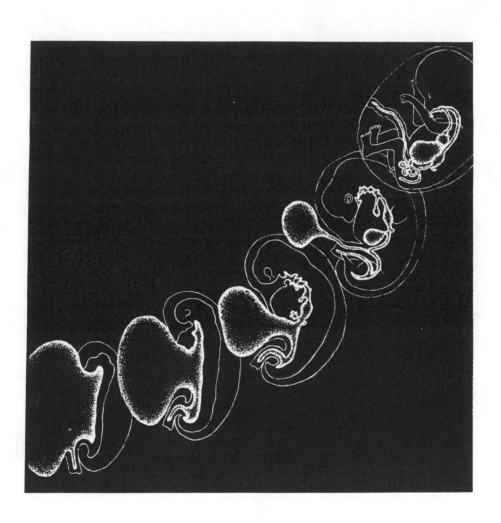

The body is not just a factory that sustains the mind; it is the reality which the brain personifies. We may act as though our consciousness were cerebralized only, but without the organs the brain would be a mathematical abstraction with no living experience. We embody spirit through our breath, and love in our heart. When we lose the ability to make contact with others the rigidity and constriction is felt in the chest. Our guts convert phosphates into thought and motion, and we remove their waste through excretory sensations.

Though it is an exaggeration to say that the individualized organs of the vertebrates are like single invertebrate animals (as many early recapitulationists did), there is a hint of validity to the analogy. The heart, intestines, lungs, and genitals all function as if they were creatures with rudimentary intelligences of their own. They are imbued with nerves and blood vessels, lined with muscle and skin, supported by bone, connected to hierarchies of ganglia, and suffused in a medium of blood and lymph cells. They are like colonies of polyps, worms, and mollusks in an internal ocean. The fibers that permeate them also connect them to one another and to a cephalic creature which experiences them not individually but as its own wholeness. Of course without the brain their collective existence would not be conscious, but they would still have rudimentary invertebrate sentience; they would eat, pulsate, and discharge waste.

The somatic therapist Stanley Keleman says: "We have been taught to be mostly cognitively aware, so much that we are out of bodily experience. The languaging self assumes it is the self but it blocks out a larger muscle-brain sense of reality. . . .

"Most of us are so geared to pay attention to the images in the form that we can recongnize them, that we, in fact, never pay attention to how those images are connected to the whole person. Images are concrete, hard. Most of us don't feel the pulsatoriness or the warmth to support that image-making. We are unaware that the patterns of sensations that arise from the muscles, from the joints, and from the viscera, *are* our images."[1]

The organs of our body are composed of clusters of cells subsumed in tissues; the forerunners of these cells carried out the same or similar functions in ancestral chordates and the vertebrates of our lineage. None of our organs could exist without a substratum in time and space. At the same time, each organ must "evolve" again within the embryo.

The three layers of tissue formed in gastrulation give rise to the organs of the body. Although individual cells within the layers may remain equipotential to a greater or lesser degree through the process of development (and, if transplanted, will form radically different structures in accordance with their new locales), they are otherwise fated by their position within the emerging organism. They will become specialized within the contexts of local fields of influence and the larger field of the animating creature itself.

Cells respond to genomic regulation in two basic ways: They can proliferate by simple mitosis, or they can differentiate from their parent cell. Proliferative cell cycles lead to aggregations of the same kind of cell. Quantal cycles differentiate cells: Such cells may continue to fission in their new state, or they may lose the ability (like red blood cells and bone) and spend the rest of their existence performing a specialized function.

Although the body is an integrated whole, the layers of embryogenesis remain throughout life. In an autopsy the skin is pulled off to reveal the separate entrails, connected in blood. The face is rubbery and pliant, frozen in its last act. The film-maker Stan Brakhage, shooting a movie in the Pittsburgh morgue, perceived suddenly that the first masks must have been the actual faces of the dead, perhaps removed with the scalp in battle and worn by the victors. He recognized, then, that not only are we actors but that the dead continue to wear their costumes and hold their postures.[2]

The ectoderm, as a layer, generates the central and peripheral nervous system, the epidermis with its hair and nails, some glands (including the mammary and pituitary), and the enamel of the teeth.

From the mesoderm is derived the connective tissue, cartilage, bone, muscles, heart, blood and lymph, kidneys, gonads, spleen, various membranes lining body cavities, and cortices of internal organs.

The endoderm is deep gastrointestinal tissue, so it forms the main esophageal and respiratory tracts, tonsils, liver, pancreas, part of the bladder, parts of the ears and tympanic cavities, and various glands (including the thyroid, parathyroid, and thymus).

These layers of tissue participate profoundly in the induction of organs in one another, and, in many cases, fuse to form organs. They are programmed by the history of their predecessors in simpler organisms so that their complex relationships in the embryogenesis of higher vertebrates are simply interpolations of systems they already contain. There are no spontaneous tissues or organs, simply more intricate structures and new combinations of functions. Hence, we find even within a single embryo that the eyes derive their musculature from the same layer of tissue as the genitals, and the kidneys and brain are reciprocally penetrated by blood vessels. The bladder is not fundamentally different from the heart; it is another bag of folds in a different position in the organismic field. From order to order within the vertebrates, the radius and ulna bones arise and articulate in a similar fashion, for instance, in a bird wing, the flipper of a whale, and a human arm. The lung and thymus gland of mammals are essentially the gill chambers of a fish closed off and integrated in a new biological field.

The three layers of the body continue in the adult as three personalities, or, more accurately, three separate expressions of the personality. Some therapists believe that all traumas become somaticized and retrace the underlying pathways of embryogenesis, following the trails of the organs through sense receptors into nerve and muscle fibers and even down into the living tissue of the bone where they sustain structural distortions (hence, the oft-misunderstood role of the chiropractor as "psychologist"). With a proclivity toward lines of stress they locate the gastrointestinal tract and flow into the heart and lungs. They may even penetrate the ancient archenteron and globalize through the rays of morphogenesis. If traumas carry enough energy to reach the mitotic level of the cell they perhaps can be transformed into tumors and dysfunctions of the immune system. Pain is always driven inward, and though it becomes more and more unconscious as it sinks, it is neither defused nor obliterated; it interrupts the natural functioning of the organs and becomes concretized much as tissue itself does embryogenically. In fact, somaticized pain is a pathological substitute for healthy growth. But if disease can be driven into soma through mentation, so then can psychic medicines, curative chants, and massages; they can form new tissue too. This is the basis of all esoteric healing.

From Wilhelm Reich's somatic analysis of character through the present florescence of "body" therapies, psychologists have attempted to reorganize personality through changing men's and women's experience of their somatic pathways. For instance, Keleman notes how, in situations of anxiety, stress, and terminating relationships, we do not want to feel the body, we suppress excitation. By inhibiting sensations of sadness, unboundedness, or rage we in some serious manner deaden them; we become incapable of their natural expression over time. Most people naively rationalize that this limitation is only a superficial layer of irritation—mere habitual images—but the response patterns are transposed into the neuromusculature of the gut and heart, and profoundly structured by the alignment of living cartilage. They create deep, intractable obstacles.

Body psychotherapists teach people how to bring pulsation back to forgotten organs. Often such therapy involves kicking, pounding, flexing, tumbling, and deep breathing. Martial arts like aikido and capoeira become medicines in which combatants give energy to each other. The exercise shatters the brittle coat of resistance around the nerves and muscles and restores some of the natural range of feeling to the organs. This feeling sweeps by as a series of images and spontaneous experiences and may excite outbursts like wails or screams. But it is not pain; it is in fact pleasure, the delight of rediscovering the unharmed source. When emotions are experienced as the actual somatic life of the body they become more than just correlates of anxious and painful episodes (or flashes of ecstasy); they become the sweet anonymous current sustaining existence itself.

In the vertebrate embryo endodermal tissue invaginates to mold the archenteron. At its frontal end it contacts the ectodermal layer to form the oral plate at the mouth, and caudally it becomes the anal primordium. From front to rear this continuous digestive tube is induced by surrounding mesoderm to form distinct sections: a foregut, midgut, and hindgut. During neurulation the ventral side of the foregut is simultaneously flattened and bent upward so that two separate cavities form. The one beneath the brain gives rise to the mouth and branchial region, whereas the posterior cavity bordering on the midgut becomes the liver diverticulum, the source of not only the liver but the pancreas and parts of the stomach and duodenum. This invagination continues to bend back and down, pulling along the rudiments from the wall of the foregut cavity into a narrow tube. In front of it the cavity flattens from top to bottom and expands sideways into a pouch with a thin epithelium. This is the pharynx. Initially it rests atop the heart which bulges into it. To have our heart in our mouth is to return to our aboriginal state. The therapist Robert Sardello describes the primordial connection between heart and brain:

"In the human embryo they are not at all distinct. In the first twenty days after fertilization of the ovum, what is to become the heart lies nestled around what is to become the brain, as if the heart is its crowning glory. As the brain emerges, the heart submerges. I suspect the heart forever remembers this intimacy: The heart attacks when the brain thinks that it no longer needs the throbbing rhythm of the life below, when it thinks it can be more productive without the interference of emotion, sentiment, feelings."[3]

The primitive pharynx reembodies a series of histories of both sea and land vertebrates; as it forms, it induces a row of paired pouches. At their slits the endoderm swells outward to contact the ectoderm, eliciting membranes. In human embryos the first pouch comes to encompass the middle ear bones as it gives rise to the tympanic cavity—widening into the pharynx through the eustachian tube. The endoderm of the second pouch fuses with the mesenchyme around it, becoming tonsillar crypts (the remaining mesenchyme spawns lymph nodules). The third and fourth pouches give rise to the thymus and parathyroid glands, respectively. Beginning as a ventral pocket in the floor of the pharynx, the thyroid is later closed off and displaced to the rear.

The tongue originates from a median triangular swelling of mesenchyme in the floor of the pharynx. Two round nodules emerge on either side of it, and the three structures fuse. Mesoderm from the branchial arch provides connective tissue, blood vessels, and some muscle fibers; additional muscle fuses from the primordial myotomes of the occipital somites.

The processes of the palate, composed of mesoderm in contact with ectoderm, arise from the jaw and project downward on the sides of the tongue, which descends as the jaw develops. The lateral palatine processes combine with each other and with the forward palate and a downward growth (the nasal septum). Subsequently, bony membrane molds the hard palate.

The nasal sacs grow up into the developing brain. A pair of regional thickenings (placodes) sprout in front of the neural plate, appropriating material from the neural fold. These become the two olfactory sacs. Similar placodes invaginate on the sides of the hindbrain and form vesicles which detach from the epidermis and twist about as the labyrinth of the inner ear. Some epithelial cells flatten to become membranes, and others elongate in patches of sensory ectoderm. The ear is like a winding cave of rivers, a soft snail secreting a hard matrix to protect its viscera. The organ forms as a whirlpool in the absence of water, a vortex which separates the waves of sound in the air into notes which register on soft membranes (its characteristic labyrinth actually occurs because its expansion is cramped in some areas and swells in others). The vesicle itself, induced by mesodermal contact with ectoderm, then induces the surrounding mesenchyme to produce a spiral capsule of cartilage.

In the lining of the ventral wall of the pharynx, internal folds fuse

to make the tube of the larynx in which the cords of the voice-box develop; these are attached to the respiratory tract by folds of mucous
membrane and articulate with the trachea, the windpipe.

This median zone of the body is very old—a primitive chordate
worm endodermalized into gut and breathing channels. In the vertebrates the worm fuses with connective tissue, muscle, and potential
bone of the mesoderm. The ectoderm spreads and thickens to enclose
the body in a protective outer layer. In those spots where the skin is
neuralized and the mesoderm or endoderm (or both) contact it, capsules are induced and large-scale sense receptors form (like the nose
and the ear).

The viscera of the lower invertebrates all lie in the same cavity, but
the coelom of the mammals is divided into two sections—the thorax,
which includes the chambers of the heart and lungs, and the abdomen, which holds the whole digestive tract, including the liver and
kidneys, and the reproductive organs. These are separated by a sheet
of tendon, the diaphragm, the muscles of which contract and expand
in breathing. This central organ can somaticize fundamental emotional crises in its critical position between the esophagus and lungs,
and beside the heart. It is a spot of holding out, of dulling feeling, of
refusing to exist. A continuous partial vacuum sucks air down the
windpipe into the lung and then expels it by relaxing. Birds, with no
diaphragm, must use muscles of the rib cage and wings for breathing;
reptiles and amphibians pump air in by their throat muscles.

The lungs grow into large spongy sacs connected to the pharynx
through the trachea. Actually, they are concretions of multiple
branched tubes ending in minute air-bags. The initial lung bud
sprouts at the caudal end of the laryngotracheal tube and splits in two,
a bipolar expression of one set of genomic instructions. As an evagination of the alimentary canal, the lung cavity is as endodermal as the
gut; digestion of food and air are particularized aspects of the same
process, the same aboriginal organ.

When the lung buds contact the surrounding mesenchyme they differentiate into bronchi and limbs. The lungs then form as bronchial
trees, vast arborescences projecting downward, forking dichotomously to the sides and backward. The unbranched section, suffused
by cartilage from the raw mesenchyme, becomes the trachea. In
lower vertebrates the lungs are mere folded sacs at the end of the
bronchi, but in mammals the bronchi diverge outward, forming as
many as ten multiple stems in the right lung and nine in the left. Air
sacs (alveoli) then develop on the terminal branches so that the total
area exposed to air is as large as a tennis court.

When this system expands into the thorax, surrounding mesenchyme supplies connective tissue and envelops the lungs in membranous bags (pleurae), which lubricate them with a coating of fluid. The bronchi grow right into the body wall of the thorax and lie adjacent to the heart as the embryo folds transversely.

In the modern fishes an organ grows similarly out of the endodermal wall of the alimentary canal, the swim bladder, which allows the animals to rise and sink by filling with air and emptying. This hydrostatic sac may have originated as a primitive double-sac in the thorax of extinct fishes, a structure which also gave rise to the lungs of the land vertebrates. It was the ability to breathe atmospheric oxygen through this sac that allowed the lungfish prototypes of the amphibians to cross dry land during droughts in search of new pools of water. Such journeys were necessarily carried out over generations, and most of the "fish" became dehydrated and died in the desert. Apparently, chance mutations transformed a few of their descendants into part-time terrestrial animals with primitive lung sacs, and *their* descendants branched off into the amphibians, reptiles, birds, and mammals.

The mammalian embryo remains at least partially a lungfish; its alveolar ducts before birth are small immature bulges siphoning amniotic fluid which keeps them half inflated. At birth this fluid is suddenly expelled, some of it into arteries and veins and a good deal of it through the mouth and nose from the pressure of the birth canal on the thorax. A millennial change of habitat is accomplished in an instant, albeit traumatically. The lungs then respire only because the alveolar capillary membrane is thin enough to allow gas exchange—a state which originates ontogenetically while still "underwater."

The lungs are more than swollen bags of oxygen exchange; their in-and-out pumping inspires sentience throughout living creatures. For us they are the dynamic jelly of image-formation. All serious meditation, yoga, and shamanism begin in the mind of the breath. Internal alchemy and tantric sex take their seeds from the same furnace—not just through the mechanics of transpiration but in its continuous discharge of feeling and intelligence throughout the body. If any molecule carries the vital force and the elixir, it is oxygen, or at least oxygen appears at the point of its materialization. The lungs transmit the medicines of acupuncture, chiropractic, and deep massage through the meridians to the viscera. But to work they must carry attention with them.

Psyche is an epiphenomenon of breath, a winged internal goddess always on the edge of the dissipation of the lungs, i.e., almost entirely unconscious (but then the lungs never completely unravel).

Every outbreath is a small death for the organism, the breaking apart of its psyche, the dissociation of its ego. Every time we expire we must trust that we will inspire, that we will restore our self. But we never know for sure. That unconscious fear is the source of much anxiety and rigidity in personalities; it manifests as shallow breathing, the subtle but continuous attempt to prevent the dissolution and restructuring that occur in full breath. The body/mind tries to force reality to conform to its limited ego-sense, but the universe wants to move on, through the lungs and heart into new realities, fresh unknown surges of embryogenesis and cells, as in conception and at birth. The East Indian and Chinese philosophers have long recognized that the lungs are the organ of mentation and epistemology. It is in the pleural chambers and branchings that inhabitable images of this reality form and trap us; so it is through a conscious cultivation and retraining of breathing that reality is changed and new universes are brought into being.

As the pharynx expands during the first seven weeks after conception the esophagus elongates within the foregut. The mesenchyme and neural-crest cells provide it with the material for smooth muscle and a visceral plexus. The caudal part of the foregut dilates as the stomach, its dorsal tube expanding more rapidly than the ventral border and forcing a gradual curvature. The stomach swells into the abdominal cavity rotating ninety degrees clockwise on its longitudinal axis so that the original left side lies toward the belly and the original right side toward the back. It is suspended from the wall of the cavity by a large mesentery that develops coalescing gaps in its surface and becomes the lesser peritoneal sac. The duodenum is a ventrally projecting loop from the end of the foregut to the beginning of the midgut.

The liver diverticulum spreads to the front of the alimentary canal, its forewall billowing into folds which enclose its slight internal cavity. The tiny posterior opening that remains joins to the duodenum as the bile duct. The folds of the diverticulum ultimately break up into strands of cells with blood vessels and sinuses.

The association of gut epithelium with mesenchymal tissue induces major clusters of specialized cells in the liver (but also the lungs, kidneys, thyroid, pituitary, and other organs). Hepatic cells in particular remove from the blood substances that are poisonous to living cells, intercepting them between the intestines and the heart and converting them into urea, which is passed on to the kidneys via the blood. They also produce bile for digesting fats, which they send into the intestines along with exhausted blood cells, and they store nutrients from digestion. These hepatic cells proliferate so rapidly that the liver fills most of the abdominal cavity. The right lobe grows faster, and the left lobe develops a partial split, giving the organ its characteristic leafy shape. The liver grows so quickly that at ten weeks it is 10% of the human foetus by weight, and its participation in embryonic blood-making gives it a bright red color (the Sumerians located the seat of consciousness in the liver rather than the brain). The cells of this organ remain mitotic and continue to renew through the lifetime of the organism. As a neutralizer of poisons, the liver must be self-regenerative, though certain herbs (such as the New World desert chaparral) help it to flush the extra toxicities of modern civilization.

Another sheet of endoderm, a division of the diverticulum, wraps around the space beneath the mesoderm as the rudiment of the gall bladder. Located under the right lobe of the liver, this organ stores bile and articulates with the bile duct.

The pancreas initially buds dorsally and ventrally from endodermal cells at the tail end of the foregut. A small later-forming protrusion develops near the bile duct, and, when the duodenum rotates to the right, it is carried dorsally with the earlier-forming elements. The major portion of the pancreas is constituted by the original dorsal bud, whereas the pancreatic duct arises from the ventral bud. Connective tissue originates mesenchymally as in the other gut organs; the loose mesodermal material apparently induces the distinctive secretory cells of the pancreas with their rough endoplasmic reticulum and swollen Golgi bodies. The spleen, which is the source of blood production until late in the life of the foetus (and the lifelong synthesizer of lymphocytes and monocytes), forms directly from masses of primordial mesenchymal stuff between the dorsal mesenterial layers.

The midgut is initially suspended from the dorsal abdominal wall by a short mesentery; the cells connecting it to the yolk sac gradually break down, leaving only an umbilical stalk. This zone elongates much more rapidly than the body itself, so the intestines are projected spirally into the umbilical cord. Only by the tenth week after conception do they return inside the body. Then they are gradually rotated almost completely around, filling the coelom while the liver and kidneys decrease in proportion to the overall enlargement of the cavity. Pushing their mesenteries against the cavity wall, the intestines fix themselves in place. Rotation of the stomach pulls along the duodenum and causes it and the pancreas to fall to the right, pressed similarly against the dorsal abdominal wall.

The hindgut is made up of the colon, rectum, the upper portion of the anal canal, and part of the bladder and urethra. Its expanded rear

portion, the cloacal membrane, is folded inward by a sheet of mesenchyme. As these folds converge, they partition the cloaca into the
rectum and upper anal canal and, in front of these, the urogenital sinus. The hindgut fuses with the anal pit, and endoderm meets ectoderm at the anus, the opening at the end of the alimentary canal.

As the inside of the gut is formed and the organs are rotated, the
embryo must experience the turmoil of its own reorganized shape.
After birth such dramatic reorganization can occur only through the
emotions. The embryo literally embodies the continuous states of
confusion before reorganization; the original twisting of the gut becomes a direct repository, throughout life, for intense feelings, often
of rage. The suppression of such passions belatedly represents an attempt to prevent the primal structuring of the soma, a holding back of
inevitable change. "The conflict is going on in the abdomen . . . the
biological conflict," Keleman says of a client. "We can see the actual struggle in the abdomen as it pulls in and lets out."

He asks him to squat down and hold his hands, rocking him
slightly on his heels: ". . . breathe . . . let your head back . . .
breathe into your belly."[4]

The urinary and genital systems arise primarily from the nephrotomes, stalks of the somites that lose their connection with the mesodermal system and separate into layers around a cavity. To a degree,
the rapid growth and transverse folding of the embryo cause this intermediate section of mesoderm to become detached and migrate
ventrally. Packed with tiny tubules, themselves interfused with clusters of fine blood vessels (the glomeruli), the kidneys are pure excretory organs. Nitrogenous waste is filtered from plasma passing
through the tubules and discharged into the bladder as urine. In the
lower vertebrates the ducts open directly into the cloaca, but in higher
vertebrates the tubules fuse and drain into a common tube.

The mammalian excretory system is divided into three sections,
each arising from discrete sections of nephrotomes. The frontal pronephros is reproduced only recapitulationally in the higher vertebrates and is nonfunctional, its cervical cells reenacting their archaic
tendency to flow into the cloaca. In amphibians the pronephros coa-
lesces from mesodermal thickenings along the second, third, and
fourth somites, and, in frogs, persists as the oviduct. A functional
kidney in the larval bony fishes, the pronephros is no doubt the original excretory organ of our lineage which induced subsequent renal
sections through mutations and natural selection—a phylogeny
reenacted ontogenetically.

Although it may have some temporary embryonic function, the

mesonephros is also nonfunctional in mammals. It is formed as mesenchymal tissue (from the dissolution of the nephrotomes) separates in clumps that subsequently develop lumina. Each one grows into an S-shaped tubule which extends laterally until it reaches the common duct. The medial end of each is flattened and pushed inward by glomeruli so that a cup forms. This phylogenetically secondary kidney joins the primary duct that induces it. Only in mammals with loose connections between maternal and foetal tissues (like pigs) is the mesonephros clearly operable prenatally.

The permanent kidney of the mammals, the metanephros, develops from a posterior section of nephrogenic mesenchyme. Adjacent to the cloaca a node germinates on the mesonephric duct and protracts rearward. This bud uniquely induces the collecting tubules, for the metanephros is not a reduplication of the other kidneys. As the bud contacts the mesoderm, its cranial end expands into calyces, and these branch dichotomously to form lineages of collecting tubules. Mesenchyme provides glomeruli at their tips. The duct formed by the extension of this bud becomes the ureter which connects the kidney to the bladder. As the embryo's body grows caudal to the kidneys they are shifted from the pelvic region to the abdomen.

Meanwhile, the endodermal cloaca is divided by a membrane into a dorsal rectum and ventral urogental sinus—a curved cavity. The rectum is continuous with the alimentary canal, and the sinus expands into the mesonephric ducts to form the epithelium of the urinary bladder. The ducts are absorbed in such a way that the ureter opens directly into the bladder. Muscle layers develop from adjacent mesenchyme. The canal from the bladder initially leads into the allantoic stalk, but this extraembryonic organ later disappears. Behind the bladder the sinus narrows into the urethra, the eventual urinary duct; it also derives its muscle from contact with mesenchyme. As the kidneys produce traction in their cranial migration, the ureters of males open to the side and cranially to form ejaculatory ducts; the rudiments of these degenerate in females.

The adrenal glands lie atop the kidneys and rival them in size through the early foetal period. They form complexly, their original gonadally derived cortex wrapped in cells from the coelomic epithelium. Their core is supplied by migrating neural-crest cells. A series of enveloping differentiated structures, the adrenal glands are the source of sudden bursts of energy, obviously critical in the primordial hunt. Their powerful impact, though mediated through the old mammalian limbic system, is often a burden to men and women in society, for the hunt has been sublimated and replaced emotionally by far subtler gradations of feelings. Adrenalization can mimic emotional

authenticity and cohesive neuromuscular function, recalling the prehistoric struggle of life and death. But, experienced within society,
such states can distort personality and cause seemingly intense moods
which burn out meaninglessly.

The male and female gonads first appear as outgrowths of the coelomic epithelium on the midline of the urogenital ridge (a region
called the germinal epithelium because it was once thought to be the
source of the primordial germ cells). The germ cells arise outside of
the three tissue layers but occur endodermally and are imbedded in
the ridge as it thickens and swells into the coelom. The germ cells are
clearly aliens; they are much larger, spherical, and contain vesicular
nuclei. In humans they migrate from the wall of the yolk sac to the
gonads, and only there do they differentiate into oogonia and spermatogonia. Mesenchyme flows into the primitive gonads and collects
in strands as sex cords.

In embryos with an XX chromosome the gonadal cortex will become the ovary and the mesenchymal medulla will not develop beyond a thin epithelial layer; in the XY (male) complex the medulla
develops as a testis and the cortex regresses. Even though sexual fate
is established genetically at the time of fertilization, gonadal sex
must be induced and developed, so until the seventh week of human
embryonic life the internal sexual organs of male and female are identical: one androgynous gonad. The Y chromosome transcribes proteins which, in the general morphogenetic field, have the effect of
inducing the medulla of the indifferent gonad, the sex cords differentiating into seminiferous tubules. In the absence of such induction in
females, an ovary forms, swelling with the expansion of the germ
cells within it. Those nearest the surface of the ovary's cortex become
the primary oocytes. Because the sexual organs border on the nephric
ducts the renal channels are co-opted for the passage of sperm and
eggs to the cloaca. As the nephric ducts split longitudinally through
induction, part of the pronephros survives as the oviduct, while the
mesonephros develops both renally (as the ureter) and reproductively
(as the sperm duct). This urogenital convergence was inevitable anatomically, and it has been internalized in our lineage as the psychosomatic channel of sexualization. Through the fusion of nerves, glands,
and ducts we experience the entire alimentary system as erotic, particularly its perforations of the ectoderm at the mouth and anus.

In Freud's version of sexuality, libido is a universal current not
confined to just the reproductive organs. The mouth is actually the
first erogenous zone, experienced by the infant in the pleasure of
sucking the breasts (or a substitute nipple), regained at the level of

adult sexuality in the deep kisses of lovers. The cutting of teeth gives oral sexuality a sporadic sadistic element. The formation of weapons in the sensitive gums is one of the events that break the trance of childhood, the fantasy of the all-giving breast. The aggressive and carnivorous history of our species is unavoidable and thus appears somatically very early in development. Through childhood development libido is partially displaced to the anal excretory zone where it is experienced as a fusion of *eros* and *thanatos*, a narcissistic obsession with the body's products and functions. Only at puberty is libido genitalized in the phallus and put at the service of the primordial germ cells and the anatomy of reproduction. There alone does it participate in the creation of society—reaching across the somatic boundaries to a pleasure objectively shared with another organism.

Sexuality is a multilayered complex of pleasures and aggressions, neither originating in nor circumscribed by the genitals. The symbolic phallus (which occurs in both males and females) embryogenically precedes the penis, clitoris, and ovary in which it has its single biofunctional expressions. The primordial phallus encompasses a kind of universal *eros*, a suckling at the body of nature herself, a desire to hold everything, to retain fluids and feelings, and—simultaneously—a bottomless generosity, an impulse to cast one's feelings and the concrete seeds of the body into the vast orgasm of nature. These warring tendencies are never resolved, even in lovemaking and procreation; their dialectic is merely extended through time and brought ready-made to the organs (notably the genitals) of each new generation of men and women.

The male seminiferous cords lose their connections with their germinal epithelium and become enclosed in a fibrous capsule. The developing testis slides away from the mesonephros and is supported by its own abdominal fold. In the ovary the germ-bearing sex cords splinter into primordial follicles, single oogonia surrounded by flattened cortices. The ovary also pulls away from the retreating mesonephros and is suspended by a new mesentery.

The testes of the foetus manufacture hormones which stimulate the mesonephric ducts of the vestigial middle kidney; they become the male genital tract. Outgrowths of the ducts germinate as seminal glands, and the urethra sprouts in the surrounding mesenchyme to produce the prostate gland. The potential female ducts and glands are suppressed, but they remain unexpressed histological remnants through the lifetime of the male and must enter his psyche at some level.

The female manifests the complementary aspect of this biphasal system. Her hormones suppress the mesonephros and stimulate the

development of new ducts, a paramesonephric network from invaginations of the coelomic epithelium parallel to the unused male ducts.
The individual invaginations fuse and run caudally into the uterovagi-
nal canal. The unfused cranial chambers become the uterine tubes.
Twin bulbs from the urogenital sinus contact the end of the canal and
form a solid vaginal plate; its median cells disintegrate leaving the lumen of the vagina.

The external genitals also originate as one organ in both males and
females. Within the first month of the human embryo's life a promi-
nence develops at the front of the cloacal membrane, and swellings
and folds form on either side. In the male, androgens secreted by the
testes accelerate the growth of this organ, and it elongates as the penis, pulling the urogenital folds forward so that they form the lateral
walls of a urethral groove underneath it. Lined with endoderm from
the urogenital sinus, the folds fuse from the rear to the tip shaping the
penile urethra. Ectoderm grows back over the tip leaving a cellular
strand which splits and meets the urethral groove. As the groove zips
closed, its external orifice is pushed to the tip, the glans differentiating around it. A foreskin grows over the glans, and the labioscrotal
swellings creep toward each other, meet, and coalesce as the scrotum. The testes and spermatic cord descend into this sac through in-
guinal canals formed from the lining of the abdominal cavity. The
nerves in this organ are connected not only to the autonomic sexual
functions but to the full symbolization of *eros* in the nervous system.
The external phallus is thus a source of erotic fantasy, but also the ve-
hicle by which general bodily imagery is translated into sexual feelings and activity.

Even though lacking androgens, a morphologically identical genital grows in the female but more slowly and without a fusion of urogenital folds except at the very front of the anus. This is the clitoris.
The labioscrotal swellings expressed as the scrotum in the male remain unfused also as the labia majora. These organs embody a different quality of *eros* in the nervous system of a woman. A unique
psyche emerges from this anatomy, a female image of nature, a female experience of the self.

Different sexual destinies originate physiologically but have a psychosomatic reality that transcends their mere hormones and folds.
The sexes are never fully separated: there is a latent maleness within
the female and an undeveloped woman in every man. The object of
lovemaking and sexual fantasy is not only the Other but the Self; for
the man, the woman is the anima, the female aspect of his own psyche, and his desire for union with her is also his desire for a lost
wholeness; likewise for the woman, the man is the animus, the part

of herself that remains male, hence the completion of her own morphology through a love object. These complexes arise from the fact of our somatic existence, but they are also archetypes, eternal objects through which the shape of sentience imprints itself on the soft clay like an image in a mirror. The Rites of Dionysius taught us that "the body is the only part of the Earth we feel from within," but it is also the only Mind.

19. Blood, Bones, and Immunity

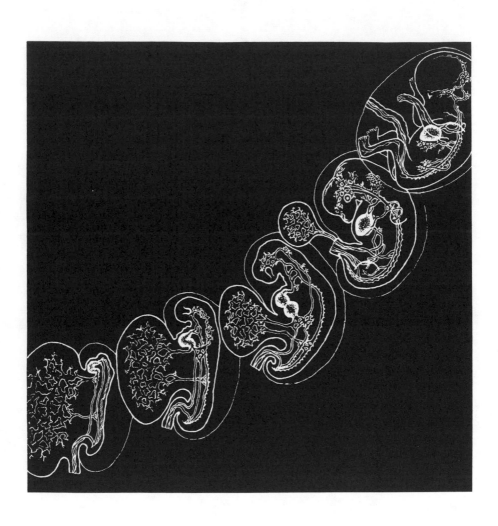

We are all miniature worlds. Our soil is supported by stony ridges and suffused with streams bearing a variety of life forms. The viscera of our bodies are wrapped in membrancs and suspended by folds and fibers. They float in a sea of ancient blood sealed in a living skin supported on branches of extracellular crystal.

Bone is the distinctive feature of the vertebrate phylum. Invertebrates are lumps of flesh supported by their own jellies or by a rigid shell, a chitinous exoskeleton; even lampreys and hagfishes are boneless chordates. Vertebrate bone provides structure for the skeleton, leverage for the muscles, and protection for the delicate viscera (particularly around the brain). The skull is essentially a bony plate, a remnant of the armor of dinosaurs which shelters the center of the nervous system and its sense organs and strengthens the jaws.

Bone consists of approximately equal volumes of inorganic crystals and collagen fibers, with the denser crystals contributing the bulk of the weight. The minute individual crystals of calcium, phosphate, and hydroxyl ions are fused to the fibers. Collagen is the tightly strung cohesive tissue of tendons, ligaments, skin, the webbing of muscles, and bones. When arranged in parallel layers in the cornea of the eye it is transparent, whereas in bone it crisscrosses to form a hard leathery substratum. In muscle the striated fibers of collagen are made up of tiny fibrils, each containing filaments of varying girth that work in opposition to generate movement.

Skeleton and muscle originate in the somites, the aggregates of mesodermal cells distributed about a central cavity through postgastrulation cell movements. As embryogenesis proceeds, they flatten and their cavity constricts into a shallow slit. They are now arranged in linings: a thick inner sheet and a very thin outer sheet. The lower half of the inner sheet, known as sclerotome, breaks up into mesenchyme, and its cells migrate into the spaces around the notochord and spinal cord, enveloping them while differentiating into cartilage. In areas of prospective cartilage the mesenchyme condenses and its cells multiply, accumulating in nodules from ten to several hundred. Actual chondrogenesis (cartilage production) depends upon the extracellular secretions of these bunched-together

265

cells. The matrix is synthesized from collagen and another complex molecule consisting of chains of sulfated polysaccharides affixed to a protein core. The collagen molecule is a left-handed polypeptide helix with three helices braided into a right-handed superhelix; held together by hydrogen bonds, the superhelices buckle into a variety of crystalline forms and are able to disassemble and form new structures under suitable morphogenetic conditions.

Meanwhile, the outer wall of the somites (the dermatome) generates a connective layer of tissue in the skin. The upper part of the inner lining (the myotome) supplies material for the vertebrate somatic muscles. Through organogensis the cells of the myotome realign themselves as longitudinal columns, each a muscle unit separated from the next by a layer of connective tissue. As the organism develops, the segments spread down between the skin and the lateral mesodermal plate. The lower vertebrates retain the primitive segmentation in their adult forms, but the segments are obliterated secondarily in the adults of vertebrates adapted to the land.

Bone develops out of existing cartilage by calcification. As individual cells of cartilage die, their thin layer of crystal is deposited around a shaft. While some of the adjacent mesenchymal cells break up into the blood-forming cells of the marrow, others become osteoblasts and deposit bone matrix on the already calcified spicules. The skeleton is a scaffold of cartilage covered with an expanding bony layer which is continuously deposited and reshaped. Like bone, dentine (ivory) consists of collagen and crystals precipitated extracellularly; the cells that lay down the grid become imbedded in the bone they form but not in the ivory.

Bone also arises directly in mesenchyme by the differentiation of some cells into osteoblasts which then manufacture calcifying substance for bony spicules. The spicules gradually thicken, producing plates of compact bone between which the mesenchyme remains spongy and differentiates into marrow.

The skull forms from mesenchyme around the emerging brain. It is historically two-sectioned—a neurocranium (or brain case) and the viscerocranium, the skeleton of the jaws. Several separate sections of cartilage fuse in the molding of the skull; subsequent ossification of mesenchyme around the brain creates the cranial vault, with fibrous areas of dense connective tissue between the skull's flat bones. These sutures allow the cranium to mold delicately in the birth canal and to accommodate the postnatal enlargement of the brain. The cranial vault grows rapidly during the first year of life and continues to expand until some time during the seventh year. The sense organs of the higher vertebrates are fortified with additional cartilaginous capsules originating in the cranial sclerotomes—so-called otic capsules around the developing ears, and nasal capsules around the sacs of the nose. Other sections of cartilage emerge in the vicinity of the forebrain, their upper edges gradually becoming wedged between the brain and the rudiments of the eyes and the nose. While neural cells migrate into their frontal zone anterior to the eye sac, their posterior section remains pure cartilage.

The original supporting system of the chordate was the notochord, which is retained by lampreys, sharks, and rays in their adult states and which exists vestigially at the core of the vertebrate skeleton, perhaps as the primary inducer of cartilage in the axial system. The migrating sclerotome of the higher vertebrates surrounds the vestige of the notochord giving rise to the intervertebral discs and centra of the vertebrae, each vertebral section developing from two adjacent sclerotomes. At this point, ontogenetically, the notochord degenerates. Sclerotome travelling dorsally covers the old neural tube and provides the material for the neural arch. A subsidiary section moving ventrolaterally into the body wall forms the costal processes which will be induced across the thorax as ribs. Underneath all of this ossifying bulk lies the primitive chordate ancestor which provided the structural symmetry for vertebrate phylogenesis and the migration of large animals out of the sea. Its relic appears briefly ontogentically within neurulation.

As terrestrial creatures evolved, their cranial plates of armor became capsules of small bones over their heads, ossified skin lining the mouth cavity. The head of a fish is continuous with its body, and the skull of a salamander is barely distinguished from its vertebrae. Gradually, through transmutation, a composite skull of skin, cartilage, and bones developed around the old chordate intersection of the opening to the gut, the seat of the nervous system, and the primary sense organs. Initially this was a great jaw (dinosaurs were mainly snapping neurons); it has since become a neurocranium too.

Bone is not dead armor; it is living matter, continuously molded by resorption of material and expanding at its open edges; it is covered with a skin, the periosteum. Arteries, veins, and nerves all work their way through minute channels to reach its cells. This disperse and extensive organ is laid down not as mortar but electrochemical crystals in a tough elastic sinew. Under stress, a crack will run into the collagenous matrix so the bone will be slightly deformed but will not crack. Between very long bones the mesenchyme separates peripherally into ligaments and vacates centrally, leaving behind tiny seams—the joint rudiments of the knees and elbows.

The tensile strength of bone is fifteen thousand pounds, and its compressive strength is about twenty-five thousand pounds per square inch, thus it has great elasticity and can resist blows and return to its original shape after intense distortion. Bone is a living fossil, a vitalized gem at the core of every organ; it is also a powerful aura, condensed and internalized, the concretizing pole of vertebrate formation. Because of the power of the whole skeleton and its viscera the *t'ai chi* master warns us not to strike with a single fist: "In push-hands the hands are not needed. The whole body is a hand and the hand is not a hand."[1]

The myoblasts of the muscles form from the myotome of the somites, with contributions from mesenchyme in the branchial arches and somatic mesoderm. The myoblasts are multinucleate tube-cells assembled on a repeating hexagonal grid. The cytoplasm of each of these tubes is saturated with myofibrils, each one containing millions of actin and myosin filaments. Actin-binding struts join myosin molecules in a symmetrical lattice. Once the process is triggered it continues as tissue self-assembly without any further DNA synthesis. The myoblasts proliferate, stretch out in parallel bundles, and then fuse. As the prospective muscle divides, each developing spinal nerve splits and projects a branch into a myotome.

The fibers of our muscles (and those of the animals) finally align in a striated pattern, with each longitudinal fiber matched to a fiber crossing it. These bundles of finely tuned string are interfused with nerves which fire their synchronized potential electrochemically.

Most myoblasts migrate away from their somites. The extensor muscles of the neck, the vertebral column, and the loins form from cells dispatched by the dorsal segment of the myogenic epithelium. Ventral myotomes meanwhile protract in sections—thoracic for lateral and ventral flexor muscles, and inferior for the pelvic diaphragm and the striated muscles of the anus and sex organs. The occipital myotomes generate the myoblasts for the tongue, whereas myoblasts from the branchial arches travel to the sites of masticating muscles, facial expression muscles, and the connective tissue of the pharynx and larynx (the ocular muscles arise from mesenchymal cells). Even ectodermal mesenchyme differentiates into sets of muscles—for the iris, the mammary, and the sweat glands. The smooth muscles of the heart and gut originate quite separately from the striated muscles, in a visceral layer of mesoderm in contact with endoderm.

The muscles of the limbs are derived from mesenchyme surrounding the developing bones. The arms and legs first appear as small mounds of tissue sheathed in ectoderm. Anatomically, they are the fins of old vertebrate fishes. The embryogenic limb of mammals reenacts aspects of its long metamorphosis, appearing first as an un-shaped nodule that swells, protrudes, narrows, and gracefully etches its own digits and sensory tips. The limb bud contains both cartilage and skeletal rudiments, and its very elongation draws nerves and ves-sels out along the same track, right to the fingertips.

At the site where a limb will form, a layer of lateral-plate meso-derm is induced to thicken. Although the mesoderm remains a contin-uous sheet, individual cells break their connections with it, accumulate outside its margins, and attach themselves to the inner surface of the ectoderm. This brings about the inductive field of the limb which proceeds outward as cells multiply within the disc. In hu-mans the arm buds germinate first, opposite caudal cervical seg-ments; a few days later the leg buds amass opposite lumbar and upper sacral vetebrae.

The ectodermal ridge maintains an inductive influence on the limb mesenchyme; it draws it out and gradually ripples its field distally into hands and feet—plates with marginal digits (between the phalan-ges, the cells deteriorate). As the limbs elongate, bones crystallize along their paths, myoblasts collect along each extremity and sepa-rate into dorsal and ventral (extensor and flexor) segments. Ulti-mately, the arms and legs rotate in opposite directions. The upper limbs turn ninety degrees laterally so that the elbows point backward, and the lower limbs rotate medially almost the same distance so that the knees point forward. Spinal nerves migrate along both the dorsal and ventral surfaces of the buds as they expand. When certain drugs (like thalidomide) are used early in pregnancy during primary limb formation, they may chemically inhibit one or more of the inducing factors. The limbs may wither partway through formation, fail to form hand and foot plates and digits, or perhaps not form at all.

The original limb is raw equipotential tissue; when half a limb bud is destroyed, the other half will still develop as an arm or leg. A split limb bud prevented from re-fusing will form two complete limbs. And two limb buds experimentally fused will merge into a normal limb, larger at first but gradually returning to scale.

Only the latent genome knows where a limb will form and which one will become an arm, which one terminate in a foot. A corre-sponding polarity of fields causes the feathered arms of the goose to diverge from its webbed claws. The appendicular skeleton is an aster-ism shaped within the pattern underlying the whole organism. From the primoridal chordate body—a Palaeozoic lump on the ocean bottom—the limbs shoot out like rays of light, energy becoming mat-ter; they are histogenic ikons of our extension into the cosmos.

* * *

All animals circulate internal fluids and disperse oxygen to their organs or organelles. The waves of the ocean are reborn dynamically in the slow internal flux of plants and animals.

"The first 'hearts' seem to have been nothing but faint waves of peristaltic motion (like the waves that nudge food through intestines), which gradually became localized and developed into swellings with a pulse. As circulation was mostly open and unconfined by blood vessels (as it still is in clams, shrimps, insects, etc.), heart action was more comparable to gently stirring soup with a spoon that to anything that could be called pumping—which may explain why the squid needs three hearts, the grasshopper six and the earthworm ten. And even when the heart evolved its valves with completely channeled blood flow, it still awaited a future history extending from the single-loop circulation of fish to the loop with a side (lung) branch of amphibians and finally to the now well-perfected double-loop circulation of mammals, which uses a two-chambered heart to pump blood first to the lungs to absorb oxygen, then to the whole body to distribute it."[2]

The human heart is a mesodermal organ composed of three separately arising structures: an inner lining (the endocardium) which is continuous with the endothelial lining of the blood vessels; a heart muscle (the myocardium); and a very tough cellular membrane (the pericardium) around the myocardium. The basic vertebrate heart originates ontogenetically in a sheet of mesoderm advancing forward from the blastopore and travelling between ectoderm and endoderm. The dorsal section is the swiftest moving; the ventral the slowest. By the time the neurula phase of the embryo is completed both the dorsal and dorsolateral segments of the mesodermal mantle have contacted the head region, creating an open forward space with a broad anterior base and engulfing the oral and pharyngeal zones. The rear of this space will be filled gradually by the formative cardiac mesoderm. (These early phases of heart-making are endodermally induced.) After neurulation the free edges of the mantle begin to converge and thicken ventrally in presentiment of an organ. The ventral cells of the mantle flow freely like mesenchyme; then they converge in a longitudinal strand which becomes a tube with a lumen, i.e., the heart cavity.

In human cardiogenesis a pair of elongated mesenchymal strands develop lumens, drift together, and fuse into the endocardial tube. A gelatinous connective tissue, the cardiac jelly, collects around it. The surrounding mesenchyme forms a mantle which will give rise to the myocardium and epicardium. The tubular heart lengthens and begins to dilate in some spots and thicken in others, forming sacs, arches, and a ventricle, including the aortic sac and arches, the atrium (the primitive pacemaker), and the sinus venus (the mature pacemaker into which thread the umbilical, vitelline, and common cardinal veins of the chorion, yolk sac, and embryo). Because some regions swell faster than others a loop develops, the bulbus cordis on one side of it and the ventricle on the other—the twin lobes of the S-shaped heart. As the head fold develops in the expanding embryo, the heart and pericardial cavity fall in front of the foregut and sink behind the membrane of the pharynx. The tube of the heart lengthens and bends, and the organ is submerged in the dorsal wall of its cavity, suspended by a fold.

As the internal stuff of the heart condenses, the atrium and ventricle begin to be partitioned. Veins and arteries interfuse the tissue, some of them developing as evaginations of the atrium wall. At the same time, ridges of tissue become hollowed out into valves controlling the flow of blood into the atrium; when the ventricle contracts, blood is pumped back out. The muscle layers of the atrium and ventricle become knitted together until they contract peristaltically, at first causing an ebb and flow between themselves and the embryo but eventually establishing a unidirectional flow wi h coordinated contractions. The embryo and uterus develop as ι mutually pulsating system.

Prior to this complex internal development, half the heart could be transplanted and still develop as a whole organ. Once the chambers and valves begin to form and the muscle fibers penetrate the ventricular walls, cardiac induction occurs primarily between internal fields within the heart itself. With the subsequent displacement of the developing organ from the pharyngeal region to the thorax, the mature heart takes shape, separated into a right half carrying blood to the lungs and a left half receiving blood from the lungs through pulmonary veins and sending it through the body. This mutant pump began in ancestors we share with the lungfish and has persisted as a functional requirement in the large active vertebrates that colonized the land.

Our internal milieu is a primeval sea of plasma, not much different from the body fluid of the amoeba and its kin. The liquid part of blood is essentially salt water—sweat and tears. Its "solid part consists almost entirely of coin-shaped red cells that are remarkably elastic, so flexible they can elongate and fold up and sneak through a capillary of barely half their own diameter."[3] This sea is warmed by muscle contractions and the metabolism of digestion and oxidation.

Cold-blooded lizards must be heated by the sun before they dart about. (We give up that solar "hit" in order to remain active at night and in the depth of winters.)

Erythropoiesis—blood-cell formation—is a regional specialization of the blood epithelium induced by an enzyme. Stem cells are potentiated to synthesize and store the metalloprotein hemoglobin, which actively captures and releases gases. Hemoglobin has the basic molecular structure of chlorophyll, but with four atoms of iron in place of chlorophyll's central magnesium atom. This geometry explains hemoglobin's strong ferrous attraction, and also suggests the far distant aeon when plants and animals were a single creature.

Once formed, red blood cells are dispatched into vessels to transport oxygen throughout the body and remove unwanted gases (such as carbon dioxide). The lack of yolk in the oocyte makes immediate circulation a mammalian necessity, for the foetus must obtain nutrients and oxygen to survive. The heart-pumping-blood mechanism is the first living system to perform in a mammalian embryo. In fact, formation of blood and blood vessels begins in the yolk sac. About the sixth week after conception erythropoiesis migrates to the embryonic mesenchyme; thereafter it wanders from side to side in the mesoderm, from the liver and spleen to the lymph nodes, ultimately taking up residency in the bone marrow where blood cells form thereafter.

The entire cardiovascular system first differentiates as mesenchymal cells swarming together in blood islands—angioblasts. Induced in the context of the yolk, their "purpose" is instantly vegetal—to bring food and oxygen to the multiplying cells. The transport network for this process is assembled in place by angioblasts forming epithelia around cavities and linking up in chains. These separate channels extend then by budding and fusing with exterior vessels. Blood cells, as pointed out, originate in the same epithelium under the influence of an enzyme. Terminally differentiated erythrocytes, they lose their nuclei and give up their reproductive capacity. As biconcave discs, they have become biochemically and physiologically specialized—with all other functions discarded in order to improve their geometry for oxygen transport and traversing the microcirculation networks.

The blood vessels ultimately permeate the entire adult body. They are "attracted" to vascularize any part of the soma in need of blood, even transplanted tissues. Spreading by budding and fusion, the vessels branch outward into long constant tubes crossing vast topographies and detouring around obstacles. Since the system is so flexible, it is irrigated in part by the amount and direction of blood flow itself, but, at the same time, it is not initially dependent on a heart and emerges even in its absence. Those branches that receive the greatest blood flow become major arteries, and those that receive too little deteriorate. The blood networks proliferate in the regions between the mesodermal and endodermal viscera, especially in the heart and the environs of the kidney where erythropoietic hormones are synthesized. Anteriorly and posteriorly, blood vessels assemble the paired continuation of the heart tube. Induced endodermally, the two posterior vessels become vitelline veins which collect blood from the surface of the gut; the anterior pair become the ventral aortae between the endodermal pouches of the pharynx. Blood vessels also connect the ventral to dorsal aorta, and interconnect the six pairs of aortic arches—the arteries from the aortic sac to the branchial arches. Although many of these arches disappear during human development, the fourth and sixth pair continue to supply blood to the back and the lungs, respectively.

The hematopoietic stem cells that generate blood have a number of other quantal derivatives including monocytes, leukocytes, and the specialized line of lymphocytes. Potential lymphoid cells are first visible in the thymus gland which forms in the branchial pouches. The stem cells that migrate to these tissues propagate as thymocytes under local induction. With prototypal surface topographies, separate clusters develop the capacities to form single kinds of antibody molecules; in this mature form they are called "T cells" and their immunity is termed "cell-bound" because they carry antigen-combining sites right at their surfaces and directly neutralize foreign bodies. Substances alien to an individual organism are called antigens for their antibody-provoking capacity. T cells regularly examine other cells for changes in their surfaces and attack molecules of unfamiliar shape—a vigilance which prevents potential tumors from developing and also causes the rejection of grafts and transplants.

Our response to actual infections and immunizations arises from plasma cells of another lineage. In humoral immunity a different type of lymphocyte (termed the "B cell") develops an antigenic reaction somewhat mysteriously, independent of the thymus. This reaction is complex and requires T cells as helpers as well as nonimmunocompetent macrophage cells which initiate contact with invaders.

Lymphatic vessels form the same way that blood vessels do; they are part of the venous system and may even arise as capillary extensions of its epithelium. Six lymph sacs emerge along the median axis of the body—first, two jugular, then two iliac, then a pair in the abdominal region. Vessels grow out from the lymph sacs and follow the main veins. Where mesenchyme encounters these sacs they dissolve into separate channels which become the lymph sinuses; some of the mesenchyme then differentiates as the capsule and connective tissue of an emerging lymph node.

The lymphatic system emerges very early in ontogenesis, as once in phylogenesis. The first stem cells originate in the yolk sac at the beginning of embryogenesis, though the generative process is eventually taken inside the foetus and continues in the bone marrow after birth in concert with the formation of red blood cells (of which lymphopoiesis is a variant if not a forerunner). Even stem cells from the yolk sac must enter the thymus to become immunocompetent. The lymphocytes later migrate from the thymus (although some apparently develop directly from mesenchyme in contradiction of the whole process and its seeming history).

The immune system is highly complex; it must recognize the chemical identity of the body at a molecular level. Mature lymphocytes are apparently able to read the intricate three-dimensional topologies of antigenic polypeptide chains—their peculiar ruts, extensions, and electronic and spatial configurations. It seems unlikely that thousands of distinct antibodies each arose separately in the evolution of animals, nor does such an explanation for immunal capacity account for the spontaneous ability to destroy new toxins—for instance, industrial poisons and imported viruses (perhaps even someday those from outer space). Immunocompetence appears to be an inheritable proficiency of uncommitted lymphocytes which develop the appropriate neutralization when antigens appear. We do not know how such information is stored and utilized in the immune system, though it would appear that DNA itself would have to be tapped in a variety of contexts to generate a matching lymphocyte for each novel antigen. Some biologists hypothesize a constant and a variable gene, each with its own polypeptide chains and joined by an RNA molecule transcribed from both of them, thus introducing contemporary sequences into an ancient formula. Immunity must be as old as life, for without this mechanism the complex circulation of fluids and other substances in animal tissues would spread any local infection throughout the body. Plants differ in this way; they maintain diseases locally and wither in sections; they have the ability to branch out again from almost any part.

A newborn creature is not born with a fixed sense of its own identity; it must learn to discriminate itself from others, its body from life itself. If grafts from another animal are made soon enough after birth, the organism may accept them (as well as all subsequent grafts from the same animal). However, the learning period must come to a quick end if the creature is to survive the onslaught of microbes and other hungry cells in the world.

It is almost as if we keep a separate record of what-is-not-us alongside our genetic identity. We understand poisons because we are potentially toxic; in fact, we carry our own extinction dangerously close to the benign embryology that it guards. We survive by suppressing not only billions of potential cancers but billions more inherited messages that might wish to use our cellular matrix to embody themselves. If the extinct animals of the Earth are stored in our immune system (or in another submerged cellular library), these "creatures" could tell us (in the language of biological specification) all they know, providing a set of references for immune reactions.

In terms of parallel polar heredities, our diseases psychosomatically represent us; they are the partial realization of our inhibited biological potential. The tremendous capacity we have to heal ourselves is another dimension of our capacity not only to become ourselves but to sustain our being through a lifetime. Similarly (on a purely somatic level), insect larvae destroy themselves in order to create an entirely new animal with its own life plan.

There are two kinds of diseases of the immune system. Whereas tumors are collective failures of the lymphocytes to recognize a malignancy (for example, in Acquired Immune Deficiency Syndrome— AIDS), in autoimmune disorders, the cells confuse the self with the "other" and attack their own body. In lupus, cells of the genome combat *themselves* as if an outsider. Antibodies destroy the DNA, with consequent damage to blood vessels and the kidney.

Diseases of the immune system could represent crises of identity on levels not known to orthodox Western doctors. Whereas so-called psychological (and even psychosomatic) ailments reflect conflicts in the formation of personality, immunal disorders point toward unresolved ambiguities in the very fact of biological becoming. A depletion of the "self" at a molecular level is different from a psychosis or split personality, for it must arise from an embryogenic disorientation of basic identity. Some essential truth of our nature and even of our psychological boundaries is maintained in the immune system, for it is there that a biological superego impinges on omnivorous cells.

Aging itself is a gradual failure of cellular cohesiveness, almost a loss of interest among the cells to stay bound to the rigorous organismic process. Whereas it may be inaccurate to say we tire of life, it may be true that the process itself tires of the distinction between us and them, a distinction written into biology by the evolution of species and guarded twenty-four hours a day throughout nature by the immune systems of creatures. A related process must occur at the other end of life, during pregnancy, blocking T cells from rejecting the foetus in maternal tissue. In a certain sense, maternity is a form of death to the mother who gives up part of her biological identity to her child and is "willing" to preserve the life of the species in a progeny at the expense of her own exclusivity.

The great cornucopia of nature threatens us from the moment of our birth until it devours our meats at death. We are protected from the universality of the biosphere by an individuation of blood even as we must be protected from multiple personalities and collective mind by the formation of an ego. Without these identities we would drown in madness and infection. But, from the perspective of the stars, this is one archetype whose imprint falls on every layer of mind and body. Some call it the Self, but we do not know it in its original form. Epidemics and crime waves represent other cosmic forces whose forms appear with the loss of some important identity: cultural, sexual, or personal. Little enough props up our singularities in the universe, and if some individuals lose their internal sense of distinction, they may carry out unacknowledged suicides. All deaths may finally be suicides, for the cosmos eventually overwhelms us. Our minds seem almost to reach beyond our bodies into images that cannot be sustained, or cannot be sustained until the very nature of "self" changes.

20. Ontogeny and Phylogeny

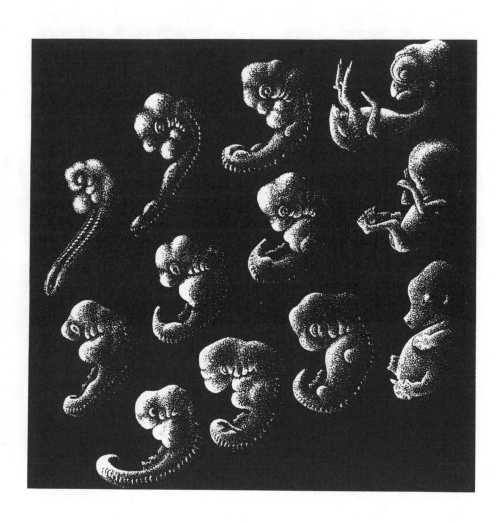

Long before our biochemical and neurological filiation with the animal world was established by scientists, there was a recurrent intuition that our lives somehow reenacted the history of our species (and perhaps of the Earth itself). From at least as early as Aristotle, the temporary gill slits and fish-like arteries in the early embryos of mammals have seemed to mark the actual reappearance of an aquatic creature. It was Ernst Haeckel who established this correspondence as a so-called "biogenetic law" in a virtual encyclopedia of writings from the 1860s to the turn of the century. To Haeckel it appeared that the same essential principle continued to push the embryo through its stages of development as once impelled its ancestors through a series of progressively more complex organisms. Recapitulation was not simply a pattern of visible resemblances; it was the driving force behind embryogenesis itself. When Haeckel pronounced, "Ontogeny recapitulates phylogeny," he meant: "Phylogeny is the mechanical cause of ontogeny."[1]

If Haeckel's "phylogenetic force" expressed itself through an actual agency (like gravity), then it would have to be a kind of embryogenic magnetism pushing anatomical traits back through stages of development, condensing some and excising others in ontogeny; the recent organs of more progressive animals would be added always at the end of the sequence. An ongoing terminal addition of new traits re-routes each embryo back through a series of its ancestors' manifestations in achieving its contemporary form. Such a sequence would explain why, for instance, a chicken begins as a tiny worm and then suddenly sprouts wings. Although Haeckel went on to identify the ostensible contemporary creatures in the lineages of other contemporary creatures, the biologist Louis Agassiz perceived that stages of ontogenesis could signify only extinct animals—the actual species in the lineage of the embryo rather than lines of divergent survivors. These "recapitulated" animals would be represented, if at all, only in fossils and, of course, would not be found among living fauna on the Earth.

Although twentieth-century biologists have uniformly rejected Haeckel's law, they accept that, in some fashion, ontogeny recapitu-

lates phylogeny. There is no other place for ontogeny to come from.
A creature could not possibly invent its own intricate blueprint in a
single generation, so it must inherit its mode of becoming from its
ancestors—not just one of them or certain key ones but the full unbroken lineage of all of them. It requires exponential generations of creatures, each incorporating existing development, to make even the
simplest modern animal.

The universe has no other way to make life except to return to the
beginning each time and to follow the same path back to the present.
It may look as though cows are born of cows, hornets of hornets, and
human beings of human beings, but they are each born only of cells.
They are created anew from something which is not them. There is
nothing in ontogeny that does not express a historical event, the organization of a mutation. The plan of an organism is assembled layer
by layer over millennial time, and ontogeny is merely the specific
process for reassembling the layers accumulated up to a given generation. Since one cannot enlarge on a pattern without including it, the
first creatures must be included within all creatures descended from
them. Although they have been replaced, protein by protein and organ by organ, their replacement has been in terms of their original
configuration, so their "erasure" lies at the basis of subsequent organisms. The genetic message may no longer include even a trace of
them, but it would not be the same message if they had not originated
it. Phylogeny is but a seamless patchwork of successive histories. It
is not a force (it does not actually exist at all except as our reconstruction of lineages). It is expressed only through ontogeny—the ceaselessly factored sum of billions upon billions of separate mutations and
the differential production of offspring.

For instance, whales and dolphins, like all other mammals, inherit
gill slits from ancient fishes and express them early in ontogeny. The
gill slits of mammals, though, are uniformly reabsorbed and do not
become functional in adults, so despite their return to the water, sea
mammals cannot use these simple filaments for breathing. They must
develop their respiratory apparatus secondarily from the clumsy air-
breathing organs of the land animals and carry them around for lifetimes underwater, breaching in order to fill them with air—despite the
fact they have not walked on the earth for millions of years (or since a
creature resembling a hippopotamus arose from an ancient forerunner
of an elephant, filled its huge lungs with air, and went fishing off-
shore). Creatures a billion years from now will probably contain the
beings of this epoch within them, including the speaking primates.
Even if lines of human descendants lose intelligence somehow, they

will experience the loss as a gradual descent (gene by gene) into unconsciousness.

Recapitulation fuses two separate systems of belief—an ancient prescientific cosmology in which the species of animals were fixed as totems (along with Sun, Moon, and stars), and a scientific view of speciation through evolution. Totemism no doubt stretches back into the Stone Age when the first human philosophers perceived the different animals as unique spirits and manifestations of numinous forces. These creatures received their energy from a cosmic realm and imparted aspects of it to lineages and clans of human beings. To the South American aborigine, the crocodile and opossum are fixed stellar entities; they do not have primitive ancestors. The jaguar could hardly be a temporary visitor to the Earth, for he is the perennial source of fire and the custodian of the cooking hearth.[2]

Claude Lévi-Strauss has demonstrated that tribal peoples *think* in plant and animal categories and that these categories are the basis of their social and religious philosophy. "How animals and men diverged from a joint stock that was neither one nor the other (and) how the black-nosed kangaroo got his black nose and the porcupine his quills"[3] are irreducible elements that explain not only the zoology and etiology of the Australian desert but also the origin of tribes, clans, and languages; as well as the reasons behind exogamy, sister-exchange, and circumcision, and why men and women must die.

Long before Aristotle constructed the rudimentary categories of logical taxonomy for the emerging Western civilizations, there were "pagan" tribes in the Aegean too. For these Stone-Age philosophers each local beast, flower, and planet signified an etiology; the specific designations were not inherited by the early scientists, but the underlying totemism was. The Kantian species are, in a sense, the last flicker of a magical system. Not until the twentieth century did it become clear that the animals (*and* the stars) were ceaselessly changing and did not *mean* anything in and of themselves.

For the eighteenth-century preformationists evolution could only be the unravelling of archetypal creatures wound into the primal germ at the beginning of time. If birds and mammals "evolved" from jellyfish, then their seeds must have already been encapsulated within the sex cells of the medusa, requiring only the maturation of the intervening seeds (species) to emerge from the chrysalis.

For the pre-Haeckelian recapitulationists the species of animals were psychosomatic stages, repeated on different scales of morphology. In the early part of the nineteenth century the renowned comparative embryologist Lorenz Oken published a sweeping

"Naturphilosophe" describing these different correspondences and homologies throughout the animal kingdom. In Oken's system the intestinal organs of the mammals represent the infusorians at one level and the rats and beavers at another. The vascular organs signify clams and sloths, but also snails and herbivorous marsupials, and finally squids and carnivorous marsupials. The higher animals pass through the permanent stages of the lower animals as they add on organs (Haeckel later adopted this as dogma). The stomach, said Oken, was once the simple vesicle of an infusorian, which became doubled in the albumen and shell of the corals, vascularized in the headless clam, and infused with a blood system, liver, and ovarium by the bivalved mollusks. Our muscular heart, testicle, and penis mark our ancient transit through the snails. "The whole animal kingdom," wrote Oken, "is none other than the representation of the several activities or organs of Man; naught else than Man disintegrated."[4] The medical anatomist Etienne Serres classified aborted foetuses by the stage of development at which they were arrested; for instance, a headless foetus was a clam (actually a brachiopod) on the basis of its apparent cutaneous respiration. The human embryo, according to Serres, must pass through every major class of animal, including the insects (in which its limbs first sprouted) and the birds (in which it initially sucked air into its lungs). This nineteenth-century science has aspects of a South American Indian creation myth in which tobacco orginated from the buried jaguar-woman, wild pigs from the fornicating humans, bats from excrement, and toads from burning sperm—not as vital seeds but as transformational categories.[5]

With Haeckel recapitulation was more than just totemism and archetypal biology; it became the one permanent law of evolution, the single concept unifying life science into a field. Though Darwin himself had little sympathy for Haeckel's writings, recapitulation advertised Darwinism to the general public as well as to the emerging practitioners of the social sciences. The chain of atavistic ancestors within became an imaginal correlate to Darwin's descent of species.

A symbolic version of the biogenetic law has served as a yardstick for human and cultural development from Haeckel on. Its uses in the twentieth century have stretched from Rudolf Steiner to Jean Piaget. What for Freud was a biological correlative for unconscious levels of mind (and an explanation for spontaneous regression) became an anthropology for James Frazer and Lucien Lévy-Bruhl, i.e., a set of natural laws whereby animism and totemism were inscribed in the genes of primitive peoples and recapitulated in the developmental stages of the children of civilization.

Whatever our opinions of psychological and cultural recapitulationism, we should remember that Haeckel intended his succession of ancestors as physical fact, not a metaphor for stages of consciousness. The human foetus is actually a worm, a clam, and a fish as it develops, and this is *why* it develops. Gastrulation occurs in each embryo only because of an ancient somatic habit which invaginated the primeval blastaea (to use Haeckel's name for this animal).

"If we now want to explain the phylogenetic origin of the gastraea (repeated, according to the biogenetic law, by the gastrula) on the basis of this ontogenetic process," Haeckel cautioned, "we must imagine that the single-layered cell-community of the sphaerical planaea began to take in food preferentially at one part of its surface. Natural selection would gradually build a pit-shaped depression at this nutritive spot on the spherical surface. The pit, originally quite flat, would grow deeper and deeper in the course of time. The functions of taking in and digesting food would be confined to the cells lining this pit. . . . This earliest histological differentiation had, as a consequence, the separation of two different kinds of cells—nutritive cells in the pit and locomotory cells on the outer surface."[6]

Ontogeny was a tendency toward motions acquired in phylogenesis, an inherited resonance: the germinal "atoms" were recorded in the nervous systems, transmitted through hierarchies of tissues to the germ plasm, and ultimately imbedded in the genitals. It was, oddly, only the Lamarckians with their theory of acquired characteristics who kept Haeckel's recapitulationist law alive into the twentieth century. According to their interpretation, progressive traits were added at the end of ontogenesis specifically because they were acquired by mature animals and then transmitted from the phenotype to the genes; dynamically, there was no other path for traits to follow.

Haeckel's image was so powerful that its influence far outweighed its applicability. From early in the nineteenth century, there were compelling alternative explanations for the resemblance of the stages of the embryo to "lower animals," and, throughout the century, most scientists saw how pure recapitulation was contradicted by abundant discrepancies from all phyla.

First of all, the whole adult ancestor is never recapitulated in ontogeny, as one would expect it to be if traits were simply added at the end of a developmental sequence; yet, gills alone do not make an embryo a fish, and limb buds certainly do not make it a fly. In addition, organs that do seem to be accelerated in relationship to other organs usually fall out of synchrony; and it would appear that some organs are accelerated faster than others. The human heart and brain both appear far earlier in the embryo than they would in a perfectly parallel recapitulation of the phylogenetic sequence.

Secondly, many foetal organs are quite obviously uterine adaptations; the placenta, for instance, does not occur in any adult ancestor. What appears to be recapitulation may simply be a series of adaptive changes in each embryo, resembling phylogenesis because of the primitive and watery environment in which the zygote spawns (especially as a free-swimming larva).

Haeckel and his followers had ready explanations for these apparent exceptions.

"All of ontogeny falls into two main parts," he wrote, "first *palingenesis* or 'epitomized history,' and second, *cenogenesis* or 'falsified history.' The first is the true ontogenetic epitome or short recapitulation of previous pyletic history; the second is exactly the opposite: a new, foreign ingredient, a falsification or concealment of the epitome of phylogeny."[7]

Cenogenesis could be caused by a number of factors: uterine adaptations; the interference of yolk cells in the differentiation of the blastula and gastrula; the displacement of cells from one layer to another, for instance, the migration of the gonads from one of the primary germ layers to the mesoderm; and disjunction in the developmental timing of organs in relation to one another. Additionally, foetal condensation would blur sequences and relationships by crowding aeons into hours.

"The development of each organ is entirely and exclusively dependent upon phylogeny," declared one of Haeckel's defenders. "But we must not expect that all the stages evolving together in a phylogenetic series will appear *at the same time* in the ontogeny of descendants because the development of each organ follows its own specific rate."[8]

However, more significant than all of these exceptions was the discovery of an apparently reverse sequence: The early embryonic features of ancestral animals seemed to occur in the adult forms of some of their descendants. Phylogenesis was "reversed." Nineteenth-century recapitulationists were confounded by the axolotl, a Central American amphibian that retained larval features in its clearly adult stages. It was a salamander that gave birth only to salamanders; the frog had been eliminated from its life cycle. Sexually mature larvae of comb-jellies and starfish were also commonly known in the nineteenth century. Yet Haeckel's "law" stuck and all these foetalized creatures were regarded as curious anomalies. In fact, Haeckel's followers confabulated an ingenious explanation: The youthful features of these animals were merely senile second childhoods caused by an overacceleration of development pushing new traits back into the most persistent juvenile traits, hence producing degenerate forms in violation of progressive evolution. So powerful was the attraction of the image that it survived even an absolute refutation of its mechanism. The supporters of recapitulation were convinced enough of its rightness that they perceived a bizarre series of vanishing cycles instead of the simple reversal that was self-evident. But the most notable instance of larval retention was also the one closest to home: The human being seemed suspiciously to have retained the traits of juvenile apes through sexual maturity: a flat face, hairlessness, small teeth, and a brain abnormally large in relation to the rest of his skull.

From the point of view of modern embryology, Haeckel's law is a sterile abstract statement, an intuition of a resemblance that does not explain how energy is transferred from living systems to other living systems and how the embryo unfolds into an organism. Even during the heyday of recapitulationism, embryologists were beginning to examine the actual development of creatures. Using elastic sheets to represent embryonic layers and rubber tubes for the brain and gut, the anatomist William His was able to imitate ontogenetic processes. He split tubes, bent them back on themselves, and stretched them with remarkable resemblance to various embryonic stages. Haeckel was appalled not because His tried to demonstrate the possible mechanical basis for tissue structure but because he showed only the immediate physical cause of morphologies and ignored phylogeny. He was an engineer, a hack, not a scientist concerned with ultimate causation. Yet Haeckel himself could never demonstrate that phylogeny was the cause for ontogeny, and, as scientists came to ask *how* rather than *why,* he was passed over, though the shadow of recapitulation continues to haunt us for other reasons, as we shall see.

The landmark contemporary work on this subject is palaeontologist Stephen Jay Gould's *Ontogeny and Phylogeny,* published in 1977. Gould reveals on the opening page that the "topic has fascinated me ever since the New York City public schools taught me Haeckel's doctrine, that ontogeny recapitulates phylogeny, years after it had been abandoned by science."[9] Gould's book is not only an account of the role that the putative relationship between ontogeny and phylogeny plays in the history of biology but also a scientific analysis of the actual (i.e., empirical) links between embryos and their ancestors. As a Darwinian palaeontologist (and a modernist) Gould summarily dismisses the psychological and anthropological parallels to recapitulation (including Freud's whole explanation of primary development), and it is needless to challenge him on this issue, for his major interest is in biogenetic timing, not psychocosmic parallels. Ultimately, he banishes universal Haeckelism from all of research biology to a realm of psychology and metaphysics, and then he demonstrates that these are spheres more of folklore and superstition than of science.

In the area of pure science Gould compares Haeckel's version of the ontogeny of the liver with that offered by the early experimental embryologists Wilhelm Roux and Hans Driesch. According to Roux "the multipolar differentiation of the liver cells . . . causes the transformation of these cells from the tubular to the framework type";[10] i.e., organogenesis lies in the cellular differentiation of basal and secreting surfaces. But, he notes, Haeckel would have searched for an ancestor that had a tubular liver in its adult state.

We now presume that genes control both absolute morphology and rates of growth through their coding of proteins and enzymes. Novel forms have arisen in two ways only: "by the introduction of new features or by the displacement of features already present."[11] Insofar as all mutations occur at discrete points in history they are expressed synchronously in ontogeny (either as a divergent creature or a lethal defect). Before a mutation two variant creatures must share a common ancestor; they are one lineage without any foreshadowing of a division. When a mutation changes the genetic message of a group, its members continue to express this new pattern as if it had been inherited from time immemorial. We must not regard such transformations as rare or abnormal for, in fact, all tissues and organisms arose once by random genetic alteration.

The late-nineteenth-century rediscovery of the work of the Austrian geneticist Gregor Mendel provided the mechanical principles for Darwin's "origin of species." Twentieth-century microbiology has since planted the physical basis of that mechanism at the heart of the cellular nucleus. As Darwin himself intuited, evolutionism and pure recapitulationism are actually in contradiction. The animals are not completed archetypes; they are transistory patterns in a flowing current. Speciation occurs only because systems cannot be frozen; in the random tumult of nature they express the energy stored in them by the chance coherence of prior patterns. Ever-changing climates, topographies, and ecosystems continually potentiate certain obscure mutants of established species, so new kinds of creatures emerge with changing environments. These transitions are often cataclysmic, such as: an alteration in the balance of oxygen and carbon dioxide in the atmosphere; massive floods and earthquakes; glaciers; volcanoes spitting out islands; comets and meteorite showers; radiation storms from distant stars (shuffling the genetic alphabet); and, of course, land bridges thrusting up between continents once separated by water; and channels cut between seas—all these may bring about the sudden arrival of giant predators like sharks, panthers, or dinosaurs. It is little wonder that the ubiquitous birds and butterflies, the marsupials and marmoset monkeys arrive only amidst chaos and often at the beginning of a modern epoch.

The living fount is unpredictable. It rushes into some habitats while ignoring others. Many of its life histories are bizzare, especially considering that far simpler creatures could have been created but were not. Pond flukes must reach the intestines of songbirds in order to breed, and they accomplish this only by infesting certain snails in a manner that makes their feelers look like "brightly colored caterpillars."[12] Cuckoos lay their eggs in other birds' nests (and these huge intruders apparently appeal to the species that must nurse them), so their species flourishes even though they seem to retain no nest-building skills of their own. Salps form hermaphroditic chains in alternate generations. One clam fashions an artificial lure from its brood pouch and outer skin, and an angler fish baits itself by somehow having a dorsal fin modified and attached to the tip of its snout. Both decoys are so perfect that they bear the precise dorsal and anal "fins" and "tails" of the tiny fish they are imitating.[13]

Generations of bee-hawk moths have been spared from predators only by their inherited resemblance to the stinging bumblebee, a creature to which they have no close kinship. Likewise, swallowtail butterflies have spread through Africa and Madagascar mimicking the foul-tasting danaids. From region to region they match them by species, varying from black and orange to black and creamy-yellow to dappled black, white, and orange, while imitating danaid flight.

Innumerable kinds of flies are avoided by birds and reptiles only because they have developed some mimicry of stinging insects—a few hairs to suggest the thick fur of the bee, antennae, or, in the case of the bee-fly, smoky-brown translucent wings with their leading edge darkened by a single vein. Of course, there cannot be too many mimics before the lesson is lost; if harmless varieties outnumber noxious ones, then both are attacked with a combined high percentage of success until the former are exterminated and the danger restored.

"Let it be borne in mind," wrote Darwin, "how infinitely complex and close-fitting are the mutual relations of all organic beings to each other and to their physical conditions of life; and consequently what infinitely various diversities of structure might be of use to each being under the changing conditions of life."[14]

If all the living creatures that ever inhabited the Earth appeared before us, we would behold a panorama of varieties almost indistinguishable from one another, with partial folds and stumps, misshapen rudiments, half-formed wings, twisted shells, creatures barely able to move, creatures drowning in their own breath—all of them surviving at least a few generations. The gaps between existing types of plants and animals are filled by life forms that have become extinct.

Evolution is blind and amoral; because it is unconscious it can be neither bloodthirsty nor nurturing except by chance. After following a narrow trail of species for a billion years the life current can suddenly abandon it, leaving only a fossil record of shells or giant wings in sandstone. It can also suddenly adopt a strategy it seemingly rejected in prior millennia. Some creatures are left eating their own children or laying an egg every half-second or secreting huge spiral matrices. That is literally their price of being born, their fate. Once the hornet and the pollen are linked in the maelstrom, they lock, and their sentence is written again and again in the gnosis of the world as fresh as spring breezes and new purple clover, each time as if it never happened before. Accidental and remote connections have become partly conscious acts, and they are sustained by the energy and *eros* of newborn creatures each generation, and cannot be exterminated as long as the cycles continue and sex cells are exchanged.

In 1828 Karl Ernst von Baer had offered an explanation for the resemblance of embryos to ancestors that was to be far more compatible with Darwinian thought than Haeckel's recapitulationism, but it was overlooked (for the most part) until the decline of Haeckelism. According to von Baer, ancestral features persist only because they were once the general organic configurations from which the specific traits common to any line of descendants developed. Prior stages of organization are always the raw material for subsequent differentiation. If they were completely eliminated, the embryo would have no history; there would be nothing from which new organs could emerge. But they are *not* the regressed and condensed replicas of adult animals; they simply follow the lineal trail of tissue assembly.

Early embryonic stages of vertebrates so closely resemble invertebrate adults because the majority of invertebrates living in the primal ocean have not departed as significantly from the last equipotential state. They remain, from a vertebrate perspective, partial embryos in an oceanic womb. We might say, simply, that ancient features return in the life of the embryo only because they were never eliminated; they represent no force, no law of development. They represent (in Darwin's words) ''a community of descent'' prior to their divergence by mutation.

For Haeckel the egg simply marched through its programmed stages, its memory traces of ancestral beings. For von Baer the egg was a germinal mass produced by earlier phases of evolution. It could no more regress than any mature animal could. It too was a "terminal adult," but at a different point in its development.

Von Baer, a vitalist, and Darwin, by comparison a materialist, provided the two poles for modern evolutionary theory. Whereas Darwin proposed the phylogenetic mechanism for speciation and survival, von Baer discerned the cosmic pattern of differentiation. The unfolding of the primal ovum applies equally to galaxies, solar systems, oceans, and the primeval cells in those oceans. Systems of energy and matter flow from homogeneity to heterogeneity, from primal density to myriad varieties of microstructure. Von Baer's laws of differentiation are the axioms of the science of embryology:

"1. The general features of a large group of animals appear earlier in the embryo than the special features.

"2. Less general characters are developed from the most general, and so forth, until finally the most specialized appear.

"3. Each embryo of a given species, instead of passing through the stages of other animals, departs more and more from them.

"4. Fundamentally therefore, the embryo of a higher animal is never like a lower animal, but only like its embryo."[15]

In August of 1971 Julian Huxley told Gould that Haeckel's law of recapitulation is "a vague adumbration of the truth,"[16] and Gould concludes that this truth must be the importance of temporal displacement of genes in evolution. Haeckel called such displacement "heterochrony," but he considered it only a distortion of palingenesis, his "true natural history." For Gould heterochrony more precisely describes the dissociation of traits from one another *in either direction* during development. The acceleration of some traits relative to others leads to recapitulation; the retardation of some traits (relative to others) leads to paedomorphism. Gould's point (in keeping with the ethos of modern biology) is that these displacements represent neither progressive nor regressive evolution and reflect no preference for either juvenile adults or recapitulated ancestors in the formation of higher phyla; they are simply different strategies of survival made possible by heterochronic mutations.

Gould reminds us that as early as 1918 the geneticist Richard Goldschmidt, in his work with geographic variation in gypsy moth populations, intuited the existence of "rate genes"—genes which caused large differences in patterns of pigmentation from small changes in developmental timing. Goldschmidt wrote:

"The mutant gene produces its effect . . . by changing the rates of partial processes of development. These might be rates of growth or differentiation, rates of production of stuffs necessary for differentiation, rates of reactions leading to definite physical or chemical situations at definite times of development [—] rates of those processes which are responsible for segregating the embryonic potencies at definite times."[17]

Without "rate genes" phylogeny could occur only by abrupt introductions of new material. Heterochrony, however, allows creatures to use their existing complexity as the blueprint for ceaselessly diverging variations at the same level of complexity. The evolution of triploblastic creatures opened an immense range of somatic paths, new morphologies temporally displaced from the one simple underlying creature. The first coelomate worms may have been highly specialized mutants, but some of their embryos, through further mutations, could retard linear "worm" aspects and produce sexually mature juveniles with radically different potentials, including the templates for billions of different kinds of insects, spiders, and crustaceans. As it is, most new organs probably began as pathologies and were lethal in all but a few creatures, in which they became functional through the discovery of new ecological niches.

Mutations causing recapitulation or paedomorphosis can enter the embryonic process at any stage, so there will always be two kinds of heterochronic potential in a genotype: one, continuing a strategy of specialization by retaining adult forms of ancestral animals, condensing them, and surpassing them by terminal modification; and, two, radiating from their partial development short of full ancestral maturation and adapting to a variety of microenvironments through different expressions of their genetic potential. Without heterochrony complex organisms would become dead ends. Random somatic mutations would simply cause monstrosities (as commonly occurs in our own species). However, heterochrony allows for the rearrangement of a whole sequence of development by a slight genetic displacement.

Recapitulation and paedomorphosis can occur by quite opposite modes with different evolutionary meanings. If some aspect of somatic development is retarded while embryogenesis proceeds at the ancestral rate, then the adult is juvenalized by neoteny. If the embryo becomes sexually mature precociously, i.e., while still a "child," then paedomorphosis has occurred by progenesis.

In a wide open environment with abundant resources some aphids apparently spawn wingless forms which mature rapidly by progenesis. Since there is ample vegetation to feed on, extra energy need not be consumed in sensorimotor mobility; they can just sit in place eating. Other progenetic forms become parasites, developing core organs so rapidly that they end up as little more than gonads and stomach; they are sex cells more than creatures: If their hosts are abundant all the rest of their physiology can be dispensed with in the "haste" to produce more offspring and fill the new environment. Or, in the language of neo-Darwinism, such activity fills the environment with paedomorphs faster than with more mature and endowed species. Dwarfism may also be adaptive, especially where tiny creatures enter otherwise untried niches; for instance, parasites in the organs of clams and fish. It is possible that whole phyla of minute metazoan creatures such as roundworms and rotifers have progenetic origins. The actual line leading to the land vertebrates could have originated from tunicate tadpoles with short larval phases. Although progenesis seems primarily to be a strategy for abundance of offspring at the expense of complexity of tissue, genetically plastic paedomorphs (like these tadpoles) might have developed new lineages if transferred by chance to suitable environments.

According to Gould, neoteny is a more promising mode for the emergence of higher taxa, for it preserves the morphological plasticity of unspecialized juvenile forms. Whereas progenetic paedomorphs may lose evolutionary potential, the more conservative neotenous paedomorphs usually gain potential when the development of subsequently crucial organs is retarded along with maturation.

The two different types of paedomorphosis arise in insects from virtually opposite mechanisms: metathetely (or neoteny) when an increase in the amounts of juvenile hormone causes childhood features in adults; and prothetely (progenesis) when the juvenile hormone is suppressed and adult traits appear larvally, often at a premature molt (relative to ancestral forms), with subsequent molts suppressed. If they are activated by opposite biogenetic mechanisms, it is no wonder that progenesis and neoteny result in different modes of adaptation despite their expression in similar appearances. Embryogenesis is paradoxical in exactly this way.

If a mutation causes an ancestral trait to be displaced backward, then recapitulation occurs by acceleration, the classic Haeckelian mode. If full somatic development continues at an ancestral rate while maturation is delayed, i.e., if only gonadal development is retarded, then another kind of recapitulation can occur, and this is called hypermorphosis and often leads to larger and more differentiated organs like the antlers of elks and giant mollusk shells. When ontogeny continues past its prior termination point, it can also lead to immense creatures like dinosaurs and whales in relatively few generations. Once the scale of growth and maturation is tipped, animals can shoot from one size range to another until the physics of sheer anatomy produces its own limitation.

Neoteny is more common than hypermorphosis or progenesis in the situation of a relatively favorable but bounded environment with a harsh and perilous surrounding terrain—classically, small ponds in arid regions without predators themselves but ringed by predators. Not only would there be little advantage to population growth by rapid maturity (or enlarged bodies), but there would be no incentive to colonize the outlying region. Thus axolotls and other such paedomorphs do not even mature sufficiently to live an ancestral life. Their development slows down, so they retain larval anatomy. Axolotls never become fully amphibious; they are able to stay in the water and reproduce without having to brave the shoreline environment. Initially such neoteny may have been an ephemeral metabolic tendency, and, in some species, it can be counteracted by experimental doses of thyroid (the animals then mature); most neotenous paedomorphs, however, have developed hereditary resistance to metamorphosing hormones and do not respond to exogenous treatment. The juvenile state *is* their permanent adult state.

The intuition that advanced human development was paedomorphic rather than recapitulationary and accelerated was disturbing to many nineteenth-century biologists. If juvenalization was the desirable characteristic for advanced status, then it was clear that the Mongoloid races were most deeply foetalized in some respects and thus capable of the greatest development. But recapitulation seemed to favor the African races with respect to other traits. The implicit contradiction ran deeply enough that man was conceived of simultaneously as a retarded and accelerated animal. To a certain degree this is accurate, for mutations occur in limited groups of tissues, not universally; and, whereas many key human traits may be paedomorphic, others are more likely recapitulationary. The growth of the brain for example, may be either.

In general, the advanced mammals have evolved through retarded development, smaller litters, and long gestations. Most simple mammals are born with survival skills; humans have become secondarily altricial, apparently because of their immense brain growth outstripping the capacity of the birth canal (but perhaps also because of certain neuropsychological advantages in a brain which develops among a diversity of external stimuli). The primates are already retarded in relation to the rest of the mammals, so the hominids merely continue the paedomorphic trend. The implication is that, if individual human beings were somehow allowed to continue growing and developing indefinitely, they would become more simian, like Aldous Huxley's Fifth Earl of Gonister in *After Many a Summer Dies the Swan*, who, by his 201st birthday, from using an extract derived from the intestinal flora of carp, had turned into a hairy muscle-bound ape.[18] As it is, only retarded development allows us our already unnaturally long life span on this planet, and it is to be presumed that if our rate of maturation could be slowed even more by heterochronic mutations, we would become more childlike, i.e., more human.

Gould argues that man is paedomorphic not because of any one trait but because an overall retardation of development led to a general selective matrix through which all aspects of human morphology were simultaneously reevaluated in nature, and later, culture. Men and women remain embryogenic even after they leave the womb: Witness the late eruption of their teeth, their body growth through adolescence, and, most notably, the postnatal expansion and convolution of their cerebral tissue—indispensable aspects of the human condition. When foetal growth rates are retained through adolescence, potentially complex genetic messages play out some of their myriad possibilities in actual configurations.

Recapitulation and progenesis still occur with regard to certain traits, even in the overall context of neoteny. Gould cites as recapitulationary the "early fusion of the sternebrae to produce a sternum; the pronounced bend of the spinal column at the lumbo-sacral border; the fusion of the centrale with the naviculare; and several aspects of pelvic shape."[19] Progenetic traits include relative loss of pigment and body hair, orthognathy, labia majora in women, loss of brow ridges and cranial crests, general thinness of the skull bones, long neck, thin nails, eye orbits under the cranial cavity, and reduced teeth. It is the underlying trend which is neotenous, i.e., extension of the life span, persistence of cranial sutures, secondary altricial dependence, and the general lengthening of the time of body growth.

Ninety-nine percent of our genes are identical to those of the apes; yet we probably could not even breed with one of our hominid forerunners if we were ever to discover a tribe of these creatures hiding from us in remote caves or forests. Only the collectivity of heterochronic mutations throughout our evolution has led to such a departure from the simian line. These were almost exclusively displacements of existing elements rather than new genetic traits. But the effect of heterochrony was dramatic and irrevocable. Hominoid, and then hominid, populations became demographically separated, and, in the context of culture and language, child-like men and women became domestic, educable, and symbol-possessed and possessing. These creatures suddenly leapt the seemingly uncrossable chasm separating nature from culture and timelessness from time. In 1926, biologist Louis Bolk wrote: "I would say that man, in his bodily development, is a primate fetus that has become sexually mature."[20]

As ancient drives become latent and are sublimated, new forms arise, psychically as physically. In combinations of recapitulation and foetalization, mental phenomena of different orders exist simultaneously. These are mere images in the record of the mind, but in the fossil record we see them as successive species, Australopithecus, Pithecanthropus, Neanderthal, followed by the various tribes and races of humanity. Gradually, the physical and the psychological come together, and retarded and accelerated features merge in a mind which obliterates their antitheses.

21. Self and Desire

Humans come into being as animals; yet there is a gulf in the animal world between "us" and "them." On our side creatures have conscious awareness of their existence in the universe; we dwell among names and symbols. The animals remain in nature among things. We forget and remember intermittently that we are natural events; the animals do not comprehend the distinction. We are caught, seemingly forever, between being and nothingness; they, equally eternally, *are*. The paradox is that we are one of them and could not have come into being without billions of their prior generations. Our difference is still an animal thing. And, apparently, they could not have quarantined forever the symbol and the name.

Any animal brings a certain amount of sentience into the world, even a protozoan amoeba. Compared to a sponge or jellyfish the many species of worms are highly intelligent; compared to any of them the snails and insects are virtual philosophers. The octopus and salmon are beginning to individuate and have the inklings of personalities. Amphibians and reptiles have become partially individuated, so in myths and fairy tales we give them voices and let them speak in their own behalf.

The thoughts of warm-blooded animals are literally in their bloodstreams. These creatures appear to dream in sleep as we do, to probe the world with a slightly detached curiosity snakes and frogs never have. Bears, seals, dogs, horses, mice, cats, and even birds are "people" by our standards—people in their own classes set off from regular human society. In a great number of cultures other mammals and birds have higher social standing than some human beings. Conversely, humans can be made into "animals," through caste systems, slavery, and the taking of prisoners in warfare.

Monkeys and apes live at the boundary of our condition. After the fact they look like unfinished replicas of human beings. Like us they live in groups; they chatter; they play; they are curious (the rest of the mammals and birds are not as spontaneous in either behavior or speech).

There is the perhaps apocryphal story of the laboratory chimpanzee raised as just another child in a family among children. When

297

funds for the experiment ran out, the chimpanzee was taken to a zoo. There he sat, behind bars, to himself a child, crying, and wondering why he had been put in a cage with apes. To himself he was a person, a child with fur who resembled an ape. On the other side of the coin is the abandoned child raised by wolves who growls at humans and refuses to speak. These contemporary legends not only reify our guilt about the animals but expose our apprehension of their nearness to us.

We are far more alone than we usually realize. Because the animals resemble us we think that they aim at our designs and fall short. We project our motivations onto their actions, judging them only by their apparent functional goals. But ants do not build cities or fight wars; squids do not dance seductively to attract mates. They respond to their own bodies and ganglia, their ruffled flesh, the ancestral habits of their lineage. We can describe a dragonfly's activities but we cannot know the wingspread it feels as it beats its helicoid cuticles. We cannot imagine the intimacy of currents against a fish or a whale's big song. We have no inner reference for the taste of honey to a bee or the magnet that pulls the eel upstream.

Long prior to language a creature experiences itself and its own existence as real; it knows exactly who else is real. Alien beings may suggest "food" or "danger," but they are not "people." The spider looking at another spider sees the same thing we see when we look at each other—another human being (to universalize our term). If we were somehow reincarnated in their bodies we would find spiders as irresistible as we now find the partners of our own species.

Peacocks and pheasants display iridescent colors and parade before their prospective mates. Other species of birds hiss, raise their tails, and spread their wings. Fighting fish open their fins like painted parasols. The male dytiscus beetle strums a rhythmic tune on his femoral ring using his hind legs. Songbirds call each other with the same haunting melodies, again and again. The female spider casts down a thread and slides partway along it; the male catches the bottom and climbs up to meet her. She still may decide to eat instead of mate with him, but he strokes and feels her body with swift jittery legs and then plunges his palp deep into her vagina. He may even wrap her in a cocoon of his silk so she is immobile while he enters her.

Crab ancestors initiate what become Panamanian carnival dances. Moths are drawn to each other by smells which, to us, resemble raspberries and vanilla. In some species of flies the males compose a dancing swarm, and this stirs the otherwise placid females from among the bushes. It would seem that the women are transfixed by the pattern of the dance; they do not see the individual males but the swarm itself.

More than any other feature of the primordial animal world, sexuality looks like language. In their foreplay even snails and clams seem to recognize their psyches, to respond to the differentiation of the Other within desire. Warm-blooded animals are dimly conscious of the festival. Sea swallows transfer fish from the beak of one to another during the dance of mating; orangutans embrace and nuzzle as they hang breast to breast by their arms from branches. Flying foxes hug similarly in the air, and beavers kiss while paddling through water.

Obviously, our body images, instincts, desires, and rituals come directly from our animal heritage (if not from any animals presently on the Earth). Still something stands between us; despite the development of the human psyche by mere degrees of cell-stuff, there is an unpassable barrier between the existential situation of the human and that of any animal. It has been called language, symbol manipulation, experience of past and future, and, of course, the soul; but, whatever it is, it is important *not* because it is "human" (it can occur in quite alien creatures on other worlds) but because it changes the nature of the universe through the creature that embodies it.

Animal behavior is almost fully unconscious and compelled. No matter how complex the notes in the song or the steps in the dance, these creatures do them because they *are* them. Their sexuality brings them into contact with beings of their own kinds and provides the thread of continued life for the species. This is what the mysterious notes in the song mean to them, but to us, they remain nostalgic in another way. Their guttural resonance *we* find pastoral; they are a pure form of a thing we experience derivatively. They seem like abandoned commitment to natural desire, however contrived and ritually constrained the acts themselves. (Of course, we may also become quite mimetic in our passions and so-called spontaneous tremblings; we are not the carefree dancers we pretend to be.)

Finally we cannot interrupt the animals, either within or without, because we cannot address them. We can only enrage them, distract them, torture them, bore them, or trick them (and ourselves) into believing we are not there. They will keep their dignity forever. The experimental scientist towers over the bee in generations of intelligence but he is puny against the fixed time of its species.

The fantasy of stealing or winning animal secrets is as old as our banishment from the kingdom. Shamans of diverse cultures use hallucinogenic plants and even cut pieces of their flesh off to transpose reality and gain access to these forbidden dimensions. Trespass is so dangerous that every act must sustain the seriousness of the goal. Don Juan Matus as moth or hawk imagines he sees his prey as spots of brightness, he detects invisible movements and forms in the land below. There is always the possibility that if we could enter the alien consciousness of other minds we might touch the secret of life, dead-reckoning its reflection through two opposed but equally interiorizing systems.

Otherwise, through the code of our own half-remembered, half-invented songs we explore the melodies which lie behind the themes of the animals through whose expression we exist. But the musician is not able to return to the pure notes of the jay, nor are we, in hearing them, able to go back to the actual desire. In the translation of sexuality through consciousness, something primary has changed. However, between the ancient bell animalcules dancing about the sessile females of their species and the strip of porn theaters with their rhythmic barkers and flashing lights, there is a link. We cannot alter aboriginal desire; we can only watch ourselves responding to eroticism at the same moment as we enact it.

Our animal/human dualism is a trap. The more we emphasize the differences and attempt to elevate ourselves to a superior realm, the more the similarities jolt us back into the animal kingdom. The new sciences of ethology, ecology, and sociobiology consistently discover that *Homo sapiens* is just another animal species and our inventive behavior originates in typical mammalian and chordate organs. On the other hand, when we minimize the differences and proclaim with egalitarian determination that we are just another animal, our very act of self-realization refutes itself: What other animal could redefine the criteria of its taxonomy?

The polar positions may reflect ideology more than zoology. Claiming our status is purely animal, we justify our wars, our social injustices, our sexual abuses, and our greed. "The territorial imperative," wrote Robert Ardrey, "is as blind as a cave fish, as consuming as a furnace, and it commands beyond logic, opposes all reason, suborns all moralities, strives for no goal more sublime than survival."[1] Or, invoking the more primitive and brutal passages of the Judaeo-Christian *Bible,* "human supremacists" argue (from the opposite position) that "man" was created to have dominion over the animals and thus may use their lives and bodies as he wishes. The conviction of this dominion (whether divine or biological) is so deeply ingrained in our species that men may apply it to women—men who are otherwise politically liberal (or even radical) often tacitly promote the subjugation of woman.

These pretexts of superiority are rather late and labyrinthine exercises in self-vindication. Animals are hardly pacifists or jurists, but they are rarely as arbitrarily ruthless as we. The tenderness of the crocodile mother with her babies stands against the rampant child abuse in our species, not in every instance of reptile and human but as a measure of their intrinsic capacity for humane behavior. We have the required neural structures for restraint and compassion, and our survival has depended on a shaky self-imposed peace even from our Palaeocene incipience. Yet we are more callous than many of the "dumb" beasts. ("Practice love first on animals, they are more sensitive," wrote Gurdjieff on the walls of the Study House at the Prieuré.[2])

The answer lies not in sterile flight from the wild. Our self-conscious attempts to escape the animal kingdom turn us into stiffened robots. The pompously "human" human being is not our model for higher consciousness; in fact such men and women seem to have lost the animal's supple inherence in the universe, its native wisdom. Their pretenses are only stereotypically intelligent.

"When man does not admit that he is an animal, he is less than an animal," wrote Michael McClure. "Not more but less."[3]

Higher functions come not from transcending the animal kingdom but from experiencing the unity of animal existence in a new way. The seal scanning blue underwater light was transformed into the child learning the names of flowers and the scientist focusing images of galaxies in mirrors. Through the biology of our becoming we have turned the pure pain (and pleasure) of the flesh into words and concepts which explain and also alter our existence. Animal personalities have been incorporated into human personalities. Pigs and leopards remain inside us. Animal muteness has given rise to human speech.

The animals still do not face us with totally dumb and brute visages. Often they look like mirrors of us—our viciousness but also our tenderness and our flights of somatic joy. The hunter focuses on this projection of his own sentience through the sight of the gun; unless he is engaged in some clearly restorative ceremony, his choice to fire is also an attempt to blast himself out of the animal kingdom, though on another level he must change places with a creature by destroying the very aspect that is most sentient. It is his own sympathy he kills along with the deer or rabbit. The men shooting at wolves from helicopters, leaving their gasping bodies in the snow, are turning into something. They are becoming less alive, deed by deed.

We have the illusion of having created a safer universe, one that holds wild animals, like diseases, at bay. We now create our own horrors and these may be far worse than any suffered by animals, or by us as animals. The fear of loneliness and the ever-present specter of death replace the dismembering of the hare by the fox. The final terror has become the Word (so we experience pain in its absence, almost continuously, and die many times before our death), but the Word is also an elixir, and we can be reborn or transformed.

The human experience can oppose and even reverse the animal through language. In culture the territorial imperative can lead also to treaties, migrations, and even pacificism. A wealthy Chinese land-holder takes all his money and treasures out into the ocean in a sailing vessel and throws them overboard as an offering to the Dragon King who lives at the bottom. A warrior studies judo for twenty years in order to learn how *not* to fight. Generations of meat eaters reflect on their heritage and adopt vegetarian diets. Monks sit naked in caves through long winters, trying to see past the dross of mind to a state of perfect mindlessness.

We inherit the brief flares of awareness that enable a crab to find food, to awaken to its hunger in the mass of organized chemicals it constellates, to extend its claw and move across the sand. We use symbols brought into being by the ancestors of both crabs and us, and from these create the alphabets and cities the animals reflect. The brilliant yellow of the fish in the aquarium, the blue and violet plum-age of the birds in the cage delight us because they are statements prior to language. Something reaches out of the darkness in these col-orations, and that in us which reaches out of the darkness too, our mystery, meets it in admiration of its sullen beauty. The wonder of consciousness is that it seems to arise from nature and to have nothing to do with it, that it is grounded in raw hunger and predation, yet gives rise to compassion and justice. Cells eat before they do any-thing else, and that eating, transferred phylogenetically through psy-che, becomes a vision of a peaceable kingdom in which the lion lies down with the lamb. What other system has encompassed a greater reversal?

Still, we should not forget Freud's warning; for us too: *Anatomy is destiny*. The struggle we see in nature is real, and we cannot get away from it by taboos, laws, or science of behavior. The silence of the animals when we address them is not simply their own con-stitutional intransigence; it is the silence of them still present in us.

* * *

Freud made a critical distinction between the pure somatic expression of instinct, which is the animal condition, and the psychological expression of drives in the personalities of human beings. Men and women experience the id, the thingness in them which is not conscious and not human; at the same time that they struggle to fulfill it, they struggle to deny and smother it. Society is born through the uneasy alliance of these opposing impulses, through the death of pure instinct. Without taboos on incest, cannibalism, murder, and hoarding of subsistence goods the entire tribal enterprise would have collapsed, aborning, into the mythical primal horde. In Freudian dogma, civilization is the collective projection of a taboo, expressed individually through the emergence of the superego in each personality. Our species pays a heavy price for consciousness, but (even Freud acknowledges) not too high a price considering the fruits.

The French anthropologist Marcel Mauss proposed that the intital act of society was the gift; the first gift was the spouse, the marriage partner, from the outlaw tribe. The peace offering of an enemy became, after the fact, a dowry exchange between kinfolk. Writing from the unexamined male perspective of his time, Mauss explained the process in this way:

"Food, women, children, possessions, charms, land, labour, services, religious offices, rank—everything is stuff to be given away and repaid. In perpetual interchange of what we may call spiritual matter, comprising men and things, these elements pass and repass between clans and individuals, ranks, sexes, and generations."[4]

So culture charms itself out of nature, but perhaps the gift already had its forerunners in the fish passed between swallows and the insect wrapped by the spider as its nuptial libation. Even the nests of songbirds and sticklebacks are tokens of an unconscious generosity.

To Freud civilization was hardly a resolution of desire in language; it was the enemy of desire, a grim obstacle to be overcome (in dream and fantasy if not waking action). In his version the primordial libido seeks such unhindered gratification that it could be satisfied only by continual sexual opportunity. But uncontested expression is impossible, not because of the laws of society and not even because of the inherent demands of the superego. From the moment the symbol entered psyche, pure expression of any instinct became impossible, and this intrusion may be universal on such a profound level that its rudiments had begun in the higher mammals if not simultaneously in all living tissue.

The distorting process of the psyche removes the actual object of desire by subtle degrees so that one must finally always have sex with a stand-in. But we achieve partial fulfillment, and as we do, we push the symbol further into history. Through the sublimation of *eros* the animal world is humanized (and, as the Marxians warn, very often capitalized too, so that the gift is taken back and libido is hoarded in goods, driving some classes of humans to the outskirts of economic survival).

This far along in the process, it hardly helps to try to become less human. For one, the passage back (as everywhere) is through psyche, which will deflect it. Our "animal" methods are always semantic, and thus, artificial. For two, animal existence is unconsicous; so the less human one becomes, the more one loses the "ego" of the desire. If one achieved total dehumanization, which is impossible, the libido might be liberated, but it would no longer be human. As dear as the ideal of the pure orgy is to the human heart it seems to lead inevitably to degradation and mutilation, even among previously humane lovers.

Wilhelm Reich took a different point of view on sublimation. He asserted that libido was far stronger and more fundamental to our nature, even in its raw state. Sexual desire connects us to the vital fluid of life itself, and, in our contact with libidinal energy, further life is given expression and form. According to Reich, and in absolute contradiction to Freud, men and women could have the whole of sexuality and still be human. Desire transcends language; desire can give the gift without a symbol; desire *is* the gift. Taboos and sexual perversions are thus diseases we have imposed on ourselves collectively over time through our submission to civilization.

Reich adopts a moral position in place of a psychological one, but it is unclear where *his* enemy came from. Culture apparently arose from nature without any extrinsic intercession. And nature, prior to man and woman, is filled with acts of sadomasochism that outdo even the most decadent human practices. If sexuality were born as love in the world of nature and then corrupted by men and women (or men alone in the feminist argument), then nature would have two remedies: either it could provide us with a means of curing ourselves (since the disease is not natal); or it could kill us off and try the experiment of consciousness again with a more humane species. However, it is possible that love does not precede us, that we are still bringing love into existence through *eros*, through language, that crocodiles, ravens, and whales (to mention just a few of the others) are equally in the process of bringing love into the world through their own glimmers of consciousness, and that this process has been going on since the origin of life.

But we are also shadowed by the unnamed hungers of beasts not in our lineage. The female spider devours her mate, the mantis dismembers and eats her lover during the act of intercourse. Are these bodily expressions of an actual psyche, or are they only the chemical activities of cells? Are we trapped in the libido of the spider, or do we merely pretend to be in order to evade the subtle *prana* of love? Twentieth-century pornographers have lost the distinction between pleasure and pain. Despite a wealth of artificial erotica, their passions draw no image, so they invoke gang rape and dismemberment to take themselves to the bottom, beyond which, they hope, there is no mystery, no allure. The animals become their projections of unexperienced lusts, but they have no idea who the animals actually are.

Rapists often smugly claim that they bring the primal sex act to their victims too. But most women intuit that the aboriginal event cannot be recovered or even sexualized satisfactorily. There is energy but no channel. Despite our inevitable retreat we are never quick enough to be superficial, and our unattended bruises become traumas.

We are now enslaved by lusts that trivialize our actual capacity for feeling. Our recreational sadomasochism is, in fact, a pathetic exaggeration of the real suffering of sentient beings. We are caught in a desperate rush to get everything, in an illusion that by getting everything in the most explicit way we will not be shortchanged or denied. But the snuff films and bondage ceremonies numb us to the joys of our actual desires. In light of the modern sexual pathology we look back through Jean Henri Fabre's eyes with a sense of portent as well as wonder:

"In the course of two weeks I thus see one and the same Mantis use up seven males. She takes them all to her bosom and makes them all pay for a nuptial ecstasy with their lives. The male, absorbed in the performance of his vital functions, holds the female in a tight embrace. But the wretch has no head; he has no neck; he has hardly a body. The other, with her muzzle turned over her shoulder, continues very placidly to gnaw what remains of the gentle swain. And, all the time, that masculine stump, holding on firmly goes on with the business!"[5]

Desire begins outside the dream of life in the mantises and continues past it into the unconscious depths of their physicality. Fear and desire run together to the point where both disappear, along with everything else.

Although they too must go where nature leads them, humans

can experience their individual *eros* through their own personalities rather than through some shadow of primordial hunger. We cannot solve the ancient mystery, so we come to value our dignity and individual expression far more than the muck of collective *eros*. Our bodies may still respond to animal darkness, but we are in the process of differentiation, and when murder becomes *eros* we are truly lost.

"A headless creature, an insect amputated down to the middle of the chest, a very corpse, persists in endeavouring to give life. It will not let go until the abdomen, the seat of the procreative organs, is attacked."[6]

But this is not pornography or rape. It simply happens. It comes into the world through the germ plasm and is expressed in the flesh. If these were crimes, the tiger and the vulture would go hungry, and the dragonfly and the shark would be imprisoned for genocide.

Mantises can do nothing about it, but we who stand facing each other armed with ballistic missiles and grotesque warheads both allow and condemn it—snuff films, S & M nightclubs, kiddie porn, piracy and rape of boat-people and other refugees, death squads, torture and mutilation in the jails and prison camps. The generals send armies of children through the fields of land mines to clear the way for the tanks. We have the weapons of gods, and we are not yet even awake.

It has grown from the mute wars of the Stone Age to the primitive bellicosities of the Caesars to the modern epidemic of the psychopathic generals, the mass murderers, the armed street warriors, the Gestapo armies firing into crowds. What is new is not the activity but the desperation. It no longer seems possible to be human, so people are willing to try anything else, anything they can think of. Or, perhaps, as the orgies and whippings of the Marquis de Sade over two centuries ago augured, the association of desire and mutilation is as old as our species (and only now are the limits, or the illusion of their absence, put on public display). It is almost as if the most venal stroke against God could alone bring Him back to life, and only the exploration of every morbid erotic fantasy could lift from us the burden of oppressive consciousness.

Pornography is not just innocent titillation or harmless erotic fantasy. In the context of epidemic psychopathology in the culture of the West it provides images of disease for those who are already sick. The full expression of *eros* will be healing and medicinal only in the context of social and economic justice, at a time when men and women live comfortably with each other in the truth of their own biological natures.

We are incarnated in a world where the slaughter of innocents has reached epidemic levels and members of our species treat other species as objects, statistics, or mere industrial machinery. Animals are hunted and tortured for pleasure, butchered for food we do not need, and maimed in useless experiments. Guinea pigs are injected with carcinogens, cats are lobotomized, and monkeys are made to run on treadmills until they drop from exhaustion, but the treadmill never stops. There is no one to intercede for their lives. They are true political prisoners. Behind the masks of self-important scientists and so-called objective experiments sits merely another legally protected pornographer. Of course, the experimenter imagines his tests are important in the context of the academic bureaucracy he serves; he loses planetary perspective: The animals are defined as different from us in a way that justifies his going as far as he wants without doubt or remorse. Does he not see that their bare liver and nerve sheaths are his liver and nerve sheaths, that they shed the same blood and weep the same tears? At least those "savages" who tortured and maimed their captives did so in a ceremony that recognized the courage and honor of the victim, and then put their own lives on the line in the same battle.

Most experiments are arrogant and useless, they do not even have to occur. We cannot disguise our brutality behind statistics and careerist papers. We are far too old for that; we have been through the executions of the tribal elders and the sacrifice of children, women, and other sentient beings to the dark spirits behind the Sun.

The mayhem of our species comes not only from misogyny, not only from the traumas of brutalized childhoods; it is also what we inherit from not being human (though not necessarily from being animal). It all happened so long ago; even the old men of the Australian Aborigine tell us it was at the dawn of time, or before time itself. The mythologized account of the crime echoes through the generations: "The men of the Dreaming committed adultery, betrayed and killed each other, were greedy, stole and committed the very wrongs committed by those now alive."[7]

In the summer of 1983 Will Baker was invited on a hunting trip in the Peruvian Amazon by two Asháninka Indians, Carlos and Cuñado. Soon after they begin tracking, a monkey couple ("their faces small and old as time,"[8] Baker writes) comes through the trees to look at them. Cuñado nocks his arrow, draws, and fires. The female starts as the shaft enters her small body; her fingers fondle its hardness, and she drags herself back and forth, uncomprehending. The unsuspecting male runs up to her and pulls at her shoulder, trying to hurry her away. Cuñado's next arrow pierces him, and he bolts from her aid and pinwheels through the branches. The hunters "hoot at this slapstick agony, this silly tale of fidelity."[9] The question is not: Why do we kill for food? (Not yet anyway.) The question is: Why are we laughing? How long can we remain in the jungle in darkness in a night of faraway stars?

Even Freud must have realized, in the remorse of his later years, that the primal act goes far beyond the genital symbolism in which it masks itself. It is a transformation rite happening neither here nor elsewhere, neither in time nor outside of time, and involving events indecipherable, as such, but fundamental to the crisis of human life:

"I saw the soul of a man. It came like an eaglehawk. It had wings, but also a penis like a man. With the penis as a hook it pulled my soul out by the hair. My soul hung from the eagle's penis and we flew first toward the east. It was sunrise and the eaglehawk man made a great fire. In this he roasted my soul. My penis became quite hot and he pulled the skin off. Then he took me out of the fire and brought me into the camp. Many sorcerers were there but they were only bones like the spikes of a porcupine.

"Then we went to the west and the eaglehawk man opened me. He took out my lungs and liver and only left my heart. We went further to the west and saw a small child. It was a demon. I saw the child and wanted to throw the *nankara* (magical) stones at it. But my testicles hung down and instead of the stones a man came out of the testicles and his soul stood behind my back. He had very long *kalu katiti* (skin hanging down on both sides of the subincised penis) with which he killed the demon child. He gave it to me and I ate it."[10]

There is no longer any name for this. There is no tribunal, no judge. There isn't even a convenient psychosis. The death camps appear on the face of history like raindrops splattered on the glass. In the Australian version it isn't malevolent; it is just a shadow masked by a nightmare itself masked by the rudiments of a ceremony. And if we hope to find the evidence or explanation for it in the embryology described elsewhere in these pages, we will be as sorry as those who look to the documents and (now) the videotapes of history in an effort to escape it. We are not likely to find our way out by the animal or the aboriginal shaman either.

In 1761 Georg Wilhelm Steller described an instance of loyalty on the battlefield among sea cows attacked with harpoons by Russian sailors:

"When an animal caught with the hook began to move about somewhat violently, those nearest in the herd began to stir also and feel the urge to bring succour. To this end some of them tried to upset the boat with their backs, while others pressed down the rope and endeavoured to break it, or strove to remove the hook from the wound in the back by blows of their tail, in which they actually succeeded several times. It is most remarkable proof of their conjugal affection that the male, after having tried with all his might, although in vain, to free the female caught by the hook, and in spite of the beating we gave him, nevertheless followed her to the shore, and that several times, even after she was dead, he shot unexpectedly up to her like a speeding arrow. Early the next morning, when we came to cut up the meat and bring it to the dugout, we found the male again standing by the female, and . . . once more on the third day when I went there by myself for the sole purpose of examining the intestines."[11]

The expedition slaughtered so many of these animals that they were extinct within twenty-seven years of their discovery. To many scientists this is the only crime—but that is too much grief for us, too little for Steller's sea cow. If animals have rights, if they exist (like us) to experience and explore the universe, to feel the ancient wonders of their tissues, then they have rights as individuals, not just as members of endangered species. But there seems little chance of remedy in the next century or even the next millennium, if we ourselves survive. Justice for animals is a cause we are still approaching at great distance.

Instead, we have in the present the acrimonious debate between those who uphold the right of the woman to abortion and those who uphold the right of the foetus to life. The battlelines are, as usual, drawn in the wrong place. To put such weight on the lives of the unborn is sanctimonious unless the plea for life is a general plea of compassion for all sentient beings (or at least all human beings). To make accusations of murder against women who abort unwanted embryos without opposing slaughter and oppression in general is not charity; it is moral duplicity. What about the millions of children and animals who die so that other creatures can maintain a higher standard of living and consume them and their goods? What about the murders of children and adults in the undeveloped world, either directly through the armies and police of the nuclear-scientific hegemonies and their client states, or indirectly through policies of economic imbalance and resource exploitation? Do the anti-abortionists oppose killing in war? Do they oppose the mass production of implements of torture and murder to be sold to the highest bidder or provided to our surrogate enforcers? Do they think the universe, or, for that matter, God Himself, feels any less pain for the slaughter of the wolf or giraffe than for the abortion of the foetus? Are not the fish and the frog foetal souls in other states, equally impregnated with nerves? Those who bomb abortion clinics as avenging angels should also be freeing animal souls from death camp laboratories and attacking weapons factories.

This nation exists through the theft of land from its previous occupants, the systematic murder of aborigines, embryos and all. But, unfortunately, conquest has been the rule rather than the exception. Some tribes kill their own young to keep families small; surely they would destroy other tribes for a spring, a valley, or hunting ground. Infanticide and genocide form a cycle over millennial time.

But to say that the embryo is not alive, or not human—not yet sentient and thus not murderable—is another form of self-deceit. It is impossible to abort the foetus without killing the person incarnating there. But we mistake rhetoric and journalism for reality. The marketplace and work force replace our inner voices, our somatic wisdom, replace even desire. Goods seem to interest us more than ourselves, consumption more than the lucid calm of the body at rest. ''Pro-choice'' means nothing if it does not include the unarticulated choice of the living soul of the embryo. Its hunger for life, its desire for unfoldment, is as strong as ours, even before it has words in which to express a personality. A court may rule that the embryo has no rights, but the cells and gathering consciousness of the creature are real and seek manifestation. We should not diminish the mystery we are by pretending that all experiences can be institutionalized or sold. Courts of law cannot redress the seeming biological injustice making women alone the carriers of zygote. Progressive technology with its massive bureaucratic governments does not understand who we are, so we should not put our identity in its hands. The role of the woman as the bearer of life is ancient beyond words; it is not just an unfair allotment or a capitalist exploitation; it is an unsolved mystery and an opportunity, as is life itself.

Even the death of a fly requires atonement, although we may push it out of mind a thousand times, as we crush the body, before we notice. The makers of weapons operate under the same anaesthetic until they are awake.

When young children are kidnapped and sodomized or mutilated by adults, otherwise merciful men and women propose the most brutal vengeance. And so the pornographic current spreads through the rage and revulsion it inspires. At the core of this disgust must lie some attraction to the acts, or they would not compel such spontaneous need for revenge and suppression. If it is realized in some of us it is nascent in all of us. Most people haven't the slightest predisposition toward pornographic and abusive deeds, but these exist on the edge of their "normal" sexual behavior, in fantasy spheres and unbidden nightmares. The desire for vengeance is always partly a self-loathing, a reaction against the failure to suppress entirely the pornography within. So the war-like aspect of our shadow self goes to war against war, or is thrilled to be a warrior, either in reality or imagination. The death penalty is meant to obliterate not only the hated murderer who expressed the atrocity but every aspect of sympathetic imagination of it in us. When we participate in the second murder, the punitive one, we give new energy to the original crime. The event cannot be resolved or concluded in any way. The murderer is never a total alien; he must be one of us.

The Gary Gilmore story, as told by Norman Mailer in *The Executioner's Song*, contains the central dilemma of our twentieth-century crisis. Other centuries have experienced mass brutality and murder, but we so little know what to do about it and how to act next. With the insanity defense on one hand and the firing squad on the other, we are paralyzed between irreconcilable intentions. We want to pardon and to be pardoned. We want to give our disadvantaged and traumatized children a second chance; yet, all too often, the killer is rehabilitated, comes through his madness to a feeling of hope and a rebirth of love, and kills again, or kills again anyway. A part of us wants to put the murderer to death so that the dead are avenged and the living feel safer. But we can't protect ourselves as long as we don't know who the real killer is, and, either way we act, we are thrown back into self-loathing and rage; we are trapped together in the same society, the same inherited bodies and nervous systems.

Gary Gilmore understood this problem better than any modern philosopher. "Kill me," he said. "For your own sake." He had already suffered the worst of America's prisons, from beatings by guards, as he stood crippled by disease and blinded by infections, to months of solitary confinement, hallucinating for weeks at a time, half-awake. He got so that he thought he could withstand anything; he was a prison-created yogi with immense diabolic powers; he was a living incarnation of America's shadow. When he was finally released he lasted nine months in Utah before committing two unnecessary murders one night in a rage that seemed to him inevitable even after it had passed. He had taken two random lives from people for whom he later had sympathy, and yet he accepted the rage as it happened, its largeness bigger than he. He was sorry that they were dead, but he expressed no remorse that he had done it.

Between the imprisonment of Gary Gilmore and his execution six months later he sat in jail meditating on the big issues of life and death in society, not as a monk in a seminary but a perfectly literal man who accepted what had happened to him so fully and profoundly that he wanted the firing squad he had earned. His love for Nicole Barrett was more excruciating to him than anything he had suffered in a lifetime of prisons, but it also illuminated his whole brutal existence as he waited to die. He was an animal suddenly made conscious of everything he had done and been.

In that state of grace Gilmore bared his being so openly and directly that everyone around him, everyone involved in the case, saw a reflection of their own hypocrisy and fear. Gilmore was beyond ideology; he did not spout pieties about justice and capital punishment; he did not pretend grief or guilt; he spoke neither for nor against the Mormons; he simply showed them the creature they held captive, and, on that basis, he met them halfway, as their human equal. He spoke for the prosecutors, the judges, the lawyers, the other prisoners, the guards, and even the humanitarians who tried to save his life against his wishes. He took away their easy assumptions and forced them to face the brute facts of this existence, their own as well as his. They realized that and respected him for it. In the end his visitors could hardly tell the difference between a condemned murderer in his cell and a holy man in his robes, except that Gilmore didn't try to be sweet or gentle. He was crude, belligerent, and aggressively disgusting; like a master warrior hardened by combat, he trained everyone who came into his presence; he changed who they were. Even the priest who tried to bless him at the moment before the fusillade found the roles reversed and himself being blessed and accepting the murderer's last blessing.

Gilmore understood the severity of the gods; he did not want to talk himself into believing he was just another "innocent victim of society's bullshit." [12] He demanded that the Mormon State of Utah follow through on its own sentence. The general outpouring of sympathy in his last weeks didn't fool him. Sentiment was cheap compared to the battles he had been through. He knew that his only possible atonement was to teach us by facing the truth without flinching, thus proving it *was* the truth. He said, in essence: You have no choice. Don't lose your self-respect by pretending to save or reform me. "Let's do it." [13] [His last words before the blessing.] But don't pretend I'm an inhuman killer who has nothing to do with you or that you've solved your problem by shooting me dead. I'm the best of you as well as the worst. And neither you nor I know what's behind any of this. Kill me, but we're still in it together.

He became a media figure because he spoke truly for a civilization that had experienced death camps and atom bombs and was nowhere near a resolution or even a catharsis:

"I'm so used to bullshit and hostility, deceit and pettiness, evil and hatred. Those things are my natural habitat. They have shaped me. I look at the world through eyes that suspect, doubt, fear, hate, cheat, mock, are selfish and vain. All things unacceptable, I see them as natural and have even come to accept them as such. I look around the ugly vile cell and know that I truly belong in a place this dank and dirty, for where else should I be? There's water all over the floor from the fucking toilet that don't flush right. The shower is filthy and the thin mattress they gave me is almost black, it's so old. I have no pillow. . . . .

"It seems to me that I know evil more intimately than I know goodness and that's not a good thing. I want to get even, to be made even, whole, my debts paid (whatever it may take!), to have no blemish, no reason to feel guilt or fear. I hope this ain't corny, but I'd like to stand in the sight of God."[14]

Aboriginal justice once sought to heal the wound by obliterating time. In some native societies, a murderer is asked to replace the victim in the victim's family. Having deprived that family of one of its offspring he must become the thing he has taken away. Murderers actually adopt the clothes, the wife, the children, and the parents of the one they have killed. In very small tribes the familiarity between members makes the group a large family. The act of homicide is dealt with not as the atrocity of an outsider but from within the understanding of the clan. Since the murderer will be reborn into the tribe again and again, he must be diagnosed and cured; he must return to humanity. The family dispels the darkness by taking in what is left of the human being in the murderer. They accept the ultimate collectivity of the species and may, even unconsciously, consent that the bond of murder is still a bond, however agonizing and degrading. The victim likewise will return in subsequent generations, so the crime must be expiated in all aspects before his soul seeks blind revenge. Among animals in general, and visibly among packs of mammals, this replacement is axiomatic; the individual is the species.

The horrors are finally integrated because there is nothing else to do with them. The mantis has its young, and they too mate, breed, and devour. The wars end, and the dead are buried. Years later, the identities of those in the cemeteries fade as the whole generation passes like some forgotten Dakotan tribe—even the memory of their existence blurred with all other existences.

As we approach the millennial turning point of Western civilization, we are obviously trapped in a number of highly polarized ideological battles. The Biblical fundamentalists are at war with the pure scientific materialists as to what religion will be imposed on the next generation (which will make its own decisions anyway). Neither side perceives that its entrenchment represents a character neurosis more than a commitment to the truth.

The nuclear activists rightly blame the military establishment for arming the Earth beyond any strategic rationality and thereby jeopardizing all life forms. War is not just glorious revolution or "politics by other means"; it is an addiction to a "power high": the war chiefs have always enjoyed the rush of the battle, whether they ride mounted with their troops or sit before the glitter of three-dimensional video units. War is also an indulgence of machismo fantasies, a denial of the "weakness" (the *tao*) and universal feminine within. This "last ditch" resistance to our humanity is disguised as heroism.

But destructive activities cannot be ended by decree, and they certainly cannot be convinced to end by the demonstrations against them. In serving to expose the collective disease, formal protests are healing to all, even to the targeted enemies, but in rhetoricizing and oversimplifying the causes of violence, indignation also becomes self-aggrandizing and false absolution of the protesters. The architects of weaponry and makers of war operate ever under the mystery of the undiagnosed causes of war. The lessons of Napoleon are submerged in Tolstoy; the texts of Tolstoy are further deconstructed by the living scripture Hitler and the Third Reich engraved into the spine of Europe. The black magic of human warfare recedes back through Machiavelli, Philip of Spain, Theodoric and the Vandals, Alexander of Macedonia to the Egyptians and Stone Age conquistadors. Hitler's unintentional warning echoes like the meditation gong through *Apocalypse Now* in Vietnam and Cambodia, the Red Guard furies, and the tribal genocide of modern Africa. Until we as a species experience the true mystery of war and peace, or until enough of us meet its old warrior gods individually to change human consciousness, it will not be possible to disarm. The current stalemate represents how far we have come and how far we have not come from unquestioned violence and the compulsion of the battle. The weapons and the generals are not themselves the causes of war; they are the symptoms of our latency and incubation; they cannot be eliminated without being replaced from within.

We embody the collective shadow as well as the psyche. The horror of the nuclear weapon is that it existed in our fantasy so long, and not just in the fantasy of the warlord and tyrant, in *all* our fantasies as a projection of both power and fear, and the ultimate horror/allure of a sleeping dragon within raw dust. The "death rays" of Zeus and Saturn are also the death rays of the Martians and the high-frontier space chiefs. They are the struggle of mind over matter, of frustrated idealism and unachieved self-knowledge let loose in the cave of the shaman and the workshop of the alchemist. The wound is self-inflicted, though we still suspect a malefic force. Witness Robert Ardrey's lyrical version of a twentieth-century credo:

"Our history reveals the development and contest of superior weapons as *Homo sapiens'* single, universal cultural preoccupation. Peoples may perish, nations dwindle, empires fall; one civilization may surrender its memories to another civilization's sands. But mankind as a whole, with an instinct as true as a meadow-lark's song, has never in a single instance allowed local failure to impede the progress of the weapon, its most significant cultural endowment.

"Must the city of man therefore perish in a blinding moment of universal annihilation? Was the sudden union of the predatory way and the enlarged brain so ill-starred that a guarantee of sudden and magnificent disaster was written into our species' conception? Are we so far from being nature's most glorious triumph that we are in fact evolution's most tragic error, doomed to bring extinction not just to ourselves but to all life on our planet?"[15]

"I have seen it done with my own eyes, and have not recovered from my astonishment,"[16] wrote Fabre of the mantises.

Robert de Ropp answers him in the words of the Marquis de Sade: "Oh, rest assured, no crime in the world is capable of drawing the wrath of Nature upon us; all crimes serve her purpose, all are useful to her, and when she inspires us do not doubt but that she has need of them."[17]

Newborn baby fish are captured instantly by underwater insects. A group of young squids experience their first puff of ink and are swallowed *en masse* by a whale. But fish and squids feed on crustaceans and snails.

When Idi Amin served a former minister's head at the dinner table,

Frank Terpil, the CIA renegade, did not balk. "How could you go on working for him?" the reporter asked.

"I don't make the rules. This is what the world is."[18]

The universe doesn't recognize me. I must start over from the beginning.

Someday, says Dostoyevsky's Inquisitor, "the beast will crawl to us and lick our feet and spatter them with tears of blood. And we shall sit upon the beast and raise the cup, and on it will be written, 'Mystery.' But then, and only then, the reign of peace and happiness will come for men." [19]

Ivan protests, in the name of Dostoyevsky and, in fact, for all of us:

"Not justice in some remote infinite time and space, but here on earth, and that I could see myself. . . . If I am dead by then, let me rise again, for if it all happens without me, it will be too unfair. . . . I want to see with my own eyes the hind lie down with the lion and the victim rise up and embrace his murderer. I want to be there when every one suddenly understands what it has all been for."[20]

The message of the twentieth century is that we can no longer evade anything (instinct or not). The nuclear bomb has even turned the Third World War into an excruciating dialogue we must maintain with daily attention despite all distractions, or it will begin. Behind closed doors the enemy diplomats continue to speak to each other because they are trapped in the same ancestral language, the same pathology, the same polar fantasies of golden cities and of ashes and toxic air. There is always the chance that, despite their opposed interests, the words themselves, the armored feelings will force them to acknowledge each other's realities. But what then?

Despite proposals for nuclear freezes and mutual destruction of weapons I suspect we may have to talk until the end of time, simply to survive. Or we had better plan for this long a dialogue, just as we must plan for the radioactive matter and other poisons we have created. We are also the species that spoke so long ago of turning swords into plowshares. If it were simple we would have done it, assuredly. And even though it is difficult—in fact, impossible—we have no other choice but to bear this hope through time along with the bombs and the holocausts. "Pure deterrence," just as Henry Kissinger and the military existentialists would have it—a deadly game of unenacted battles and unwritten treaties and their partial violations until the end of time.

We now approach Mars in all our nakedness. We stand alone in the heart of the battle with only the Word. Olive branch in one hand, grenade in the other (as usual), we stand always for revolution, for justice, for truth; inevitably we stand for the Earth. And the terms of our survival till now remain a mystery.

As an emerging current of conscious desire, our faint incipient ego encounters the unexperienced desires of the cells, their hunger for substance, their unceasing differentiation which consumes old cells in the birth of new ones. Computers notwithstanding, we cannot create mind from a sterile liquor, and we cannot reform nature by a rational attack upon its perversions. Our task is more difficult. We can reclaim the darkness only by conducting its radiance through our lives.

And there is no rule of thumb: One person can nurse the sick in Bangladesh while another person irrigates a small farm in pre-Columbian Arizona. Some dispel evil through kung fu or aikido, while others spread discord through the same arts. No one is totally diabolical; every person enacts some quantum of photosynthesis, but likewise, a portion of the darkness sticks to any life.

The Sioux Indian prays for his game and lures it to come to him by the beauty and integrity of his chant, the clarity of his attention. Animals may have the power to call to other animals to become their food, or they may collaborate across species in remorse and understanding at the moment of the kill. An unexamined telepathy could join the hunted and hunters throughout the planet, but, as the first shamans intuited, only when the hunt obeys the great ceremony, not when slaughter goes beyond scrupulous attention. The closeness of our cells to one another, animal to animal, ensures a sympathy of some sort—a sympathy expressed in *eros*, a force of love not necessarily separate from the required cannibalism of species within nature. All thoughts are already universal on some level, even without planetary telepathy.

It is perhaps the *eros* of the hunt that rapists and sadomasochistic pornographers abuse. To portray human bodies with guns and knives around erotic organs is to proclaim the alienation of not only desire but power itself, and to proclaim as well, the loss of the ceremony in which all of these lusts and aberrations are transformed and sated without degradation of wasteful violence.

Even the great healer Paul Tillich kept pictures and accounts of women beaten and crucified. "Nature, society, and soul are subject to the same principle of disintegration," he wrote. "They all are possessed by demons, or, as we should say, by psychic forces of destruction."[21]

But his wife, upon discovering these artifacts after his death, added: "I was tempted to place between the sacred pages of his highly esteemed lifework these obscene signs of the real life that he had transformed into the gold of abstraction."[22]

The evidence appears in the congenital dysfunction of our race. We are joined through the animal kingdom to some original strand, call it sexuality or virus, foreshadow it in parasitic worms or sharks. The clergy sheds no light at all, wanting only proof of their apocalyptic myths. It was bloody from the beginning, but that does not mean it will stay that way forever. We don't in fact know.

We cannot be carnivores without being killers too. From the point of view of plants, we are just another mutant that has lost the ability to feed directly from the Sun. What if this ability were regained and transmitted through the cells? This would be remarkable, considering the thousands of years of predation our metabolism embodies. Our *apologia* for the whole animal kingdom is based on the circumstantial evidence that there is no other path to knowledge. Yogis still promise we can someday materialize the right chords to draw our sustenance from vibrations of air, without killing even plant life, to drink from the Sun and the psychic field around us, but if that's where we're headed, we obviously have a long way to go.

No diagnosis of the problem or possible solution is any longer too extreme. We miss the point when we act only conscientiously, so even our humanitarian gestures are confounded. This media-conscious, high-technology civilization creates the mirage of progress against famine, tyranny, and disease. But, in self-protection, we are blind. Between the unnamed forces of voodoo and the quantum dance of atoms lie untold universes of suffering and redemption. We may pretend to heal the Earth, but what about whole civilizations destroyed on other planets, individuals in pain on worlds around other suns? If we were to accomplish a lasting peace on our world, would we then have to worry about other planets that perhaps do not even exist? But if they do, they are part of the universe, part of consciousness; and, ultimately our sympathy must be extended through eternity to those victims too, creatures we could never know. Not because it does any good but because it forces us to view the crisis in its actual bigness while at the same time reminding us that we do not know who and where we are and thus what powers we have. If we could bring peace to this planet, we could probably bring peace to the universe.

We do not know. But that is not the problem. The problem is that we pretend we know, or think we should pretend. We think we know what we are and what the world should be. But nothing about cell life, or DNA, or the self-assembly of tissue suggests we have any idea of whence we come into being. Our whole culture and technology might be an evasion of our natural condition, our true dormant power. We could be avoiding our own natures, missing the solutions to our crises; worlds without end more fulfilling than this one might be within our grasp. But these lie in the margins of an unconscious inner life we presently flee in all our ideologies and institutions. It would seem now to have to begin in silence again. We must drop all our expectations and see just how quiet and observant we can be—keeping our eye perfectly on each thing as it arises. We have created so much noise; yet the thing we are is so soundless and perfect it might simply become.

22. Spiritual Embryogenesis

In a real world the physical and spiritual cannot be separate. There is but one universe, a universe that appears to change only in our changed perceptions of it through millennia. There is no fixed arrangement of props in the laboratory, only an unending cyclone through which waves of energy pass, revealing and obscuring while phenomena elapse.

We can never know first causes or reconstruct the beginning of things. Creation does not even originate in time as we understand it. Since access to data is limited to a nervous system formed inside nature, we are condemned to follow circles, half of them dwindling until they disappear in the subatomic forerunners of matter, and half of them stretching across the stars to infinity; circles all the same, and, in effect, defining the modern plight.

We could hardly underestimate the physical effects of the cosmos. We are at the mercy of gravity, heat, and the chemical activity of molecules. The weapons and industrial machines we have invented are absolute dictators of cause and effect within their realms; power politics and economic exchange rates appear just as irrevocable. It is no wonder that a system of knowledge based exclusively on measurable phenomena is so important to us. What we learn from strict scientific inquiry may be disturbing (as regards our mortality and minor standing in the universe) and even perilous in its applications (poisoned rivers, nuclear warheads, epidemic psychopathology), but, abstractly, it *always* clarifies our condition and teaches us what and where we are. Like birth and death science cannot be fooled, and it cannot be cheated.

Spiritualism, on the other hand, develops from the invisible aspects of the universe. Although we cannot demonstrate that such things even exist, men and women from the historic beginning of consciousness have intuited intelligent forces shaping our world— gods, ghost-shadows, and archetypes (to cite just a few of the ways in which these entities have been named). From the evidence of the assembly of vital animals and the certainty of their own egos within, humans have imagined some sort of drive behind nature. Else why should we experience this carefully crafted body/mind, these won-

323

ders and desires? Why did it not all lie still forever, from the beginning?

Many of the founders of modern science believed in this supersensible agency, whether they called it God, vital force, or Qualities (in the Platonic tradition). Some (Kepler, Newton, Darwin, and even Driesch) addressed it directly, and assigned it the ultimate destiny of planets and creatures. Because these scientists could never square supernatural forces with the daily flux of data, and yet spoke in terms of both astrology and astronomy, occult and secular biology, their works have (to modern sensibility) a schizophrenic ring, and historians of science pay attention only to their paradigms that have survived modern disciplines. Others (Galileo, Pasteur, and Haeckel among them) were pure experimentalists, but the unarticulated background of their work still acknowledges the hidden agency. In either instance, something large and cosmogenic remains: Ultimate matter comes into the world through extrasensory forms, but it then must obey the vectors of mechanical forces (which are themselves sums of archetypal effects).

The custodians of science have rightly attempted to rid it of this mystical baggage, for invisible abstractions can never be proven or disproven. If science did not purge the ghosts it would not be able to work its way through the inexhaustibly complex integers of quantity. When we attach spirits to inexplicable effects we lose the opportunity to decipher their actual physical workings. We would hardly have discovered the gene if we had remained loyal to the vital force.

The mystics and vitalists had gotten lazy by the nineteenth century. The mere fact that mechanists had always failed at a full description of causes and effects in nature was enough to justify their faith. They did not have to demonstrate or even experience for themselves the existence of supersensible forces. Authoritative mysticism had lost its ancient wellspring, and as much as science, it needed fresh empirical data.

During the nineteenth century, mechanistic science made startling and unprecedented headway. Equipment and techniques of investigation improved, and whole areas that had been conceded *in perpetuity* to the vitalists were reclaimed, one by one, as physicochemical causes were revealed. In 1828, the German chemist Friedrich Wohler synthesized the organic molecule urea, demonstrating, ostensibly, that even complex substances could be imitated in a mere laboratory; the modern-day gene-splicers are his intellectual descendants. In 1859, in *The Origin of Species,* Charles Darwin translated the basic premises of the conservation of energy into living systems. The same universal force that held the Moon in its orbit and caused water to run downhill was shown to be responsible for the myriad diversity of plants and animals, with no exterior agency or energy. In 1893, Max Rubner applied that same law of conservation of energy in a strict and functional manner to animal tissue and its metabolism.

Jacques Loeb, a German biologist who believed that so-called human will was merely a form of chemical tropism, shocked nineteenth-century vitalists by artificially fertilizing sea-urchin larvae so that plutei formed without the participation of sperm. After the rediscovery of Mendel's principles of genetic transmission and phenotypic expression of traits, turn-of-the-century biologists could explain the physical diversity and behavior of all living organisms as the result of differential fertility and mortality, cell mutation, and the molecular properties of organs. Twentieth-century scientists have since restored the old esoteric connection between remote suns and the stuff of life on Earth, but only by making the stars lifeless and their relationship to life statistical and random. The *élan vital* was hardly a flame passed down by spirits and angels; it was a thermodynamic variant, an accident in a regime of ultimate lifelessness. We no longer require gods because we have been exposed as not being truly alive. We are now molecular ripples mimicking something grand (but something that cannot happen in a universe such as this). The mechanist position is typified by J.D. Bernal's smug witticism:

"It is difficult to imagine a god of any kind occupying himself creating, by some spiritual micro-chemistry, a molecule of deoxyribonucleic acid which enabled the primitive ancestral organism to grow and multiply. The whole hypothesis has now come to its natural end in absurdity."[1]

But it is equally incomprehensible that this gossamer world of thought and design came into being through a thermodynamic accident and the chance properties of amino acids in blind interactions. Though nearly every scientist rotely professes belief in this epistemology, none of them behave as though they are mere jumbles of chemical concatenations. They act like official spokesmen for the gods. On the one hand, it is the only reasonable explanation; on the other hand, it is utterly ludicrous. Everyone knows it is the only reasonable explanation; yet, everyone knows that it is utterly ludicrous. The modern sensibility has also come to its natural end in absurdity.

Even as the success of mechanical science took most of the ground from traditional occultism the failures of that same science gave the vital and hermetic traditions surprising new ground. Physical experimentation still has not solved the mystery of life let alone the secret of the mind, and it has not even adequately explained inanimate matter (Clerk Maxwell's nineteenth-century electromagnetic hypothesis foretold the coming downfall of pure progressivism). If the first wave of eager scientists assembled a machine, the second wave has gradually dismantled it (though some still hope it will turn out to be a computer). In the atom, there was no machine. Substance and motion were different forms of the same substratum; even light and matter were interchangeable, and time and space turned out to be regional jargon for something quite different in the large.

The mysteries of mechanical science were once again the mysteries of spiritualism, though in an entirely new way that suggests these ties will be dissolved only to be reestablished as aspects of a larger mystery for as long as *we* exist. A generation after scientists were certain that a mechanism would be found for all motion and form Werner Heisenberg and Albert Einstein were once again addressing the agency behind nature and attempting to characterize the invisible Shaper. But this is not the spiritual universe in place of a mechanical universe; it is the basic mystery of all phenomena. Like yogis, scientists must return to their wonder and sense of creation within, for that is where the impulse for inquiry and intelligence arises. The awareness of being in this vast and magical manifestation of things, of being one of them and, at the same time, in the slipstream of a singular flow of mind, transcends any theory, law, set of symbols, or gods. "The frog never lies," the old laboratory biologists taught us. And now, in a millennial crisis of identity, we are the frog.

"We do not know what anything *is,*" Da Free John warns. "We are totally mindless, and totally beyond consolation or fulfillment, because there is no way to know what anything *is.* The only thing you can know about anything is still *about* it. But you do not know what it *is* . . . *is* . . . *is* . . . or why it happens to be. You have not the slightest knowledge of what it *is.* And no one has ever had it. Not anyone. Not Jesus, not Moses, not Mohammed, not Gautama, not Krishna, not Tukaram, not Da Free John, no one has ever known what a single thing *is.* Not the most minute, ridiculous particle of anything. No one has ever known it, and no one will ever know it, because we are not knowing. . . . The summarization of our existence is mystery, absolute, unqualified confrontation with what we cannot know. And no matter how sophisticated we become by experience, this will always be true of us. . . .

"No matter what sophisticated time may appear, no matter when, in the paradox of all of the slices and planes of time, any moment may appear in which men and women consider the moment, no one will ever know what anything *is.* "[2]

In the end we have only our own reality to go by—our very existence as a means for probing reality, not only as objectifiers and sorters of data but as the objective basis itself of information and its discrimination. For millions of years as a species we have implicitly understood that our life itself is the meaning of life and the source of our identity in the universe. The most significant truths cannot be demonstrated and, in many cases, cannot even be articulated as questions. Likely, all of us are aware (at heart) of who and what we are, but we cannot even acknowledge it internally. Yet those who most fiercely dispute the transcendent world are able to deny it only because they experience their own immersion in it without sacrificing their cynicism. They are, more accurately, disputing someone else's transient and ideological version of a reality they experience in their own way. If we did not know implicitly who we were, we quite likely could not exist at all; we would be a series of robot-like gestures strung together by unexperienced commands.

The names given to things represent exclusive clubs and social milieus, not real systems. Because one person is committed to "hard science" and another to "astral projection" does not mean that the scientist does not intuit the astral realm or the occultist does not accept the inviolable laws of matter. They are both playing temporary roles using stock images. Instant conversions, both ways, remind us of the true secrecy in which people guard their final positions.

(An American writer who returned from years of study in India to come upon the program "Cosmos" on television perceived that Carl Sagan's strained hyperbole—depicting the material universe in ornate personal tapestry with repeated professions of wonderment—was, in fact, his resistance to his own spiritual "coming out of the closet"; in muffling a lifelong impulse to acknowledge gods, he disowned them with all the fierce righteousness of his "space scientist" credentials, but always reassuring us—and himself—with fake naive charm, that he stood in awe nonetheless.[3])

But those who adopt the mystical and fill their lives with its symbols and accoutrements are not necessarily more spiritual than those who do experiments in laboratories. Each has chosen a certain concrete imagery for the internal mystery and a way to pass the hours on the Earth. The most asserted atheism, the most devoted experimentalism must come from the gods too if there are gods at all.

And yet nothing so divides the modern world as the split between those who believe all things arise from spirit and those who believe that all things are matter only—even if we do not know who is who behind the personae of Church and State. The physical Sun that burns down into twentieth-century beehives and batteries is real enough, working its photosynthesis through the cosmic history of hydrogen, and so is the bloody infant who comes into this world through the cellular lattices of embryogenesis. The universe could have a spiritual history too, without ever telling itself outwardly in sun-stars and stones. A newborn frog acts and responds, beyond time and space.

Even an amnesiac, who has forgotten every factual detail of her life, acts from the essential truth of her being; loss of memory is not loss of self. There is likely only one intelligence in the universe, and what we hear in our heads is also what the universe hears echoing through its galaxies and what the darkness heard in the birth of atomic form and living crystal. We are the crack in reality, the metaphysics of creation.

Spiritual theories of embryogenesis provide their own meanings for cell division and morphogenesis. Where biologists propose hierarchical fields and homeostases, occultists refer to a transcendent archetypal force that precedes matter. The spiritual agency is the string that holds the beads, though the beads are already held to each other, from their origin, by electrochemical forces. Shamans and occultists have always viewed cause and effect in the microcosm as a reflection of cause and effect in the macrocosm: Men and women are the shadows of angels, and the different plants and animals are hermeneutic replicas of spirits. Under this influence some early biologists interpreted the biological species as fixed steps in a ladder or chain flowing ever upward from lower, less sentient forms to the higher, more intelligent organisms. In the mid-eighteenth century the French lawyer and scientist Charles Bonnet described the likely future state of our globe:

"Man, then transported to a dwelling place more suitable to the eminence of his faculties, will leave to the ape or the elephant that first place which he occupies among the animals of our planet. In this universal restoration of the animals, therefore, it will be possible to find Leibnizes and Newtons among the apes or the elephants, and Renaults and Vaubans among the beavers. The more inferior species, such as the oysters, polyps, etc., will be in comparison with the more elevated species in this new hierarchy what birds and quadrupeds are in comparison with man in the present hierarchy."[4]

When Haeckel advanced the concept of ontogenetic recapitulation, occult biologists immediately incorporated it into their theory of archetypal species. Not only does ontogeny recapitulate phylogeny (as Haeckel singly proposed) but, they added, both ontogeny and phylogeny (on different scales) recapitulate cosmogony (the history of the universe and the migration of souls). The occult anatomist Hermann Poppelbaum specifically rewrote the biogenetic axiom to read: "Microcosmogony is a reflection of macrocosmogony."[5] The stages of the embryo recapitulate both the physical and spiritual processes leading to the human state.

This has become a seminal vision for twentieth-century spiritualism, providing material for some bizarre cosmic melodrama (which we shall see). As Gould (quoting Huxley) remarked, recapitulation contains the germ of a hidden truth. To occultists who had inherited millennia of teaching that the stars rule our organs, that hidden truth could only be the imprint of the cosmos itself on the embryo and the expression of a supernatural history through its physical unfolding. Probably the most complete adaptation of Haeckel's proposition in a universal occult system was accomplished by Rudolf Steiner (under his own caution that Haeckel himself would have "unmistakably declined this dedication").[6]

In Steiner's creationary tale human beings pass through (and recapitulate foetally) four ancient universes. Although he gives these realms the names of ordinary planets in the Solar System he is actually proposing zones in other dimensions on which our ancestors incarnated.

The whole of this level of creation began in a spiritual germ as large as the present-day universe. Primeval specters descended from the monad onto (or into) the world of Old Saturn. At the moment of fertilization the human embryo briefly reembodies the seed of that long-vanished universe.

On Old Saturn the human ancestors were asleep and unaware; those souls who incarnated directly from there onto the Earth appeared in the Azoic era, still unconscious, as minerals. In the beginning they were capable of becoming human, but they were too precipitous, and their consciousness remains yet in the higher spiritual worlds. Other souls that were not human progenitors awakened fully on Saturn and passed into angelic realms.

The Saturnian world was entirely physical but not in the sense we think of; it contained no atoms, no molecules; it was a physicality only of heat effects. Through the aeons of Saturnian time the human ancestors acquired the possibility of incarnating in new bodies, thus bringing the epoch to an end. After passing through an interval of cosmic night these creatures were reborn on the Sun. First they recapitulated their sap in the new medium, while the Sun remained dark. Then envelopes of ether sprang up around them and ignited their world with flame.

The etheric substance of the Sun traced the germinal organs of an animal body in the heat effects inherited from Old Saturn. Even today ripples of this ether sustain the living carapaces of creatures on Earth;

without this aura they would return to the mineral realm and disintegrate. Our physical heart is a replica of an etheric heart, and our physical brain is an etheric brain imprinted in protoplasm.

By occult chemistry the liquid of our bodies is already a "living" thing, an intricate global sense organ. Through water the life body of the Sun responds to the appearance of matter on Earth and transmits soft material patterns into protoplasm. Water responds to every breeze over its surface, every slight change in temperature and point of contact, from extraterrestrial fluxes of gravity to the interruption of a vine or bug. To any snag it responds rhythmically, sending out a precise scale of ripples or waves.

Water is not simply a miraculous chemical; it is an intermediary between the rocky planet and the invisible ether, recording the impressions it receives from both poles and distributing them in currents. Note the effect of rain on a parched valley; prehistoric seeds spring from dormancy and all life becomes alert to its own nature. The moisture the desert sage distills from the night atmosphere fills the air with medicine. Even a garden hose on a hot afternoon brings astral refreshment. In the rivers and seas of this world and in the watery clouds of our Saturnian atmosphere we can observe the organs of actual creatures becoming visible and then receding. The fins of sting-rays appear in the intersecting waves of the oceans; the umbrellas of jellyfish form where springs surge into surface waters; the paired vortices of the heart are foreshadowed where two streams meet underwater. The worm is already a river, and the feathers of birds are currents of air embodied. Even bone preserves the criss-crossing etheric ripples.

On the Sun the potential man/woman was not completely asleep, but lay in a dreamless sleep, occasionally stirring to observe the brilliant flutter around him through a sense organ which would be incarnated as the pineal body. Overwhelmed by the harmony and completeness of the creation around it, this hermaphroditic being watched the external world flow by in bright soul pictures, semiconscious reflections of its own being; but it did not recognize them or comprehend the significance of its objective identity. Such is the state of consciousness of those beings who incarnated directly onto Earth from the Old Sun—the plants; their egos remain behind in the lower spiritual zone.

Our ancestors inhabited the esoteric kingdom of this era, the Hyperborean; its realm of sunlight is recapitulated in the yolk sac (the occult embryologist Karl König has remarked that the yolk sac of a young aborted human still holds a fluid bright as gold).

The world of the Sun was too brilliant and overwhelming for the human ancestors to evolve any further there; thus it came to an end, and, after a passage as seeds through the cosmic womb, the Sun beings evolved onto the Old Moon. First they recapitulated the sap of Old Saturn and the etheric signature of the Sun; then they secreted a new carapace, an astral body, a liquescent envelope that drew them from their eternity of slumber. Without a lunar double the life body would have remained permanently unconscious. Even today the astral shroud alone keeps animals awake to an exterior world.

The astrum expressed itself initially in physicalization, a gradual condensing of the human ancestors and their fellow beings into faint willowy creatures. Our predecessors wavered here between their nostalgia for the Sun and the excitement of their independent nature. They felt the first pangs of longing and desire.

As the human initiate received its physical nervous system its revels now truly ended. No turning back, no exit, no bluff this time! It perceived the vastness of things around it and began to fathom its plight. Only then did our ancestors understand, dimly, that they were a copy of the cosmos and not the cosmos itself, that something belonged to them alone. It has been a vision and a prayer ever since.

This was the Lemurian age, corresponding atemporally to the Mesozoic on Earth. The pharyngeal slits of the embryo date from this time, for they are not only the gills of Mesozoic fish but the organ through which Lemurian ancestors received the form-giving vibrations of the cosmos, sound-tones from beyond even the stars, recoiling through the watery astral atmosphere. Speech does not only originate in the grunts and bellows of beasts but is also an imprint of the symbols of higher realms within astral receptors. As the ancestors swam, their dream-like propulsion through these tones became language in the throats of some animals, while other hastily incarnating souls awoke with gill slits in the Earth's primeval ocean.

During the Lemurian age the physical universe was solidifying. Galaxies coalesced; stars separated from nebulae and cast planet webs. The actual Sun was incarnated with a retinue of worlds including the Earth, each of them with astral sheaths many times their physical girth. Since the astral world is not structured by gravity or space-time, creatures were able then to move freely without energy sources or receptacles. The higher astral beings danced through the worlds of the universe drunk on cosmic tones. That is, they are still dancing.

Those creatures who jumped from the astral onto the Earth became invertebrates. Because the Moon is a hardening and drying system, jellyfish, worms, crabs, insects, and starfish dominated the Mesozoic. In Moon consciousness there is a perfect correspondence between image and object, so the wisdom of these animals is in their organs not their minds: in claws and stingers, phosphorescence and shells. Cold-blooded vertebrates follow the invertebrates out of the old Moon. They have no sense of their own presence; they live in their ancestors. They are dreaming, Poppelbaum says, as if it were always the day before yesterday. Spiders arrive spinning a web before they can personify it; glands and spinner hooks are their intelligence. Crabs dance about each other waving their claws and changing colors, but they are asleep. The earthworm follows the undulation of its own body through the grains of soil. The frog and the grasshopper sing. The tiny tardigrade goes back to sleep.

"All animals," said Steiner, "live as it were under the surface of the sea of color and light."[7] For life on the Moon, existence is a single photograph, snapped once, at birth. The clam lies within its shell sucking the astral liquid, a copy of the universe.

Though their wisdom is trapped in their organs, the invertebrates, like all other creatures, contain a soul and a spark of the creative power of nature itself. Carl Jung wrote: "If the glow-worm could be transformed into a being who knew that he possessed the secret of making light without warmth, that would be a man with an insight and knowledge greater than we have reached."[8] The secrets of the universe are distributed through all levels of conscious and unconscious matter and not just mind. The first prokaryote and eukaryote cells, and even the winds and waves stirred by gravity, were seeds of intelligence seeking embodiment through the physical transformation of matter.

Look at the charmed countenance of the newborn wren, the wildebeest foal struggling to its feet, they seem almost to be trying to recall another place and time. "Animals are fixed ideas incarnate,"[9] wrote Henrik Steffens in 1822; they are souls that incarnated abruptly, leaving their egos behind on the Old Moon. Each one of them is thus the physiognomic expression of its astral body.

When the world of the Old Moon came to an end, the advanced Lemurian beings departed for higher spheres and a new world emanated from the cosmic forge, a radiant body. The original Earth appeared first as Old Saturn, a shell of heat and air. Then its aura coalesced and merged elementally with the rains of the physical planet. This moment recurs ontogenetically as the circulation of blood through the placenta. Then the astral bodies of the planets were recapitulated; they continued to condense and harden past the stage of the Old Moon. Souls that found solidification too painful swiftly vacated; the others were attracted to the planets. At the point of fullest animal incarnation the inner skeleton is mineralized on Earth, and at the opposite pole the aura persists, a vestigial gateway into the astrum during sleep.

Just as the newborn Earth was about to ossify entirely, spiritual beings interceded and dislodged its heavy Moon. Steiner took this act of rescue to mean that *man cannot incarnate himself*. The withdrawal of the Moon reflected immediately in the division into sexes of all subsequent animals. Outwardly less perfect, less complete, these creatures now had space in which to develop internally.

With the lifting of the Moon the human ancestors assumed erect posture and gradually became visible. Their hands and feet were differentiated, and nutritive and reproductive organs sprouted nodules for their voices. Restrained from rigidification in the lunar body, men and women acquired the spark of an ego, an objective consciousness. This would distinguish the Earth from other incarnations.

During the Mesozoic and Tertiary epochs many souls hastened onto the terrestrial plane, becoming animals there and leaving behind their bodies in the higher realms for subsequent souls. The ancestors of men and women lingered in Lemuria and then Atlantis; correspondingly, their embryos are retarded in the womb, remaining flexible and uncommitted. Crocodiles, chickens, and mice all start like humans, as mere coiled knots of flesh with a backbone and nerves radiating from the primitive streak like meteor showers, but they are suddenly trapped in hard artifactual organs. The more quickly they hasten to migrate (cosmogonically), to incarnate (ontogenetically) the less consciousness they retain and the more their wisdom lodges in organs.

At the very end of the Atlantean era, the esoteric eleventh hour, the last prehuman creatures rush into bodies: apes followed by prehistoric tribes of hominids—Australopithecines, Pithecanthropines, and Neanderthals. They struggled seemingly to the very end to avoid solidification, but they incarnated still an instant too soon. The embryonic gibbon bears the stamp of this prehuman embrace of flesh; it is human but human like an old man, wrinkled and hardened, dead to the possibility of ego while yet unborn. In esoteric tradition man and woman are not descended from the apes, nor, in fact, from any of the life forms on the Earth. They are the force holding together the branches of creation; other genera are the side effects of *their* formation.

Spiritual evolution descends through Saturn, Sun, and Moon incarnations to meet physical evolution ascending through star debris, minerals, bacteria, plants, invertebrates, fishes, amphibians, reptiles, mammals, primates, and hominoids. It is the same evolution reflected in different mediums—a single confluence of two streams, one bringing physical organs into mind and ego, and the other radiating spiritual essence through the world of molecules and organisms. Together they form the tree of life. The bodies of life forms are but shaggy replicas of their souls.

The present-day kingdoms of plants and animals are reminiscent of long-ago partings, says Steiner; they are the equivalents of what we were on other worlds, not this one. There are no fossils of our ancestors because they were not material until they were human; thus, there is no physical evidence for our evolution.

In Steiner's vision, spirit rushes toward matter. What we see in matter is merely the physical shadow of a spiritual event. The physical event is self-contained in its own kingdom, but it is directed by the spiritual entelechy. Thus, where scientists see physical evolution progressing by random degrees, complexifying to sustain intelligent life forms, Steiner saw spiritual evolution descending into matter and organizing it. Where spirit sinks too quickly and suddenly, its whole force is absorbed in matter and it is frozen in minerals, in inorganic crystals. These are spiritual beings too, but so deeply imbedded in physical substance they have no flexibility. As more spiritual force is withheld, body becomes softer and more receptive, less fixed by its transfiguration. Plants spring from a quantum of restrained spirit, but they are predominantly frozen in incarnation. All plant life on the Earth is equivalent to the story of one man or woman, Steiner says.

Animal life follows, from amoeba to earthworm to snail to salmon to antelope, awakening bit by bit. Except for the markings that distinguish species from species, nothing separates one crab from another, one albatross from another. Their individuality is subsumed in their ancestral type. "The biography of one man," wrote Steiner, "corresponds to the history of a whole species in the animal kingdom."[10] And each animal biography, the flying squirrel for instance, is phenomenologically an eternity.

Occult biologists refute our descent from the apes on this basis. The generalized ancestors of men and women never slid that far into matter; they were present in human form even during the heyday of the mollusk and the dinosaur, not here but in Lemuria. In their paedomorphic development humans reenact the spiritual prehistory of the Earth, a legacy that single life histories strive to illumine. We are all heroes seeking a grail, though few realize it. As we age, however, we harden, and, if we lose touch with our astral and etheric envelopes, we become more fully materialistic.

Although the wisdom of human beings can never enter their peripheral organs, it has been replicated in artificial organs—machines, weapons, and even money. People obsessed with these things become mockeries of animals. They barter the possibility of spirit for the shadow of power on this one plane. A similar fate befalls the elders of technological cultures if they become trapped in endless fascination for material novelties. Instead of joining with collective wisdom and evolving into seers, they wish to individualize each new object in their psyches. Their deaths actually bring wisdom back into the world.

Cold-blooded animals have no mind and no voice; they speak only by abrasion of exterior body parts. The chirping frog and hissing snake exist at the boundary of this condition. The higher animals perceive something of the tragedy that has befallen them; they have a voice but no words. They try to express their strangeness and isolation, their loneliness, but the words come out as cries and groans. Even when triumphant they are plaintive. In odd moments the eyes of other mammals may meet ours, and their expression is one almost of bewilderment. They are trapped in the dream of the astral body, and their faces reflect either the sadness or horror of the missed opportunity.

"It is not the ox that bellows," writes Poppelbaum, "not the dog that barks—a bellowing comes from the ox, a barking from the dog—*through* the dog. From the land of dreams it pours into the world of man. . . . An unredeemed being is striving for expression in it. . . . The voice of an animal is wrung out of it as though by a nightmare; not produced in the free course of breath—it is full of the destiny that is suffered but not understood!"[11]

The Indra lemur and howler monkey seem almost to force their lungs out, but they still screech from the hollow center of being. The turtle bears an ancient mask-face of pain.

"It is as though something *veiled* were living behind the physiognomy. Something that *craves* to shine through but is withheld by the body's rigidity! This impression becomes positively grotesque and horrible in the case of an insect. Looking at the head of a wasp or a butterfly, perhaps through the magnifying glass, we cannot help almost shuddering. The merciless rigidity and hardness of the casing, out of which the eye, an immobile point, its surfaces walled in, stays there lidless and ever open; that fearful leverwork of the parts of the mouth working mechanically; the hurried jerk and cramped groping of the proboscis; the antennae, always trembling, and yet not looking truly 'alive'—it is as if one saw a ghost, a phantom suspended by invisible threads, that pretends to be alive but in reality is only a moving mechanism."[12]

The truth vaguely adumbrated by Haeckel's theory of recapitulation may have been more than just the significance of heterochrony in speciation; it may also have been the esoteric link between embryogenesis and cosmogenesis. That is, Haeckel helped to create post-Darwinian spiritualism by providing a mechanism for the invisible operation of the macrocosm within the microcosm. He became an unwitting mage, cited as religiously by twentiety-century occultists as Hermes was by the Mediaeval alchemists. Recapitulation is in fact the modern version of "As Above, So Below." It is the basis for the miniaturization of astronomical events in human tissue.

With the separation of the first polar body from the oogonium, says the anthroposophical doctor Karl König, the removal of the Old Sun is recapitulated. As the second polar body is shed in the formation of the mature oocyte the human ancestors depart from the Old Moon. That is, primordial episodes of creation leave their signature in the germ plasm, and the cells *must* depict them even as they must (at the same time) obey their morphogenetic situation. Because of the cataclysmic withdrawal of the Moon the small globe of the yolk sac is drawn inexorably out of the yolk-sac vesicle. Then the billions of souls who streamed to the new Earth become millions of spermatic beings swarming about each egg.

The theosophical scholar Manly P. Hall describes the cell as a microcosm of planetary fields and forces: the cytoplasm is the Sun; the centrosome, which activates mitosis, is, of course, lunar; the nucleolus, the seed of the mind, is the planet Venus. Mars occurs as the archoplasm, an emotional sheath and the astral body of the cell. Saturn is the nucleus, the emerging animal nervous system. Jupiter and Mars do not form separate orbs but are dissolved in the nucleolus with the human aura.

In Hindu lore the egg is the "natural abode of Vishnu in the form of Brahma; and there Vishnu, the lord of the universe, whose essence is inscrutable, assumed a perceptible form. . . . In that egg, O Brahman, were the continents and seas and mountains, the planets and divisions of the universe, the gods, the demons, and mankind. And this egg was externally invested by seven natural envelopes; or by water, air, fire, ether, and Ahamkara, the origin of the elements, each tenfold the extent of that which it invested; next came the principle of Intelligence; and, finally, the whole was surrounded by the indiscrete Principle; resembling, thus, the cocoa-nut, filled interiorly with pulp, and exteriorly covered by husk and rind."[13]

After fertilization (in König's version) the trophoblast forms first, the dwelling place of the soul. The chorion recapitulates the infinitude from which individual worlds manifest; Hall interprets it as the atmosphere of the cell, the outer crust corresponding to the corona of the Sun or the shell of a chicken's egg. König's amnion is the massive astral sea in which the embryo floats; for Hall the amnion is composed of sidereal ether, and the amniotic fluid is a body flowing from "the ten divinities of the sun."[14] Hall's allantois (the outpouching of the caudal surface of the yolk sac) is a ripple in the etheric current through which vital energies pass onto the world. According to König, Lemuria itself is recapitulated in this protrusion, whereas the yolk sac is the Earth, the sandy beach onto which Palaeozoic animals are washed in the first waves.

The midwife Jeannine Parvati remembers a dream the night one of her babies imbedded in her own endometrium:

"I am being initiated into a secret medicine society—a cluster of women are together, watching as I burrow into soft, rich earth. We are chanting, 'I am entering the sacred circle of mugwort.' "[15]

Viewed from beyond its biological mechanics, conception is a spiritual episode, a ceremony in the life of the organism as surely as any rite of puberty in old Australia or Israel. It is the passing of the sacred paraphernalia from an aging priest to his or her successor.

The blood force of the initiate grounds itself in the trophoblast. Stem cells sow generations of corpuscles. Angioblasts, erythrocytes, and lymphocytes stream into circulation as replicas of the first galaxies and stars. This is not only a mineral blood; it is a universal etheric force combining chlorophyll, hemoglobin, pure sunlight, and vibration. The equipotential liquor of plants, cold invertebrates, and warm-blooded birds and mammals occurs as the life-bearing liquid in the human embryo.

In occult anatomy, blood is a sacred vapor, the essence of the soul. Ties of blood are not simply metaphors; they are karmic bonds of the deepest sort. When the Comanche chief slices his hand with a blade and holds it out to meet the similarly cut palm of an ally, their pact is signed in blood, i.e., in the magnetic field of the Earth. (Would that arms negotiators took their bodies as seriously!) Sorcerers and shamans have always sought the blood of their targets: To gain access to the blood is to hold power over the soul. The blood mysteries go back to Isis and then to the gods and goddesses of the Old Stone Age. Through the "smoaky" vapors and emanations of this arcanum all forms of disincarnate entities and ghosts may be summoned, hence, the sacrifice of animals at the altars of the invisible world. Necromancers, likewise, invoke satanic entities and demons from the fumes of blood.

On the seventeenth day after conception (in Steiner's cosmology) the soul enters the body (the Hindus say twenty-two days, the Sikhs forty, but of course "all are describing *aspects* of consciousness/soul—not a fixed *adherable* commodity"[16]). König proposes that the etheric element of the soul merges with the amnion, its physical being with the yolk sac and its astral layer with the allantois. The soul penetrates the embryo through the chorion. But there is a paradox here: The soul is present from the beginning of the Earth, or evolution would not have occurred. Only because its cosmic history has *already* been recapitulated (phylogenetically) can the creature recur (ontogenetically). The human archetype imbues the trophoblast and then the embryoblast, both of which it has already formed through the evolution of the mammals.

Haeckel's confusion of acceleration with retardation, fatal to his ultimate position in science, becomes not only meaningless but facilitative in cosmogony, for as the more perfect spiritual realm works forward in time toward incarnation, it must go backward toward Primal Intelligence. Carnality is a primitive ancestral quality and at the same time the evolutionary path of life on Earth. Thus, the human being is a paedomorph for retarding solidification but, contrarily, a recapitulation of the maximum number of prior universes. In fact, the genius of Haeckel's error is that it accurately describes the dilemma of the soul. The only reason it is sent on the long transmigration of lives is to be able to recapitulate past universes while successively embodying more and more of its objective spirituality in each one. Phylogenesis can go only forward in time, encumbered as it is with the actual physical baggage of history. Cosmogenesis goes backward and forward at the same time: at one pole toward the celestial sphere of its origin and at the other toward the imprint of its ego on the Earth.

The force of the ego first appears in the primitive pit and then spreads through the agency of the streak. Its presence drives inward as a groove; then it folds over into a vessel—the neural tube; objective consciousness molds genetically active flesh. The forward thrust of this force locates in the head, and its peripheral awareness disperses to the torso, limbs, and tributaries of the nervous system. The finger of God reaches out to the finger of Adam in a painting by Michelangelo. The neural groove sinks into the mineral world, and the notochord is detached; with the separation of the Moon the primordia of the vertebral column ascend one by one, the Atlantean child rises. This act of raw and unabashed magic should be almost familiar by now; as William Blake told us in a former age: "Man has no Body distinct from his Soul."[17]

Materialist scientists now seem to locate more and more of our psychology in purely mechanical activity. Moods and mental disorders are routinely ascribed to chemical imbalances and genetic predispositions—no need for psychotherapy. Dreams are said to be nothing more than the biocomputer clearing its cells of useless data in sleep; their emotional and psychic contents are declared meaningless. But these hordes of mechanists were born yesterday. The body *is* the soul; it is the soul's replication in flesh and blood, the simultaneous impact of all its prior incarnations.

In gastrulation endoderm forms as a recapitulation of Saturnian sap; the raw mineral body of the plants and animals lays down an aboriginal gut. Ectoderm is Moon tissue, the source of thought. The separation of our lunar double is recapitulated again and again through the multiple sense organs. Ectoderm withdraws from itself and reflects back against its interiorized surface; the lens of the eye and the labyrinth of the ear are induced by polarities from a primal embryogenic field, the Earth. The segregating tendency of the Moon reflects a creative principle in the ectoderm; the outer skin is impelled to form a sensual inner layer, a receptor for exterior stimuli. Simultaneously, a third germ layer, unknown in the simplest animals, develops along the contact zone of ectoderm and endoderm. This mesodermal tissue is a long erogenous ridge upon which actual reproductive organs form. Where mesoderm penetrates endoderm, meso-endoderm forges volition, will, desire. Where lunar nerves pierce the mesoderm, feelings sprout along filaments and channels. The kidneys, gonads, muscles, and bones all occur in mesodermal pairs under the bilateral influence of the Moon in ectoderm.

Hall points out that the extra-embryonic membranes are in fact the aura of the foetus, the solidifying aspects of supernatural bodies. When the infant is discharged onto the planet it is flung from these visible placental shrouds into the magnetic field of its aura in which it is likewise suspended.

"The placenta is a being," writes Parvati. "It is of its own ac/cord. The placenta would have a related (but not identical) aura to the mama and the baby. How we treat it and its cord will affect the baby and the mother. . . . Why do we want to cut and get rid of the placenta?

"The reason, in part, most want to dispose so rapidly of the placenta is the simple fact that it is dying. It's hard to watch a being die. As you watch the newborn lotus baby come fully into life, the placenta goes fully into death. We appreciate watching the old ones pass on as the new ones come in. . . ."[18]

The allantois and the yolk sac disappear once their images have been impressed—the Lemurian and Hyperborean epochs. The amnion continues to develop until, by the end of the seventh week, it is pulsating on its own like an external heart and lung. As the neural network swells through the organism, the waters of the amnion are most active. The celestial message passes in ripples through the astral medium, and the embryo hears them through the pineal gland and learns about its coming incarnation. The late human embryo recapitulating his Atlantean incarnation in the womb is still clairvoyant, tuned to the higher spheres; he listens to the clatter and chatter of the "lesser" world he is descending into while his body forms the armament of that very world around him.

"Babies are pure love," says Parvati. "A spiritual midwife vows to welcome all babies *gently*. . . . She is the guardian for babies, and does all she can to end oppression and suffering by sado-medical rituals of O.B./pediatric religions. . . .

"Each birth is a ceremony for all time. When we attend a birth without making a claim, we are healed."[19]

Gut branches into tiny channels, liver and spleen sprout, and the first chambers of the kidney appear. The era of Atlantis is recapitulated; it is the Tertiary on Earth; the unborn child is submerged in the astral waters of the flood; then the waters burst and a new being is floated out into the cosmos like an ark.

The visible shards of creation are strewn to the ends of time and space, unbelievably far beyond our limitation of mind. Trillions upon trillions of stars surround the Earth, but radiant embryo orbs also penetrate the Earth, awakening through its atmosphere, in its oceans and caves. The young owl hoots, the sow bug scurries through leaves, the fish pulsate in their schools. The newborn is a planet separating from a sun. It does not so much leave the womb as enter another womb: the cosmos woven into the night sky.

The ornate spiritual universes of Steiner and Hall are not the only possible ones. In fact, we do not know what is possible. We must finally accept, in light of the harsh reality of being born and dying, that what we are is a continuation of what the universe is, so all our hopes and fears could not be irrelevant to cosmic process; else how could they have occurred? Our wild hopes for eternal rebirth, our dread of hell and extinction are part of the universe too.

The inevitability of death is the same as the inevitability of birth. It is a part of our essence. The same forces that brought us here, that embrace and cling to life, are the forces that will take us from here. If we shun and vilify our sure deaths, then we must in some way deny the fact of our life. We are in the hands of the gods anyway, and if they are not able captains, we were in trouble long before dying; we were in fact in trouble before being born. Of course we must be exactly as good as they have made us. Their thoughts must be our thoughts, flitting across the outer edge of dreams (but we must not overspiritualize the universe, for we do so always at risk to the psyche; the universe is quite spiritual enough without our decadent gods and gaudy symbols).

The embryo forms as light reflected unto itself. This body of flesh which produces fluids and feelings and with which people seduce each other and from which more come—is made out of cells as a light show is made out of light.

23.   Cosmogenesis
and Mortality

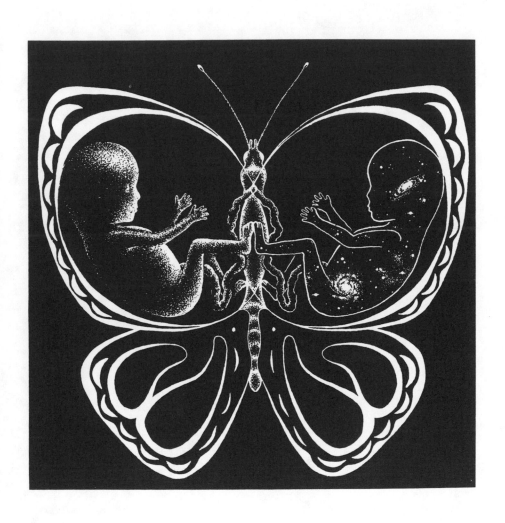

According to Chinese cosmogony, the body is shaped and sustained by two counteracting forces. A *yang* centripetal energy, Heaven's force, originates in the depths of infinite space and flows down through the uppermost spiral of the heads of both mother and embryo. Meanwhile, the rotating Earth discharges a *yin* centrifugal force upward through the genitals. Embryogenesis begins in the ceaseless antagonism of *yin* and *yang* drawing helices out of the undifferentiated unity. Through continued vibrations, ever more complex spiral chains condense, forming atoms, then molecules, and, finally, genetic fibers. The physical embryo is a realization of the structuring processes in matter provided proximally by the parental cells. The energized patterns originating in sperm and ovum are attracted to each other, and their intersection-paths move back and forth forming tissues in layers. In a sense each organ is an interference pattern of waves through each other, a vortex produced by streams of liquid "interweaving from manifold directions,"[1] all redistributed through the global whirlpool of the gastrula. The currents which curl and fold within themselves finally push out as limbs, feelers, vesicles, and delicate surfaces of skin and nerves.

Heaven's force charges the brain cells as transmitters of consciousness; continuing downward it energizes the uvula, heart, stomach, genitals, and other organs. Where it encounters Earth's force *chakras* are formed—reservoirs with unique spiritual as well as physical functions. The basal chakra locates around the bladder and genitals; subsequent ascending ones emerge at the ovary and intestines, solar plexus, heart, throat, brain, and crown (or cerebral cortex).

As they swirl outward, the internal spirals of the organism intersect the streams of Heaven and Earth. Centrifugal motion molds hollow organs such as intestines, stomach, bladder, and gall bladder. Centripetal force condenses as lungs, heart, spleen, liver, and kidneys. The currents are also distributed through twelve primary electromagnetic channels, the meridians. Anglicized as Lung, Large Intestine, Stomach, Spleen, Heart, Small Intestine, Bladder, Kidney, Heart Governor, Triple Heater, Gall Bladder, and Liver, these meridians

343

become the pathways of *yin/yang* energy and the loom of the life body on which the physical body is assembled.

Currents of invisible *chi (ki)* energy flow between these channels and are assimilated in the organs, but they are not fully represented or concretized by any other circulatory system such as the blood or the nerves (though they must underlie these and determine the courses of their materialization). Insofar as the currents continue to carry vital energy through the organism for a lifetime, embryogenesis never ends; creatures are continually supplied with the original currents of which they are fashioned. Food and breath merely supplement the *chi*: the body is primarily sustained by the embryogenic flow of spiral helices. The *chi* provides strength and intelligence, and, through years of study and discipline, can be tapped directly as in the charges given off by a *t'ai chi ch'uan* master who jolts an opponent fifteen feet in the air by the slightest touch.

Two other meridians form on the dorsal and ventral midlines of the body. The electromagnetic energy streaming up along the front of the embryo from a point between the anus and genital to the tip of the mouth establishes a Conception Vessel which furnishes deep tissue during embryogenesis. A connecting channel (the Governing Vessel) enters the mouth, flows down the inside of the body through the digestive system, and comes back out in the region between the genital and the anus. From there it runs up the dorsal surface of the body, over the head, and in through the mouth, distributing energy to the body's periphery.

The formation of the body by polarizing currents leaves a complex network of tissue homologies and organ relationships. The brain and intestines are complementary spirals generated at opposite poles of the organism, joining to produce, at midstream, the double-vortex of the heart. The nervous system contracts and concentrates in the brain, whereas the digestive system expands and disperses through the intestines. The brain is centripetalized intestines; the intestines are centrifugalized brain. The formative energy of the *yin* digestive system is also complemented by the *yang* contracting spiral of the respiratory system. The circulation of blood and flushing of waste proceed in twin eddies from the heart.

The arms and legs are formed by secondary waves of tissue. The basic meridians follow the surfaces of the limbs outward, winding around the elbows and knees, the wrists and ankles, and flowing to the tips of the fingers and toes, interacting in complex morphology. The outer surfaces of the limbs and phalanges originate at the cores of these spirals and thus correspond to the inner aspects of the organs. Conversely, the inner surfaces of limbs and joints concentrate as the outward flow from the peripheral zones of inner organs. Because of this etiology needles and massages can cure internal conditions when applied even at points remote from the disorder. An entry point on the arm may draw greater medicinal power to the heart or lungs than one on the chest because the former activates the entire channel leading to the organ and traverses a zone of original formative energy. The traditional Chinese physician uses needles, subtle herb potions, and burning herbs in cones in attempts to retune the body/mind's currents to its embryogenic harmonies.

In some practices, the face is taken as a microcosm of the body; thus, the cheeks become complements of the lungs, the tip of the nose corresponds to the heart, the bridge of the nose to the stomach and pancreas. The ears counterpoint the kidneys, the mouth the anus, the teeth the vertebrae, and both sides of the forehead the reproductive organs. In iridology, the eye is interpreted as a miniature of the organism and used diagnostically, with parts of the iris corresponding to specific internal sites. In one style of acupuncture, the ear is treated as a partially aborted twin, a small replica of the embryo, and is used like a voodoo doll to treat corresponding organs.

The meridians begin at points of intersection between the embryo and the electromagnetic layer around it and travel inward through the body. In gastrulation and subsequent organogenesis the original peripheral surface of the organism orients dorsally. Thus the major Entering *(Yu)* points are on the back, girding the spine. Continuing to receive energy from the atmosphere after birth, these remain important activating nodes in acupuncture treatment. After contributing to the formation of organs along their paths the currents culminate in Gathering *(Bo)* points. They are then discharged through channels leading to the Well *(Sei)* points on the tips of the fingers and toes. Thus hands and feet can be treated to reach all the other organs by reflexology.

Each of the Entering, Gathering, and Well points, as well as points between them, have unique roles in balancing the energy of the body (as they did in molding the body initially). The life current is either bubbling up, bubbling down, spreading out, eddying, fluttering, oscillating, or carrying out some other formative motility as it flows through each duct. In similar fashion, electromagnetic forces and ley lines passing through the Earth generate volcanoes, waterfalls, springs, mountain ranges, buttes, islands, and river valleys, while galactic forces charge single and double star systems and planets of various size, as well as moons, dust clouds, comets, and so on. The body is a topographic replica of the planet and the cosmos; it contains the same complex streams and topological functions, but working in a different medium and animated by the specific qualities of sentient life.

Herbert Guenther writes: "The very fact that one images what may well be a process occurring on a cosmic scale in humanly intelligible terms suggests that the energy involved itself has an intelligent character which can easily be construed as a mind informing, as it were, its own material—the water of the ocean (the energy, *thig-le*) assuming a distinct shape (wave, *rtsa*) through the very dynamics of the ocean's energy (the surging, *rlung*)."[2]

Matter and consciousness are the two counterpoints of creation. They begin at opposite poles but stretch out in space and time and engulf each other. The infinitude of the universe is actually the appearance of mind, like a sun dispersing darkness and homogeneity—a ceremony repeated in the birth of each living thing. So we always approach cosmogenesis through the waking dream of our unknown desire.

When God is personified in *Genesis* and other etiological myths, He is a being like any other: He must confront a basic inertial resistance to the creativity of His own psyche. The psychologist, Charles Poncé, points out that Gods become Earthmakers because They are lonely:

"Ultimately, loneliness is something or someone representative or symbolic of an interior component of ourselves that longs to be united with us, or that we long to be united with. Expressed psychologically, this yearning is for something that is not known to us consciously, that is hidden deep within our souls."[3]

Jehovah, Brahman, even the Winnebago Earthmaker are powerful but sterile demiurges unless They can gain access to the spontaneous shapes of Their unconscious minds.

"In an attempt to discover the hidden aspect of Himself with which He wished to be united, God allowed his unconscious to create an image of the material attempting to come into the light of consciousness. In much the same manner that the unconscious in human beings creates dreams reflective of our hidden nature, so too did God spin out of His unconscious that aspect of His hidden nature with which He was not reconciled. Whereas our unconscious lives are ephemeral and without material substance, His unconscious life took on substance. . . . His images were to live themselves forward regardless of His commandments in much the same manner that the process of our unconscious lives live themselves forward regardless of our conscious and rational commands. . . . In other words, Adam symbolized . . . the unknown portion of God's psyche that constituted His blind spot."[4]

A ray of creation originates outside time and space, said the Russian occult scientist G. I. Gurdjieff. Entering the cosmos at a speed faster than light, it crashes down into the galaxies, splintering (without diminishment) into the star systems it spawns. From there it explodes into individual suns, and from these nodes into single worlds, such as the Earth-Moon system. Creation is unavoidable, for in order to cross the gaps between domains the ray must fill enormous intervals with shocks, each one equivalent to the jump from one musical note to another. The shock of the first octave occurs between the Absolute and the Sun, and its effect is beyond our knowing. Subsequent shocks are filled by the spontaneous manifestations of Sun, Earth, and organic life.

For Gurdjieff, atoms are cosmic notes vibrating within and between scales. Each atom has chemical, cosmic, and psychic properties which take on discrete shapes in different zones. For instance, there are hydrogens on levels from 6 to 12,288, with hydrogen 384 defined as water, hydrogen 192 as breathable air, and hydrogen 96 as rarefied gases. Hydrogens 48, 24, 12, and 6 are identified as "matters unknown to physics and chemistry, matters of our psychic and spiritual life on different levels."[5]

Consciousness not only arises from matter, it *is* matter, transmuted across intervals by living organisms. Molecules of food and air are metamorphosed into molecules of mind and spirit. Subtle impressions of the surrounding world are also digested and, thus vitalized, transmit motion and energy to animals; without images they would not survive an instant.

It is matter which evolves, not the personalities of man and woman. Human beings have the special capacity to translate molecules to a higher spiritual octave. In fact they *must* do so, or their souls perish. Only the upper realms are stable. All other substances, even the psyches of people, are broken up molecularly in the world of nature and become the changing chaff and debris of worlds and the light of stars. Souls are literally transcendent molecules. By breaking with compulsive patterns of behavior and habitual thought men and women can see their actual situation in the universe; through such self-awareness they can transmute the molecules of their being to the next octave.

The difficulty of awakening is immense, for our actual situation is so uncomfortable we do not want to experience it. "Man *can* awaken," Gurdjieff proclaimed.

"Theoretically he can, but practically it is almost impossible because as soon as a man awakens for a moment and opens his eyes, all the forces that caused him to fall asleep begin to act upon him with tenfold energy and he immediately falls asleep again, very often *dreaming* that he is awake or is awakening. . . .

"A man may be awakened by an alarm clock. But the trouble is that a man gets accustomed to the alarm clock far too quickly, he ceases to hear it. Many alarm clocks are necessary and always new ones."[6] (Otherwise we will surround ourselves with alarm clocks that keep us asleep.)

Gurdjieff accepted Darwin's system of natural selection, but only as a shadow of cosmic evolution—genetic mutations become one method of shock. For a planet to maintain life at all, some species in its biosphere must make souls. If consciousness on a world fails (perhaps through mechanization), then the ray of creation will fall from that zone and all things in it will perish. However, as long as the process of soul-making continues, all molecules (even the oxygen in the atmosphere and the phosphorus in the crust) can one day become souls.

Gurdjieff's cosmogony is Islamic and Zoroastrian at its roots, with Hindu and Buddhist elements. Pierre Teilhard de Chardin, a French priest and archaeologist, independently proposed a similar mingling of matter and spirit in cosmic evolution but with heavily Christian overtones and quite different consequences. In 1955, in *The Phenomenon of Man,* Teilhard described how the spiritual fire of the Sun endowed its planets with occult interiors. When these orbs were dispersed into the unredeemed darkness of space they sought once again to become stars, but lacking the gravitational force, they reconstructed the solar world in a new medium—as chemical reactions, crystals, plants, animate creatures, and, ultimately, mind.

"When I speak of the 'within' of the Earth," Teilhard wrote, "I do not of course mean (the) material depths . . . only a few miles beneath our feet. The *'within'* is used here . . . to denote the 'psychic' face of that portion of the stuff of the cosmos enclosed from the beginning of time within the narrow scope of the early Earth. In that fragment of sidereal matter . . . as in every other part of the universe, the exterior world must inevitably be lined at every point with an interior one. . . . By the very fact of individualisation of our planet, a certain mass of elementary consciousness was originally emprisoned in the matter of the earth."[7]

Billions of individual infinitesimal centers germinate on the Sun and transmit their energy to the cooled elemental lattices on worlds. As they coil from within and bind together in archetypal patterns, cells and nucleic acids form, and creatures emerge from these vortices, miniature stars— first, plants which reach back to their original home in the Sun and then, animals who translate the etheric fire into thought and build solar monuments and compose odes and ikons of the Sun. The interior face of matter ignites in a replica spiritual sun. This expansion of consciousness will ultimately penetrate and transform the entire planet, adding a layer to the biosphere even as the biosphere formed atop the geosphere. In true Christian fashion, this mind-sphere (noosphere) will provide for the salvation of not only the self-aware creatures contributing to it but "life itself in its organic totality."[8] The Earth will return to a solar consciouness, but to one of thought not of fire.

The emerging noosphere is "yet another membrane in the majestic assembly of telluric layers. A glow ripples outward from the first spark of conscious reflection. The point of ignition grows larger. The fire spreads in ever widening circles till finally the whole planet is covered with incandescence. . . . Much more coherent and just as extensive as any preceding layer, it is really a new layer, the 'thinking layer,' which, since its germination at the end of the Tertiary period, has spread over and above the world of plants and animals."[9]

Even if all humankind is one, even if the cosmos is contained in a single wave, we still must be individually born. Our hearts go out to other specific men and women, individuals like ourselves. Only the great saints see the same universal being in all faces, and, even then, they attend starvation and suffering that is individuated, never universalized. *"Cosmic embryogenesis in no way invalidates the reality of . . . historic birth,"*[10] Teilhard warns us. We live, and then we die, though the cosmos may go on evolving forever. From our perspective, cosmogenesis *is* mortality; our destiny is birth and death. Michio Kushi imagines us on a journey across all of time, impelled by love:

"The reproductive cells which are active in the mother's womb are the result of a journey of life-transmutations of hundreds of billions of years, which began from the infinite ocean of the universe an almost unknown time ago. They have reached the stage of organic life as the primary constitution of animal life, highly charged electromagnetic vibration. . . .

"When these antagonistic and complementary reproductive cells, the *yang* egg and the *yin* sperm, combine and fuse with each other, it is the beginning of the returning course of a hundred billion years toward the infinite ocean of life. Each of them carries its past memories and the vision of its future. When they fuse with each other—though their recent journeys separated them into different species of vegetables, different kinds of molecules and different blood cells of the father and mother—their memory that they have come from the same origin, one infinity, and their vision toward the future when they shall become one infinity, has never been forgotten.''[11]

Two lovers share billions upon billions of years of cosmic history beside which their separation is a transient flicker. Their kidneys and lungs are parts of the same body, now twained by mitosis. They have embraced millions of times before, as spores, polyps, zooids, mudfish, frogs. They have shared the blue sky and deep winter snows through cells in other bodies, as furry moles and lemurs hanging in trees. The synthesis of their germ cells is made possible by this fundamental identity. Their kiss really lasts forever. Their love is all we have.

''The memories and visions, which were carried by the parents' reproductive cells and fused into one by fertilization, are distributed to each of the rapidly growing individual cells. Each cell carries the same memories and visions, and all cells comprehensively carry the same memory and vision as a whole.''[12]

Paracelsus traces the image of man and woman to the *liquor vitae,* the ethereal life fluid which tinctures the *aura seminalis* (the physical sperm), droplet by droplet, each one an emanation of the complete human being. Da Free John notes that in some Hindu versions too the spirit incarnates not in the egg of the zygote but the male seed. The spermatic beings are living creatures, their ''little heads bright with energy.''[13] The egg is merely an assemblage of genetic material through which the sperm undergoes a metamorphosis.

''During the nine months of fetal life, the living being is not active or mobile but is rather in a state of deep rest. It is passing through a profound transformation whereby it becomes capable of functional associations that will define it as something quite different from a sperm. In other words, the individual that emerges at the point of birth from a woman does not represent the same kind of functional consciousness that the sperm did. But although its functional consciousness is profoundly transformed, the essential living being is exactly the same one.''[14]

The theosophical scholar Manly P. Hall describes the moment of conception:

''In the ovum is the plastic stuff which is to be molded by the heavenly powers. In the ovum is the sleeping world, awaiting the dawn of manvantaric day. In it lurk the Chhaya forms of time and place. Suddenly above the dark horizon of the ovum appears the blazing spermatic sun. Its ray shoots into the deep. The mother ocean thrills. The sperm follows the ray and vanishes in the mother. The germ achieves immortality by ceasing of itself and continuing in its progeny. . . . The One becomes two; unity is swallowed up in diversity. Fission begins; by cleavage the One releases the many. The gods are released. They group around the Poles. The zones are established. Each of the gods releases from himself a host of lesser spirits. The germ layers come into being. The gods gather about the North Pole. The shape is bent inward upon itself.''[15]

Madame H. P. Blavatsky continues:

''Then the embryonic creature begins to shoot out, from the inside outward, its limbs, and develops its features. The eyes are visible as two black dots; the ears, nose, and mouth form depressions, like the points of a pineapple, before they begin to project. The embryo develops into an animal-like foetus—the shape of a tadpole—and like an amphibious reptile lives in the water, and develops from it.''[16]

Hundreds of thousands of separate events clutter our lives, but a singleness of vision, an intuition of our original wholeness bonds us to the seed. We do not *remember* our spirituality as such: It impels us forward; it does not beckon us back to the womb or the cosmos. We do not necessarily even long remember the moments of actual connection from our waking life, though we do not forget their fact. In her novel *Dinner at the Homesick Restaurant* Anne Tyler describes how an elderly Pearl Tull has her son Ezra read her own diary entries from childhood to her. She is gradually dying, and she has him go on, day after day, week after week: *''purchased ten yards of heliotrope brilliantine* and *made chocolate blanc-mange for the Girls' Culture Circle.''*[17] It was 1908; she was fourteen. He continues through *''a flaxseed poultice on my finger . . . some gartlets of pale pink ribbon.''*[18] Another day (another year) he reads: *''Washed my yellow gown, made salt-rising bread, played Basket Ball.''*[19] Finally he comes to the entry she was seeking (1910):

''Early this morning I went out behind the house to weed. Was kneeling in the dirt by the stable with my pinafore a mess and the perspiration rolling down my back, wiped my face on my sleeve, reached for the trowel, and all at once thought, Why I believe that at just this moment I am absolutely happy . . .

''The Bedloe girl's piano scales were floating out her window, and a bottle fly was buzzing in the grass, and I saw that I was kneeling on such a beautiful green little planet. I don't care what else might come about, I have had this moment. It belongs to me.''[20]

Her adult life had been filled with the abandonments of loved ones and other broken dreams. She had stood most of the years behind a cash register in a grocery market. She was thoroughly disappointed by what had become of her; yet her existence, the soul of her being was not diminished.

"It is a good day for dying," said the old Indian warrior supine in the tipi of his son.[21] "Dying and living again," the Tibetans tell us in the West through their classic, the *Bardo Thötröl:* We are in an unbroken cycle of cycles, even the condemned killer realizes as he awaits the denounement of this existence:

". . . . . just me Gary Gilmore thief and murderer. Crazy Gary. Who will one day have a dream that he was a guy named GARY in 20th century America and that there was something very wrong. . . ."[22]

But then Mailer himself moved from the dismemberment of Gary Gilmore in a Utah morgue to the mummifcation and second life of the corpse, the ka-soul, in nineteenth-dynasty Egypt; after all, nothing that matters has changed in the brief epoch since:

"How could my mind continue to think while they pulled my brain apart? . . . . some wretched mixture of lime and ash . . . poured in by the embalmers to dissolve whatever might still be stuck to the inside of my skull."[23]

The Tibetan Book of the Dead is a narrative account of the passage of souls between incarnations through the *Bardo* realm where it experiences the same emotional projections as it did when embodied, hence the psychogenesis of fear and desire. The emotions are an aspect of the soul's inherent nature, not of the exterior situations through which it passes. The individual is reincarnated by the pull of his or her own illusions and valuations of reality. The embryo is not just an archetypal spirit corporealized; it is a manifestation, in genetic terms, of the hungers that draw creatures back into life, a materialization of what-we-already-are. That is why birth is irresistible.

The "dead" person is attracted to the flesh when he or she sees projections of men and women making love. These holograms draw the spirit toward the partner of the opposite sex. By identifying with that person's lover the spirit takes on its own sex—male if it is embracing a female, female if it is embracing a male. The boy, from the beginning, is his mother's lover, the girl her father's. If this were an actual paraphysical perception of the potential foetus's mother and father making love, we would not be able to explain artificial insemination; but it is a mythic representation of the compulsion through which spirit is embodied.

Tibetan cosmogony emphasizes choice through visualization. If strong emotions overwhelm the spirit, it could be born as a dog, a bird, a horse, or even an ant in an ant hill—to suffer the consequences of attachment until the attraction is understood and fulfilled. This stage is pictorialized in the text:

"There will be projections of males and females in sexual union. If you are going to be born as a male, you will experience yourself as a male and feel violent aggression toward the father and jealousy and desire for the mother. If you are going to be born as a female you will experience yourself as a female, and feel intense envy and jealousy of the mother and intense desire and passion for the father. This will cause you to enter the path leading to the womb, and you will experience self-existing bliss in the midst of the meeting of sperm and ovum. From that blissful state you will lose consciousness, and the embryo will grow round and oblong and so on until the body matures and comes out from the mother's womb."[24]

The assembly of our being is the central event of our lives. The radical and ultimate experience is incarnation. We begin as a mere mindless spark, and almost hourly our organs change; over weeks they are shifted and rearranged to a degree that would be sheer agony to a waking man or woman. The tumult within the embryo is worse than any war wound or systematic torture. Organs are torn apart; nerves sear through flesh; arms and legs swell outward and fragment into joints and digits; our bones are all broken; the brain throbs with new cells and nodules. It is more than an operation; it is once-in-a-lifetime primal surgery. And yet the seemingly fragile babe within the womb experiences this turmoil in silence and seemingly without pain. Some would even claim it is an ecstatic ride. The laying down of tissue is visionary and cosmic to a degree that no subsequent state of consciousness can recapture. Time and scale disappear, panoramas change from within, and the wall of shaping flesh becomes the universe.

Far from helpless, the esoteric body of the embryo should be tuned into the formative intelligence of more perfect worlds—entertained by an eternal choir of voices. Whether we identify this directionless ringing with blood and cells or with archangels and etheric spirits, the embryo is archetypally wiser than at any point in its existence to come, knowing everything as nothing in the state before birth. We can sometimes see this god-like figurement in the face of the newborn.

These roughly 280 days in the womb, says the Gurdjieffian astronomer Rodney Collin, are a full third of our life. Cell division and differentiation move a thousand times faster than events in the cities and villages of our planet. Everything we are is constructed there, and each instant of the embryo's perception is packed with breathtaking events. Without reference points in language and thought we cannot remember them as what they are, so we awake at birth remembering them as what *we* are, and the rest of our life becomes footnotes to this primal identity. In the embryo we become what life on Earth is; we become animal and then we become human. We share this collective consciousness or unconsciousness with every other human being, and we share parts of it with all the other life forms on the Earth. Billionfold cell junctions and gut, axons and hemoglobin are what we are all about, and from this gnosis we come together separately to build a world.

The planets are also involved, Collin tells us: At critical stages of embryogenesis glands and other organs are activated by cosmic resonance from the Solar System. The Sun/heart is the center of a progression of spirals which winds through the planets (glands) and then, in each organism, uncoils world by world (organ by organ) like a spring. Every endocrine function is a receiving set and transformer for a discrete planetary influence. Specifically, each gland (or the nerve-plexus associated with it) responds only to the electromagnetism of one planet, a message which is strongest when that planet is at its zenith relative to this world and gradually dwindles as it approaches and sinks below the horizon. Since the heavens have a unique configuration at the moment of fertilization, the influences of the planets on the glands follow a distinctive mathematical progression based on their orbits. The invisible ''long body'' they would spin in the heavens if they trailed luminescent threads behind them they in fact weave in each organism through the glands. Here embryogenesis explicitly reflects cosmic movement, or, esoterically, the planets are ruled by the evolving creatures of their worlds.

The symbolic relationship between the heart and the Sun (between the bloodstream and light itself) has long been discerned. One force operates in two mediums: warming and nourishing the remotest planets and every organ to the periphery of the body. One force synchronizes cosmic and microcosmic metabolism. Just as the planets diffuse the material of the Sun back into the Solar System—not only as primary radiation but plurally back and forth between orbs—so do the glands and organs refract each other's modulations back and forth through the bloodstream. The Earth recovers its reflected sunlight in decreasing microdoses from the Moon and other planets, and every organ is subject to the altered polypeptides of the enzymes it disperses. One universal force carries oxygen, hydrogen, nitrogen, and organic molecules of carbon back even to the Sun itself. The conviction of such a confluence of bloodstreams lies at the core of contemporary astrological thought.

Depending on which planet is strongest, first at the moment of conception and then at the moment of birth, different types of personalities will arise. The first three glands are endodermal: the thymus affecting growth leads to the solar personality; the pancreas affecting digestion and assimilation is lunar; and the thyroid governing respiration is mercurial. The next circuit of the helix is mesodermal: the parathyroid with its role in blood circulation is Venusian; the adrenals which infuse the cerebrospinal and voluntary muscle systems are martial; and the posterior pituitary which influences sensation and the sympathetic muscles is jovial. The third spiral is ectodermal: the anterior pituitary awakening the mind is saturnine; the gonads *(eros)* are Uranian; and the Neptunian pineal organ has an unknown function associated with galactic energy faster than light.

Like any concretized system of occult correspondences, Collin's astro-embryogenesis has its drawbacks and virtues. It biggest drawback is that it is arbitrary, artificially wedged between the Solar System and the body, and thus may have no physical relevance at all. Its virtue is that it creates images of our internal development and correspondence to the universe. Like all spiritual systems, it suggests that the subjective experience of an invisible (and quite possibly spurious) event, if accurate to some aspect of our hidden nature, points us through a glass darkly to the unknowable truth.

The second third of human life, Collin says, begins with birth and lasts 2800 days, or to the seventh birthday. This is the time of childhood when the organism comprehends its individual existence, learns how to respond to its habitat, and acquires language. Once a child speaks he enters also a dialogue with himself. He awakens from the shadows, whether they be bliss or terror, and his desire must become human. This initiation marks the completion of the second ''embryogenesis.'' Without language a creature is wild and free, as unborn, in a sense, and as pagan as the foetus. Writing has served a similar function in the development of civilization: Tribal peoples often fight to remain illiterate because once they read and write they can be counted and taxed.

The last third of human life stretches from 2800 to 28,080 days, when for a brief moment the planets sit in the rough configuration they had when the sperm penetrated the ovum. Saturn-return in the prime of life has been widely publicized as the astrological counterpoint to the midlife crisis, but in fact all the planets return, some of them once, most of them many times. As we may have intuited when we looked back in sadness through the clarity of mind into our ''lost'' years, the whole of adult life is no more than seven years of childhood or 280 days in the womb.

"Thus between conception and death man's life moves faster and faster until at the end the hours and minutes pass for him a thousand times faster than they did in the hours of his conception. This means that less and less happens to him in each hour as life progresses. His perception spreads over a longer and longer period, but in fact this longer period is only an illusion since it may contain no more than did the infinitesimal fraction of a second of his first sensation.

"He thinks to tame time by measuring its passage in years, but time cheats him by putting less and less into them. So that when he looks back over his life and tries to calculate it by the scale of birthdays, he is in a strange way foreshortening his existence, like a man looking at a picture which elusively curves away from him. In another figure, we can say that man falls through time as solid objects fall through air—that is, gaining momentum, or passing faster and faster through the medium as he goes."[25]

What then, does life amount to? The throbbing of a cardiac muscle? The rolling layers of tissues and nerves? The primal breath of blood and air? The radiance of the first light in the brain? We live long past the psychic wholeness of the beginning, but we can never add to its luster. The sky of childhood is still blue, the apples are red, and the flowers in the faded field are yellow. In the end life will vanish into the darkness that apparently persists past the end of time. But these too are just words.

Kushi says that one ontogenesis simply leads to another. Just as the foetus was born from water into air, so we will pass from this body into the world of vibration. From the uterine womb we enter the womb of the Earth, and from this transforming matrix we pass into another dimension, and then another, until we reach our nature. The old man, the old woman are really babies, about to be born—their wrinkled flesh but a husk, a veil. Each process of embryogenesis is to equip us for a mode of existence unimaginable before it occurs. Such apotheoses are easily proposed but not necessarily reassuring. Our mortality startles us again and again. After all, we give everything to this condition—our memory, our plans, our identity, our body, our mind, our children. What is left if these are taken away? And yet we continue to eat, and take care of children, and make love, and study and fight to the death for what we think we believe in. But death has no ideology. Its fullness is the same as its emptiness.

After we are no longer alive, all of eternity will happen without us.

We wake suddenly in the middle of the night, and perceive infinite time exploding beyond us; we stare through the dark room at a universe that has existed for trillions of years before we were conceived.

So macabre, so astonishing, so unlikely! In the context of all this our life seems a worthless thing, a trick. But who is there to trick?

Why go through all that trouble to put us here for so short a time? Why equip creatures with the *dharma* on a flight so skewed and swift?

"The great message of the universe is not that you survive," Da Free John told his disciples. "It is that you are awakened into a process in which nothing ultimately survives."[26] But it is not, either, that everything is destroyed. Nothing, in fact, is.

"Yes—it is true that we are not smothered, ended, murdered. The universe is not a machine of death."[27]

But can such knowledge help us? If a great teacher shows us, in a moment of vision, that we are immortal, does it change our situation? He replaces our few remaining seconds with billions upon billions of years of spiritual transit and unfolding, but nothing changes. The life we are living, in its ordinary sense, is what we are. The universe will never tell us anything definitive about our fate. It will continue to contradict itself and foil any grasping after certainty. We can survive only as what we are, and what we are must continuously and ultimately change.

When people seem to recall prior incarnations as Egyptian priestesses, crusading knights, and other romantic characters, they are responding to something inside themselves, though it is unlikely they were ever those people. Intimations of black holes within will seize any colorful imagery to concretize and assuage themselves. It is doubtful we could even recite "Three Blind Mice" again in an existence beyond this one.

We may have lived other lives before, but they will not be found locked away in our brain like a series of Gothic novels. They are obscure in the way this life is, when it is done. Our memories from beyond birth, if they exist at all, are like tiny crevices in thought, hollow moments that pass in a fraction of a second, for they are condensed, each one of them, to the size of an atom, expanding into lifetimes only beneath the surface of what is falsely called the unconscious but is nothing more than another archetype set to mark the boundary of existence.

The embryo teaches us how we are mortal. By the time we are born, just about everything of importance has been done. We are well on our way to death. The embryo teaches us that we are in a process of change that cannot be stopped. Everything in the universe is becoming, enveloping structure in new structure. The embryo teaches us that we are incarnating through substance—that our mind *is* substance, and a mind-like intelligence shapes it, whether that be in the genes, morphogenetic fields, or an etheric body. What we think we are is what we *are* thinks.

Much optimistic spiritual jargon is so much whistling in the dark. If a spiritual bigness surrounds us and floods down through our becoming, it would not necessarily appear as an image of rebirth or life everlasting. It might not even be sweet.

We might well experience it first as the surety of our mortality, as the nightmare into which we awake. When we are most nihilistic, most sure of the finality of death and the meaninglessness of life, we may be closest to a deathless psyche. That would be the *shadow* cast by the soul as it passes into the resistance of this world, i.e., consciousness itself.

Through the sheer terror of becoming we experience the first intuition of spirit. Our mortality experienced may be the basis of any immortality we have, and this takes precedent over any tales of heaven or, for that matter, of hell. All this discussion of spiritual embryogenesis becomes a flutter of sanctimonious fancies when we have to turn inward to the darkness and to our own nightmares for the birth of anything real. But then this is where we turned in the first place in order to become. Only from the spontaneity of our hearts or the difficulty of spiritual exercise do astral bodies and vibrations make any sense, and then only because we connect their lyrics to the joy of simply being. Certainly there is no way out of cosmogenesis, whatever it is, and there is no way out of the body. And it may finally be true that we will see our life only in the darkness between lives. Otherwise the lights are too bright.

Notes

The information in this book has been assembled from a diversity of sources, all of which are synthesized in the running narrative. Specific citation would be futile and would make little sense in a nonacademic work, so only direct quotations are documented. General sources for each chapter are listed in the approximate order of importance. The following code is used to distinguish sources from each other: double asterisks ** indicate a major source used throughout a whole chapter or extensively in a large part of it; a single asterisk * indicates a major source for a section within a chapter; and a title listed without an asterisk represents a source used for a small amount of information in a chapter. Location for all North Atlantic Books publications are cited as Berkeley (California), the current address, no matter where they were originally published.

Chapter 1.   Introduction to Embryogenesis

Murchie, Guy. *The Seven Mysteries of Life: An Exploration in Science and Philosophy.* Boston: Houghton Mifflin Company, 1978.

Balinsky, B. I. *An Introduction to Embryology.* 5th ed. Philadelphia: Saunders College Publishing, 1981.

1.   Frederick Gowland Hopkins; quoted in Donna Jeanne Haraway, *Crystals, Fabrics, and Fields: Metaphors of Organicism in Twentieth-Century Developmental Biology* (New Haven: Yale University Press, 1976), 102 f.n.
2.   John Keats, "Ode on a Grecian Urn"(1819), John Keats, *Selected Poems and Letters,* ed. Douglas Bush (Boston: Houghton Mifflin Company, 1959), 208.
3.   John Keats, "Ode to a Nightingale" (1819), Keats, *Selected Poems,* 208.

Chapter 2. The Original Earth

Hanawalt, Philip C., and Robert H. Haynes, eds. *The Chemical Basis of Life: An Introduction to Molecular and Cell Biology.* San Francisco: W. H. Freeman and Co./*Scientific American,* 1973.*

Bernal, J. D. *The Origin of Life.* Cleveland: World Publishing Co., 1967.*

Gibor, Aharon, ed. *Conditions for Life.* San Francisco: W. H. Freeman and Co./*Scientific American,* 1976.*

Wendt, Herbert. *The Sex Life of the Animals.* Translated from the German by Richard and Clara Winston. New York: Simon and Schuster, 1965.*

Crick, Francis. *Life Itself: Its Origin and Nature.* New York: Simon and Schuster, 1981.*

Shklovskii, I. S. and Carl Sagan. *Intelligent Life in the Universe.* Shklovskii's part translated from the Russian by Paula Fern. New York: Dell Publishing Co., 1966.*

Gilluly, James, A. C. Waters, and A. O. Woodford. *Principles of Geology,* 2d ed. San Francisco: W. H. Freeman and Co., 1959.

Jantsch, Erich. *The Self-Organizing Universe.* New York: Pergamon Press, 1980.

Olson, Charles. *The Maximus Poems [1950–1970].* Berkeley: University of California Press, 1983.

Duchesne-Guillemin, Jacques. *Zoroastrianism: Symbols and Values.* New York: Harper and Row, 1966.

Griaule, Marcel. *Conversations with Ogotemmêli: An Introduction to Dogon Religious Ideas.* Translated from the French by Ralph Butler, Audrey I. Richards, and Beatrice Hooke. London: Oxford University Press, 1965.

1. Aristotle, *History of Animals* (4th century B.C.) quoted in Wendt, *Sex Life of Animals,* 16–17.

2. A. I. Oparin, "The Origin of Life" in the appendix to Bernal, *Origin of Life*, 201.

3. Ibid., 203.

4. Crick, *Life Itself,* 141 ff.

5. Oparin, *Origin of Life.*

6. J. B. S. Haldane, "The Origin of Life" in the appendix to Bernal, *The Origin of Life*, 246.

7. Charles Darwin, quoted in Bernal, *Origin of Life,* 21.

Chapter 3. The Materials of Life

Hanawalt, Philip C., Robert H. Haynes, eds. *The Chemical Basis of Life: An Introduction to Molecular and Cell Biology.* San Francisco: W. H. Freeman and Co./*Scientific American*, 1973.*

Bernal, J. D. *The Origin of Life.* Cleveland: World Publishing Company, 1967.*

Gibor, Aharon, ed. *Conditions for Life.* San Francisco: W. H. Freeman and Co./*Scientific American*, 1976.*

Murchie, Guy. *The Seven Mysteries of Life: An Exploration in Science and Philosophy.* Boston: Houghton Mifflin Company, 1978.*

Bennett, J. G. *Gurdjieff: Making a New World.* New York: Harper and Row, 1973.

Dixon, Dougal. *After Man: A Zoology of the Future.* New York: St. Martin's Press, 1981.

1. Bernal, *Origin of Life,* 167–68.

2. Robert Kelly, "Staccato for Tarots"(1962), Robert Kelly, *The Alchemist to Mercury: An Alternate Opus, Uncollected Poems (1960–1980),* ed. Jed Rasula (Berkeley: North Atlantic Books, 1981), 29.

3. Bernal, *Origin of Life,* 8.

4. Francis Crick, *Life Itself: Its Origin and Nature* (New York: Simon and Schuster, 1965), 87.

Chapter 4.   The First Beings

Bernal, J. D. *The Origin of Life*. Cleveland: World Publishing Co., 1967.**

Crick, Francis. *Life Itself: Its Origin and Nature*. New York: Simon and Schuster, 1981.*

Hanawalt, Philip C., and Robert H. Haynes, eds. *The Chemical Basis of Life: An Introduction to Molecular and Cell Biology*. San Francisco: W. H. Freeman and Co./*Scientific American,* 1973.*

Thomas, Lewis. *The Lives of a Cell: Notes of a Biology Watcher*. New York: Viking Press, 1974.*

Murchie, Guy. *The Seven Mysteries of Life: An Exploration in Science and Philosophy*. Boston: Houghton Mifflin Co., 1978.*

Berrill, N. J. and Gerald Karp. *Development*. New York: McGraw-Hill Book Co., 1976.

Stanley, Wendell M., and Evans G. Valens. *Viruses and the Nature of Life*. New York: E. P. Dutton, 1961.

Shklovskii, I. S., and Carl Sagan. *Intelligent Life in the Universe*. Shklovskii's part translated from the Russian by Paula Fern. New York: Dell Publishing Co., 1966.

Haraway, Donna Jeanne. *Crystals, Fabrics, and Fields: Metaphors of Organicism in Twentieth-Century Developmental Biology*. New Haven: Yale University Press, 1976.

1.   Crick, *Life Itself,* 103.
2.   Bernal, *Origin of Life,* 146.

Chapter 5. The Cell

Wendt, Herbert. *The Sex Life of Animals*. Translated from the German by Richard and Clara Winston. New York: Simon and Schuster, 1965.*

Russell-Hunter, W. D. *A Life of Invertebrates*. New York: Macmillan Publishing Co., 1979.*

Berrill, N. J., and Gerald Karp. *Development*. New York: McGraw-Hill Book Co., 1976.*

Balinsky, B. I. *An Introduction to Embryology*, 5th ed. Philadelphia: Saunders College Publishing, 1981.*

Thompson, D'Arcy. *On Growth and Form* [1917]. London: Cambridge University Press, 1966.

Murchie, Guy. *The Seven Mysteries of Life: An Exploration in Science and Philosophy*. Boston: Houghton Mifflin Co., 1978.

Thomas, Lewis. *The Lives of a Cell: Notes of a Biology Watcher*. New York: Viking Press, 1974.

1. Anton van Leeuwenhoek; quoted in Murchie, *Seven Mysteries of Life*, 82.
2. Ibid., 83.
3. Matthias Jakob Schleiden; quoted in Wendt, *Sex Life of Animals*, 27.
4. Donna Jeanne Haraway, *Crystals, Fabrics, and Fields: Metaphors of Organicism in Twentieth-Century Developmental Biology* (New Haven: Yale University Press, 1976), 20.

Chapter 6. The Genetic Code

Crick, Francis. *Life Itself: Its Origin and Nature*. New York: Simon and Schuster, 1981.*

Balinsky, B. I. *An Introduction to Embryology,* 5th ed. Philadelphia: Saunders College Publishing, 1981.*

Portugal, Franklin H. and Jack S. Cohen. *A Century of DNA: A History of the Discovery of the Structure and Function of the Genetic Substance.* Cambridge: MIT Press, 1977.*

Berrill, N. J. and Gerald Karp. *Development.* New York: McGraw-Hill Book Co., 1976.

1. Harvey Bialy, "The I Ching and the Genetic Code,"*Ecology and Consciousness,* ed. Richard Grossinger (Berkeley: North Atlantic Books, 1978).
2. Crick, *Life Itself,* 66.
3. John Todd, "An Interview Conducted by Richard Grossinger and Lindy Hough," November 1982 (this section did not appear in the published version of the interview in *Omni,* New York, August 1984).

Chapter 7. Sperm and Egg

Berrill, N. J., and Gerald Karp. *Development.* New York: McGraw-Hill Book Co., 1976.*

Balinsky, B. I. *An Introduction to Embryology,* 5th ed. Philadelphia: Saunders College Publishing, 1981.*

Wendt, Herbert. *The Sex Life of Animals.* Translated from the German by Richard and Clara Winston. New York: Simon and Schuster, 1965.*

Glass, Bentley, Owsei Temkin and William L. Straus, Jr. *Forerunners of Darwin: 1745–1859.* Baltimore: Johns Hopkins University Press, 1959.

Gould, Stephen J. *Ontogeny and Phylogeny.* Cambridge: Harvard University Press, 1977.

Murchie, Guy. *The Seven Mysteries of Life: An Exploration in Science and Philosophy.* Boston: Houghton Mifflin Co., 1978.

1. Sir Charles Sherrington, *Man on His Nature* (London: Cambridge University Press, 1963), 95.

2. Andrew Marvell, "To His Coy Mistress," *Seventeenth Century Poetry: The Schools of Donne and Jonson,* ed. Hugh Kenner (New York: Holt, Rinehart and Winston, 1964), 457.

3. Da Free John, "The Mystery of the Spermatic Being," *The Laughing Man,* Vol. 3, No. 1 (Clearlake, California, 1982), 15.

4. Ibid.

5. Marvell, "To His Coy Mistress," Kenner, *Seventeenth Century Poetry,* 458.

6. Frank Herbert, *Dune Messiah* (New York: Berkley, 1969); *God Emperor of Dune* (New York: Berkley, 1981).

7. Anton van Leeuwenhoek, quoted in Wendt, *Sex Life of Animals,* 57.

8. Immanuel Kant, quoted in Glass, Temkin, and Straus, Jr., *Forerunners of Darwin,* 186.

9. Louis Moreau de Maupertuis, quoted in Glass, Temkin, and Straus, Jr., *Forerunners of Darwin,* 68.

10. Ross Harrison, quoted in Donna Jeanne Haraway, *Crystals, Fabrics, and Fields: Metaphors of Organicism in Twentieth-Century Developmental Biology* (New Haven: Yale University Press), 44.

Chapter 8. Fertilization

Wendt, Herbert. *The Sex Life of Animals.* Translated from the German by Richard and Clara Winston. New York: Simon and Schuster, 1965.**

Berrill, N. J., and Gerald Karp. *Development.* New York: McGraw-Hill Book Co., 1976.*

Balinsky, B. I. *An Introduction to Embryology,* 5th ed. Philadelphia: Saunders College Publishing, 1981.*

Ferenczi, Sandor. *Thalassa: A Theory of Genitality.* Translated from the German by Henry Alden Bunker. New York: W. W. Norton and Co., 1968.*

Neumann, Erich. *The Great Mother: An Analysis of the Archetype.* Translated from the German by Ralph Manheim. Princeton: Princeton University Press, 1963.*

Russell-Hunter, W. D. *A Life of Invertebrates*. New York: Macmillan Co., 1979.

Jantsch, Erich. *The Self-Organizing Universe*. New York: Pergamon Press, 1980.

1. Oscar Hertwig, quoted in Wendt, *Sex Life of Animals,* 69.
2. Balinsky, *Introduction to Embryology,* 338.
3. J. H. C. Fabre, *The Life and Love of the Insect* (London: A. C. Black, 1918) quoted in Robert S. de Ropp, *Sex Energy: The Sexual Force in Man and Animals* (New York: Dell Publishing Co., 1969), 40–41.
4. Ibid., 41–2.
5. Da Free John, "The Mystery of the Spermatic Being," *The Laughing Man,* Vol. 3, No. 1 (Clearlake, California, 1982), 15.
6. Wilhelm Reich, *Ether, God and Devil—Cosmic Superimposition,* translated from the German by Therese Pol (New York: Farrar, Straus and Giroux, 1972).
7. Arthur A. Tansley, "The Deadly Ninja: Agents of Death," in *Fighting Arts Magazine,* Vol. 5, No. 3 (Liverpool, England, 1983), 19–20.
8. Heinrich Zimmer, "The Indian World Mother," in *The Mystic Vision* [Papers from the Eranos Yearbooks, 6] (Princeton: Princeton University Press, 1968), 74.
9. Géza Róheim, *The Riddle of the Sphinx, or Human Origins,* translated from the German by R. Money-Kyrle (New York: Harper and Row, 1974), 271.

Chapter 9.   The Blastula

Balinsky, B. I. *An Introduction to Embryology,* 5th ed. Philadelphia: Saunders College Publishing, 1981.**

Berrill, N. J., and Gerald Karp. *Development*. New York: McGraw-Hill Book Co., 1976.**

Haraway, Donna Jeanne. *Crystals, Fabrics, and Fields: Metaphors of Organicism in Twentieth-Century Developmental Biology*. New Haven: Yale University Press, 1976.*

The Yellow Emperor's Classic of Internal Medicine. Translated from the Chinese by Ilza Veith. Berkeley: University of California Press, 1972.

1. Michio Kushi, *The Book of Do-IN: Exercise for Physical and Spiritual Development* (Tokyo: Japan Publications, 1979), 27.
2. William Faulkner, *Absalom, Absalom!* (New York: Random House, 1936), 142.
3. Ibid., 143.

Chapter 10. Gastrulation

Balinsky, B. I. *An Introduction to Embryology,* 5th ed. Philadelphia: Saunders College Publishing, 1981.**

Berrill, N. J., and Gerald Karp. *Development.* New York: McGraw-Hill Book Co., 1976.**

Saunders, John W., Jr. *Patterns and Principles of Animal Development.* New York: Macmillan Co. 1970.*

Moore, Keith L. *The Developing Human: Clinically Oriented Embryology,* 2d ed. Philadelphia: Saunders College Publishing, 1977.*

Russell-Hunter, W. D. *A Life of Invertebrates.* New York: Macmillan Co., 1979.*

1. Carl Sagan, *Cosmos* (New York: Random House, 1980).
2. Joseph Allen and Thomas O'Toole, "Thoughts of an Astronaut," *Washington Post* Syndicate (Washington, D.C., 1983).
3. Janet Frame, *Living in the Maniototo* (Auckland, New Zealand: The Women's Press, 1979), 117–18.

Chapter 11. Morphogenesis

Berrill, N. J., and Gerald Karp. *Development*. New York: McGraw-Hill Book Co., 1976.**

Balinsky, B. I. *An Introduction to Embryology,* 5th ed. (Philadelphia: Saunders College Publishing, 1981.**

Saunders, John W., Jr. *Patterns and Principles of Animal Development,* 2d ed. New York: Macmillan Co., 1970.*

Davidson, Eric H. *Gene Activity in Early Development,* 2d ed. New York: Academic Press, 1976.*

Thompson, D'Arcy. *On Growth and Form* [1917]. Cambridge: Cambridge University Press, 1966.*

Haraway, Donna Jeanne. *Crystals, Fabrics, and Fields: Metaphors of Organicism in Twentieth-Century Developmental Biology.* New Haven: Yale University Press, 1976.*

Russell-Hunter, W. D. *A Life of Invertebrates.* New York: Macmillan Co., 1979.*

1.   William Blake, "The Tyger" (1794) *The Portable Blake,* ed. Alfred Kazin (New York: Viking Press, 1946), 109.
2.   Berrill and Karp, *Development,* 1.
3.   Sven Horstadius, *Experimental Embryology of the Echinoderms* (Oxford: Clarendon Press, 1973), 1.
4.   Thompson, *On Growth and Form,* 85–86.
5.   Ross Harrison, quoted in Haraway, *Crystals, Fabrics, and Fields,* 99.
6.   Thompson, *On Growth and Form,* 172.
7.   Stephen Black, "Determination of the Dorsal-Ventral Axis in *Xenopus laevis* Embryos," Ph.D. thesis (Berkeley: University of California).
8.   George Oster, seminar given at University of California at Berkeley, January 26, 1982.

Chapter 12. Biological Fields

Haraway, Donna Jeanne. *Crystals, Fabrics, and Fields: Metaphors of Organicism in Twentieth-Century Developmental Biology.* New Haven: Yale University Press, 1976.**

Berrill, N. J., and Gerald Karp. *Development.* New York: McGraw-Hill Book Co., 1976.

Balinsky, B. I. *An Introduction to Embryology,* 5th ed. Philadelphia: Saunders College Publishing, 1981.

1. Paul Weiss, quoted in Haraway, *Crystals, Fabrics, and Fields,* 186.
2. Sir Charles Sherrington, *Man on His Nature* (Cambridge: Cambridge University Press, 1963), 107.
3. Ross Harrison, quoted in Haraway, *Crystals, Fabrics, and Fields,* 74.
4. Berrill and Karp, *Development,* 281.
5. Paul Weiss, quoted in Haraway, *Crystals, Fabrics, and Fields,* 177.
6. Ross Harrison, quoted in Haraway, *Crystals, Fabrics, and Fields,* 100.
7. Ibid., 88.
8. Ibid., 87.
9. Joseph Needham, quoted in Haraway, *Crystals, Fabrics, and Fields,* 113.
10. Haraway, *Crystals, Fabrics, and Fields,* 185.
11. Paul Weiss, quoted in Haraway, *Crystals, Fabrics, and Fields,* 185.
12 Haraway, *Crystals, Fabrics, and Fields,* 50.
13. Joseph Needham, quoted in Haraway, *Crystals, Fabrics, and Fields,* 136.
14. Paul Weiss, quoted in Haraway, *Crystals, Fabrics, and Fields,* 173.
15. Ibid., 147.
16. Haraway, *Crystals, Fabrics, and Fields,* 59–61.
17. Rupert Sheldrake, *A New Science of Life* (Los Angeles: J. P. Tarcher, 1982).
18. Ibid., book jacket.
19. Plotinus, *The Six Enneads* (ca. 263 A.D.), translated from the

Latin by Stephen MacKenna (Chicago: *Encyclopedia Britannica,* 1952), 117.

20. Michael McClure, "Wolf Net," in *Ecology and Consciousness,* ed. Richard Grossinger (Berkeley: North Atlantic Books, 1978), 206.

Chapter 13. The Origin of the Nervous System

Bullock, T. H., and G. A. Horridge. *Structure and Function in the Nervous System of Invertebrates.* San Francisco: W. H. Freeman and Co., 1965.**

Russell-Hunter, W. D. *A Life of Invertebrates.* New York: Macmillan Co., 1979.*

Woodburne, Lloyd S. *The Neural Basis of Behavior.* Columbus, Ohio: Charles E. Merrill Books, 1967.

Rose, Steven. *The Conscious Brain.* New York: Alfred A. Knopf, 1974.

Murchie, Guy. *The Seven Mysteries of Life: An Exploration in Science and Philosophy.* Boston: Houghton Mifflin Co., 1978.

Sherrington, Sir Charles. *Man on His Nature.* London: Cambridge University Press, 1963.

Berrill, N. J., and Gerald Karp. *Development.* New York: McGraw-Hill Book Co., 1976.

Chomsky, Noam. *Aspects of the Theory of Syntax.* Cambridge: MIT Press, 1965.

1. Stanley Keleman, "Professional Colloquium: 29 October 1977," in *Ecology and Consciousness* ed. Richard Grossinger (Berkeley: North Atlantic Books, 1978), 17.
2. Ibid.

Chapter 14. The Evolution of Intelligence

Bullock, T.H., and G.A. Horridge. *Structure and Function in the Nervous System of Invertebrates.* San Francisco: W. H. Freeman and Co., 1965.**

Russell-Hunter, W. D. *A Life of Invertebrates.* New York: Macmillan Co., 1979.*

Friedlander, C. P. *The Biology of Insects.* New York: Pica Press, 1977.*

Alexander, R. McNeill. *The Chordates.* Cambridge: Cambridge University Press, 1975.*

Rose, Steven. *The Conscious Brain.* New York: Alfred A. Knopf, 1974.*

Wendt, Herbert. *The Sex Life of Animals.* Translated from the German by Richard and Clara Winston. New York: Simon and Schuster, 1965.

Smith, Eric, Garth Chapman, R. B. Clar, David Nichols, and J. D. McCarthy. *The Invertebrate Panorama.* New York: Universe Books, 1971.

Woodburne, Lloyd S. *The Neural Basis of Behavior.* Columbus, Ohio: Charles E. Merrill Books, 1967.

Buschsbaum, Ralph. *Animals Without Backbones,* 2d ed. Chicago: University of Chicago Press, 1948.

Borror, Donald J., Dwight M. DeLong, and Charles A. Triplehorn. *An Introduction to the Study of Insects.* Philadelphia: Saunders College Publishing, 1981.

1. Frank Herbert, *Dune* (New York: Berkley, 1965).
2. Maurice Maeterlinck, *The Life of the Bee* (New York: Dodd, Mead, and Co., 1936), quoted in Wendt, *Sex Life of Animals,* 179–80.
3. Ibid., 180.

4. Eugène Marais, *The Soul of the White Ant* (London: Methuen and Co., 1937).

5. Russell-Hunter, *Life of Invertebrates,* 453.

Chapter 15. Neurulation and the Human Brain

Woodburne, Lloyd S. *The Neural Basis of Behavior.* Columbus, Ohio: Charles E. Merrill Books, 1967.**

Moore, Keith L. *The Developing Human: Clinically Oriented Embryology.* Philadelphia: Saunders College Publishing, 1977.*

Rose, Steven. *The Conscious Brain.* New York: Alfred A. Knopf, 1974.*

Restak, Richard M. *The Brain: The Last Frontier.* New York: Warner Books, 1979.*

Berrill, N. J., and Gerald Karp. *Development.* New York: McGraw-Hill Book Co., 1976.*

Balinsky, B. I. *An Introduction to Embryology,* 5th ed. Philadelphia: Saunders College Publishing, 1981.*

Alexander, R. McNeill. *The Chordates.* London: Cambridge University Press, 1975.

Russell-Hunter, W. D. *A Life of Invertebrates.* New York: Macmillan Co., 1979.

Freud, Sigmund. *The Interpretation of Dreams.* Translated from the German by James Strachey. New York: Basic Books, 1955.

1. Lord Byron [George Gordon], "The Dream" (1816) in *Selected Poetry and Letters,* ed. Edward E. Bostetter (New York: Rinehart and Co., 1958), 25.

2. Paul MacLean, quoted in Restak, *The Brain,* 36.

3. Ibid., 41.

4. Ibid., 52.

5. Andrew Weil, *The Marriage of the Sun and Moon: A Quest for Unity in Consciousness* (Boston: Houghton Mifflin Co., 1980), 257.

6. Sir Charles Sherrington, *Man on His Nature* (London: Cambridge University Press, 1963), 105.

Chapter 16. Mind

Montagu, M. F. Ashley, ed. *Culture and the Evolution of Man.* New York: Oxford University Press, 1962.** The following essays from this book were the main sources used: Oakley, Kenneth P. "A Definition of Man"; Washburn, Sherwood L. "Tools and Human Evolution"; White, Leslie A. "The Concept of Culture"; Haldane, J. B. S. "The Argument from Animals to Men: An Examination of Its Validity for Anthropology"; Etkin, William. "Social Behavior and the Evolution of Man's Mental Faculties"; Dobzhansky, Theodosius, and Montagu, M. F. Ashley. "Natural Selection and the Mental Capacities of Mankind"; Hallowell, A. Irving. "The Structural and Functional Dimensions of a Human Existence" and "Personality Structure and the Evolution of Man"; Montagu, M. F. Ashley. "Time, Morphology, and Neoteny in the Evolution of Man"; and, Brace, C. Loring. "Cultural Factors in the Evolution of the Human Dentition."

Spuhler, J. N., ed. *The Evolution of Man's Capacity for Culture.* Detroit: Wayne State University Press, 1965.** The main source in this book was Spuhler's own essay "Somatic Paths to Culture."

Le Gros Clark, W. E. *The Antecedents of Man: An Introduction to the Evolution of the Primates.* New York: Harper and Row, 1963.**

Kelso, A. J. *Physical Anthropology.* Philadelphia: J. B. Lippincott Co., 1970.*

Moore, Keith L. *The Developing Human: Clinically Oriented Embryology.* Philadelphia: Saunders College Publishing, 1977.*

Campbell, Bernard. *Human Evolution: An Introduction to Man's Adaptations.* Chicago: Aldine Publishing Co., 1966.*

Romer, A. S. *Man and the Vertebrates*. Baltimore: Penguin Books, 1954.*

Berrill, N. J. and Gerald Karp. *Development*. New York: McGraw-Hill Book Co., 1976.*

Freud, Sigmund. *An Outline of Psychoanalysis*. Translated by James Strachey. New York: W. W. Norton and Co., 1949.

Reich, Wilhelm. *The Function of the Orgasm*. Translated from the German by Vincent R. Carfagno. New York: Farrar, Straus and Giroux, 1973.

Jung, C. G. *The Archetypes and the Collective Unconscious*. Translated from the German by R. F. C. Hull. New York: Pantheon Books, 1959.

Marshack, Alexander. *The Roots of Civilization: The Cognitive Beginnings of Man's First Art, Symbol and Notation*. New York: McGraw-Hill Book Co., 1972.

de Santillana Giorgio, and Hertha von Dechend. *Hamlet's Mill: An Essay on Myth and the Frame of Time*. Boston: Gambit, 1969.

Stanner, W. E. H. *The Dreaming*. Indianapolis: Bobbs-Merrill Reprint Series in the Social Sciences, A-214, 1956.

Wilson, Robert Anton. *Cosmic Trigger*. Berkeley: And/Or Press, 1977.

Clarke, Arthur C. *2001: A Space Odyssey*. New York: New American Library, 1968.

Anderson, Edgar. *Plants, Man and Life*. Berkeley: University of California Press, 1967.

1. Erich Neumann, *The Great Mother,* translated from the German by Ralph Manheim (Princeton: Princeton University Press, 1963), 12–13.
2. Claude Lévi-Strauss, *The Elementary Structures of Kinship,*

translated from the French by James Harle Bell, John Richard von Sturmer, and Rodney Needham (Boston: Beacon Press, 1969).
   3.   James Joyce, *Ulysses* (New York: Random House, 1934).

Chapter 17.   Birth

1.   David B. Chamberlain, *Consciousness at Birth: A Review of the Empirical Evidence* (San Diego, California: Chamberlain Communications, 1983), 12.
   2.   Ibid., 9.
   3.   Ibid., 12.
   4.   Ibid., 35.
   5.   Ibid., 37–39.
   6.   Ibid., 4.
   7.   Ibid., 5–6.
   8.   Jeannine Parvati, "Notes on Grossinger's *Embryogenesis*," unpublished manuscript, 1984.
   9.   Jeannine Parvati, "Prenatal Care Guidelines," unpublished manuscript, 1982, 14.
   10.   Ibid., p. 13.
   11.   Jeannine Parvati, "Psyche's Midwife," unpublished manuscript, 1984, 1–2.

Chapter 18.   Soma

Moore, Keith L. *The Developing Human: Clinically Oriented Embryology.* Philadelphia: Saunders College Publishing, 1977.**

Balinsky, B. I. *An Introduction to Embryology,* 5th ed. Philadelphia: Saunders College Publishing, 1981.**

Berrill, N. J., and Gerald Karp. *Development.* New York: McGraw-Hill Book Co., 1976.**

Freud, Sigmund. *An Outline of Psychoanalysis.* Translated by James Strachey. New York: W. W. Norton and Co., 1949.

Schwenk, Theodor. *Sensitive Chaos.* London: Rudolf Steiner Press, 1965.

  1. Stanley Keleman, "Professional Colloquium: 29 October 1977," in *Ecology and Consciousness,* ed. Richard Grossinger (Berkeley: North Atlantic Books, 1978), 16, 18.
  2. Stan Brakhage, "Interview, 7 Jan. 72," in *Imago Mundi* (*Io*/14), ed. Richard Grossinger (Berkeley: North Atlantic Books, 1972), 362.
  3. Robert J. Sardello, "The Suffering Body of the City," in *Spring,* 1983 Annual Issue (Dallas, Texas), 153.
  4. Stanley Keleman, in Grossinger, *Ecology and Consciousness,* 23.

Chapter 19. Blood, Bones, and Immunity

Moore, Keith L. *The Developing Human: Clinically Oriented Embryology.* Philadelphia: Saunders College Publishing, 1977.**

Balinsky, B. I. *An Introduction to Embryology,* 5th ed. Philadelphia: Saunders College Publishing, 1981.**

Berrill, N. J., and Gerald Karp. *Development.* New York: McGraw-Hill Book Co., 1976.**

Matthews, L. Harrison. *The Life of Mammals,* Vol. 1. New York: Universe Books, 1969.*

Wintrobe, M. M. *Clinical Hematology.* Philadelphia: Lea and Febiger, 1981.*

Murchie, Guy. *The Seven Mysteries of Life: An Exploration in Science and Philosophy.* Boston: Houghton Mifflin Co., 1978.*

Alexander, R. McNeill. *The Chordates.* Cambridge: Cambridge University Press, 1975.*

Duddington, C. L. *Evolution and Design in the Plant Kingdom.* New York: Harper and Row, 1970.

1. Benjamin Lo, Martin Inn, Robert Amacker, and Susan Foe (translators), *The Essence of T'ai Chi Ch'uan: The Literary Tradition,* translated from the Chinese (Berkeley: North Atlantic Books: 1979), 95.
2. Murchie, *Seven Mysteries of Life.* 120–21.
3. Ibid., 121.

Chapter 20. Ontogeny and Phylogeny

Gould, Stephen Jay. *Ontogeny and Phylogeny.* Cambridge: Harvard University Press, 1977.**

Montagu, M. F. Ashley. "Time, Morphology, and Neoteny in the Evolution of Man," in *Culture and the Evolution of Man,* edited by M. F. Ashley Montagu. New York: Oxford University Press, 1962.*

Friedlander, C. P. *The Biology of Insects.* New York: Pica Press, 1977.*

Wendt, Herbert. *The Sex Life of Animals.* Translated from the German by Richard and Clara Winston. New York: Simon and Schuster, 1965.*

Gould, Stephen Jay. *The Panda's Thumb.* New York: W.W. Norton and Co., 1982.

1. Ernst Haeckel, quoted in Gould, *Ontogeny and Phylogeny,* 78.
2. Claude Lévi-Strauss, *The Raw and the Cooked: Introduction to a Science of Mythology,* Vol. I, translated from the French by John and Doreen Weightman (New York: Harper and Row, 1969).
3. W. E. H. Stanner, *The Dreaming* (Indianapolis: Bobbs-Merrill Reprint Series in the Social Sciences, A-214, first published 1956), 54–55.
4. Lorenz Oken, quoted in Gould, *Ontogeny and Phylogeny,* 45.
5. Claude Lévi-Strauss, *The Raw and the Cooked.*

6. Ernst Haeckel, quoted in Gould, *Ontogeny and Phylogeny,* 172.

7. Ibid., 82.

8. E. Mehnert, quoted in Gould, *Ontogeny and Phylogeny,* 175.

9. Gould, *Panda's Thumb,* 1.

10. Wilhelm Roux, quoted in Gould, *Ontogeny and Phylogeny,* 195.

11. Gould, *Ontogeny and Phylogeny,* 214.

12. Fritz Kahn, quoted in Wendt, *Sex Life of Animals,* 87.

13. Gould, *Panda's Thumb,* 35–37.

14. Charles Darwin, *On the Origin of Species by Means of Natural Selection, or Preservation of Favoured Races in the Struggle for Life* (London: Murray, 1859).

15. Karl Ernst von Baer, quoted in Gould, *Ontogeny and Phylogeny,* 56.

16. Julian Huxley, quoted in Gould, *Ontogeny and Phylogeny,* 267.

17. Richard Goldschmidt, quoted in Gould, *Panda's Thumb,* 192.

18. Aldous Huxley, *After Many a Summer Dies the Swan* (New York: Harper and Row, 1965), 238–40.

19. Gould, *Ontogeny and Phylogeny,* 383.

20. Louis Bolk, quoted in Gould, *Ontogeny and Phylogeny,* 361.

Chapter 21.   Self and Desire

Wendt, Herbert. *The Sex Life of Animals.* Translated from the German by Richard and Clara Winston. New York: Simon and Schuster, 1965.*

de Ropp, Robert S. *Sex Energy: The Sexual Force in Man and Animals.* New York: Dell Publishing Co., 1969.

Lacan, Jacques. *Écrits.* Translated from the French by Alan Sheridan. New York: W. W. Norton and Co., 1977.

Freud, Sigmund. *An Outline of Psychoanalysis.* Translated from the German by James Strachey. New York: W. W. Norton and Co., 1949.

Freud, Sigmund. *Civilization and Its Discontents.* Translated from the German by James Strachey. New York: W. W. Norton and Co., 1962.

Reich, Wilhelm. *The Murder of Christ: The Emotional Plague of Mankind.* New York: Farrar, Straus & Giroux, 1970.

Lederer, Laura, ed. *Take Back the Night: Women on Pornography.* Bantam Books: New York, 1982.

1. Robert Ardrey, *African Genesis: A Personal Investigation into the Animal Origins and Nature of Man* (New York: Dell Publishing Co., 1963).
2. G. I. Gurdjieff, *Views from the Real World: Early Talks of Gurdjieff* (New York: E. P. Dutton, 1975), 274.
3. Michael McClure, "Wolf Net," in *Ecology and Consciousness,* ed. Richard Grossinger (Berkeley: North Atlantic Books, 1978), 217; more complete version in *Biopoesis (Io/20),* ed. Harvey Bialy (Berkeley: North Atlantic Books, 1974), 144.
4. Marcel Mauss, *The Gift* (1924), translated by I. Cunnison (New York: Free Press, 1954), 11–12.
5. J. H. C. Fabre, *The Life and Love of the Insect* (London: A. C. Black, 1918), quoted in de Ropp, *Sex Energy,* 42–43.
6. Ibid., 43.
7. W. E. H. Stanner, *The Dreaming* (Indianapolis: Bobbs-Merrill Reprint Series in the Social Sciences, A-214, first published 1956), 55.
8. Will Baker, "Tsítsi, the Faithful," from "Three Monkeys, I am Father," in *Nuclear Strategy and the Code of the Warrior: Faces of Mars and Shiva in the Crisis of Human Survival (Io/#33),* ed. Richard Grossinger and Lindy Hough (Berkeley: North Atlantic Books, 1984), 230.
9. Ibid.
10. Géza Róheim, *The Riddle of the Sphinx, or Human Origins,* translated from the German by R. Money-Kyrle (New York: Harper and Row, 1974), 23–24.
11. Georg Wilhelm Steller, quoted in David Day, *The Doomsday Book of Animals: A Natural History of Vanished Species* (New York: Viking Press, 1981), 216–17.
12. Norman Mailer, *The Executioner's Song* (Boston: Little, Brown and Co., 1979), 306.
13. Ibid., 984.
14. Ibid., 305.
15. Ardrey, *African Genesis,* 318.

16. J. H. C. Fabre, quoted in de Ropp, *Sex Energy*, 43.

17. Marquis de Sade *(Juliette)*, quoted in de Ropp, *Sex Energy*, 43.

18. From notes taken while watching "Sixty Minutes" television show, 1982.

19. Fyodor Dostoyevsky, *The Brothers Karamazov*, translated from the Russian by Constance Garnett (New York: Random House, 1950), 306.

20. Ibid., 289.

21. Paul Tillich, *The Meaning of Health: The Relation Between Religion and Health* (Berkeley: North Atlantic Books, 1981), 15.

22. Hannah Tillich, *From Time to Time* (New York: Stein and Day, 1973).

Chapter 22. Spiritual Embryogenesis

Steiner, Rudolf. *An Outline of Occult Science,* 1909. Reprint. Translated from the German by Henry B. Maud and Lisa D. Monges. Spring Valley, New York: Anthroposophic Press, 1972.*

König, Karl. *Embryology and World Evolution.* Translated from the German by R. E. K. Meuss. In *British Homoeopathic Journal* 57 (1968):1–62.*

Poppelbaum, Hermann. *Man and Animal: Their Essential Difference.* Translated from the German. London: Anthroposophical Publishing Co., 1960.*

Schwenk, Theodor. *Sensitive Chaos: The Creation of Flowing Forms in Water and Air.* Translated from the German by Olive Whicher and Johanna Wrigley. London: Rudolf Steiner Press, 1965.

Haraway, Donna Jeanne. *Crystals, Fabrics, and Fields: Metaphors of Organicism in Twentieth-Century Developmental Biology.* New Haven: Yale University Press, 1976.

Hall, Manly P. *Man, Grand Symbol of the Mysteries: Thoughts in*

Occult Anatomy. Los Angeles: The Philosophical Research Society, 1972.

1. J. D. Bernal, *The Origin of Life* (Cleveland: World Publishing Company, 1967), 140–41.
2. Da Free John, *Easy Death* (Clearlake, California: The Dawn Horse Press, 1983), 88–89.
3. Elwyn Chamberlain, personal communication, 1983.
4. Charles Bonnet, quoted in Bentley Glass, Owsei Temkin, and William L. Straus, Jr., *Forerunners of Darwin: 1745–1859* (Baltimore: Johns Hopkins University Press, 1959), 204.
5. Poppelbaum, *Man and Animal,* 74.
6. Rudolf Steiner, *Outline of Occult Science,* xxxii.
7. Rudolf Steiner, quoted in Poppelbaum, *Man and Animal,* 106.
8. C. G. Jung, Seminar Report on *Nietzsche's Zarathustra,* X, P51F. (privately mimeographed), quoted in James Hillman "Senex and Puer: An Aspect of the Historical and Psychological Present (1967)," in *Puer Papers* (Irving, Texas: Spring Publications, 1979), 44.
9. Henrik Steffens (1822), quoted in Poppelbaum, *Man and Animal,* 85.
10. Rudolf Steiner (1904), quoted in Poppelbaum, *Man and Animal,* 136.
11. Poppelbaum, *Man and Animal,* 150–51.
12. Ibid., 149–50.
13. *The Vishnu Purana,* quoted in Hall, *Man,* 104.
14. Hall, *Man,* 103.
15. Jeannine Parvati, dream (1983) sent as part of "Notes on Grossinger's *Embryogenesis,*" unpublished manuscript, 1984 (she adds that "mugwort is Artemis' herbal ally").
16. Ibid.
17. William Blake, "The Marriage of Heaven and Hell," *The Portable Blake,* ed. Alfred Kazin (New York: Viking Press, 1946), 250.
18. Jeannine Parvati, "Prenatal Care Guidelines" unpublished manuscript, 1982, 2–3.
19. Ibid., 4.

Chapter 23. Cosmogenesis and Mortality

Kushi, Michio. *The Book of Do-IN: Exercises for Physical and Spiritual Development.* Tokyo: Japan Publications, 1979.*

Collin, Rodney. *The Theory of Celestial Influence.* London: Stuart and Watkins, 1958.*

Ouspensky, P. D. *In Search of the Miraculous: Fragments of an Unknown Teaching.* New York: Harcourt, Brace & World, 1949.*

1.   Theodor Schwenk, *Sensitive Chaos: The Creation of Flowing Forms in Water and Air,* translated from the German by Olive Whicher and Johanna Wrigley (London, England: Rudolf Steiner Press, 1965), 33–34.

2.   Herbert Guenther, personal communication, 1983. ("In an [unpublished] paper . . . I stated it in the following way—and here [see quote] I add the three Tibetan terms relevant to embryo formation.")

3.   Charles Poncé, *Papers Toward a Radical Metaphysics: Alchemy* (Berkeley: North Atlantic Books, 1984), 66.

4.   Ibid., 67.

5.   G. I. Gurdjieff, quoted in Ouspensky, *Search of the Miraculous,* 175.

6.   Ibid., 221.

7.   Pierre Teilhard de Chardin, *The Phenomenon of Man,* translated from the French by Bernard Wall (New York: Harper and Row, 1959), 71–72.

8.   Ibid.

9.   Ibid., 182.

10.   Ibid., 78.

11.   Kushi, *The Book of Do-In,* 27.

12.   Ibid., 28.

13.   Da Free John, "The Mystery of the Spermatic Being," *The Laughing Man,* Vol. 3, No. 1 (Clearlake, California, 1982), 13.

14.   Ibid.

15.   Manly P. Hall, *Man, Grand Symbol of the Mysteries: Thoughts in Occult Anatomy* (Los Angeles: The Philosophical Research Society, 1972), 84–85.

16.   Madame H. P. Blavatsky, quoted in Hall, *Man, Grand Symbol,* 85.

17.   Anne Tyler, *Dinner at the Homesick Restaurant* (New York: Berkley, 1983), 270.

18.   Ibid., 274.

19.   Ibid., 279.

20.   Ibid., 284.

21.   *Windwalker,* a motion picture, producers, Arthur R. Dubs and Thomas E. Ballard, director, Kieth Merrill.

22.   Norman Mailer, *The Executioner's Song* (Boston: Little, Brown and Co., 1979), 360.

23.   Norman Mailer, *Ancient Evenings* (Boston: Little, Brown and Co., 1983), 26.

24.   Francesca Fremantle and Chögyam Trungpa, translators, *The Tibetan Book of the Dead: The Great Liberation through Hearing in the Bardo* (Boulder, Colorado: Shambhala Publications, 1975), 84.

25.   Collin, *Theory of Celestial Influence,* 156.

26.   Da Free John, *Easy Death* (Clearlake, California: The Dawn Horse Press, 1983), xxii.

27.   Ibid., xxiii

BIBLIOGRAPHY

Alexander, R. MacNeil. *The Chordates*. Cambridge: Cambridge University Press, 1975,

Allen, Joseph, and Thomas O'Toole. "Thoughts of an Astronaut." Washington D. C.: *Washington Post* Syndicate, 1983.

Anderson, Edgar. *Plants, Man and Life*. Berkeley: University of California Press, 1967.

Ardrey, Robert. *African Genesis: A Personal Investigation into the Animal Origins and Nature of Man*. New York: Dell Publishing Co., 1963.

Baker, Will. "Three Monkeys, I Am Father."In *Nuclear Strategy and the Code of the Warrior: Faces of Mars and Shiva in the Crisis of Human Survival,* edited by Richard Grossinger and Lindy Hough. Berkeley: North Atlantic Books, 1984.

Balinsky, B. I. *An Introduction to Embryology,* 5th ed. Philadelphia: Saunders College Publishing, 1981.

Bennett, J. G. *Gurdjieff: Making a New World*. New York: Harper and Row, 1973.

Bernal, J. D. *The Origin of Life*. Cleveland: World Publishing Co. 1967.

Berrill, N. J., and Gerald Karp. *Development*. New York: McGraw-Hill Book Co., 1976.

Bettelheim, Bruno. *Freud and Man's Soul*. New York: Alfred A. Knopf, 1983.

Bialy, Harvey. "The I Ching and the Genetic Code." In *Ecology and*

Consciousness, edited by Richard Grossinger. Berkeley: North Atlantic Books, 1978.

_____, ed. *Biopoiesis (Io/20).* Berkeley: North Atlantic Books, 1974.

Black, Stephen. *Determination of the Dorsal-Ventral Axis in* Xenopus Laevis *Embryos.* Ph.D. thesis. Berkeley: University of California, 1983.

Blake, William. "The Tyger" and "The Marriage of Heaven and Hell." In *The Portable Blake,* edited by Alfred Kazin. New York: Viking Press, 1946.

Borror, Donald J., Dwight M. DeLong and Charles A. Triplehorn. *An Introduction to the Study of Insects.* Philadelphia: Saunders College Publishing, 1981.

Bullock, T. H., and G. A. Horridge. *Structure and Function in the Nervous System of Invertebrates.* San Francisco: W. H. Freeman and Co., 1965.

Buchsbaum, Ralph. *Animals Without Backbones,* 2d ed. Chicago: University of Chicago Press, 1948.

Campbell, Bernard. *Human Evolution: An Introduction to Man's Adaptations.* Chicago: Aldine Publishing Company, 1966.

Chamberlain, David. B. *Consciousness at Birth: A Review of the Empirical Evidence.* Chamberlain Communications, 5207 Benton Place, San Diego: California 92116 (1983).

Chomsky, Noam. *Aspects of the Theory of Syntax.* Cambridge: MIT Press, 1965.

Clarke, Arthur C. *2001: A Space Odyssey.* New York: New American Library, 1968.

Crick, Francis. *Life Itself: Its Origin and Nature.* New York: Simon and Schuster, 1981.

Da Free John. *Easy Death.* Clearlake, California: The Dawn Horse Press, 1983.

_____. "The Mystery of the Spermatic Being." In *The Laughing Man,* Vol. 3, No. 1. Clearlake, California: The Dawn Horse Press, 1982.

Darwin, Charles. *On the Origin of Species by Means of Natural Selection, or Preservation of Favoured Races in the Struggle for Life.* London: Murray, 1859.

Davidson, Eric H. *Gene Activity in Early Development,* 2nd ed. New York: Academic Press, 1976.

Day, David. *The Doomsday Book of Animals: A Natural History of Vanished Species.* New York: Viking Press, 1981.

de Ropp, Robert S. *Sex Energy: The Sexual Force in Man and Animals.* New York: Dell Publishing Co., 1969.

de Santillana, Giorgio, and Hertha Von Dechend. *Hamlet's Mill: An Essay on Myth and the Frame of Time.* Boston: Gambit, 1969.

Dixon, Dougal. *After Man: A Zoology of the Future.* New York: St. Martin's Press, 1981.

Dostoyevsky, Fyodor. *The Brothers Karamazov.* Translated from the Russian by Constance Garnett. New York: Random House, 1950.

Duchesne-Guillemin, Jacques. *Zoroastrianism: Symbols and Values.* New York: Harper and Row, 1966.

Duddington, C. L. *Evolution and Design in the Plant Kingdom.* New York: Harper and Row, 1970.

Fabre, J. H. C. *The Life and Love of the Insect.* London: A. C. Black, 1918.

Faulkner, William. *Absalom, Absalom!* New York: Random House, 1936.

Ferenczi, Sandor. *Thalassa: A Theory of Genitality.* Translated from the German by Henry Alden Bunker. New York: W. W. Norton and Co., 1968.

Frame, Janet. *Living In the Maniototo*. Auckland, New Zealand: The Women's Press, 1979.

Fremantle, Francesca, and Chögyam Trungpa, translators. *The Tibetan Book of the Dead: The Great Liberation through Hearing in the Bardo*. Boulder, Colorado: Shambhala Publications, 1975.

Freud, Sigmund. *Civilization and Its Discontents*. Translated from the German by James Strachey. New York: W. W. Norton and Co., 1962.

_____. *The Interpretation of Dreams*. Translated from the German by James Strachey. New York: Basic Books, 1955.

_____. *An Outline of Psychoanalysis*. Translated from the German by James Strachey. New York: W. W. Norton and Co., 1949.

Friedlander, C. P. *The Biology of Insects*. New York: Pica Press, 1977.

Gibor, Aharon, ed. *Conditions for Life*. San Francisco: W. H. Freeman/*Scientific American*, 1976.

Gilluly, James, A. C. Waters, and A. O. Woodford, *Principles of Geology*, 2d ed. San Francisco: W. H. Freeman and Co., 1959.

Glass, Bentley, Owsei Temkin, and William L. Straus, Jr. *Forerunners of Darwin: 1745–1859*. Baltimore: Johns Hopkins University Press, 1959.

Gould, Stephen Jay. *Ontogeny and Phylogeny*. Cambridge: Harvard University Press, 1977.

_____. *The Panda's Thumb*. New York: W. W. Norton and Co., 1982.

Gordon, George (Lord Byron). "The Dream." In *Selected Poetry and Letters*. Edward E. Bostetter, ed. New York: Rinehart and Co., 1958.

Griaule, Marcel. *Conversations with Ogotemmêli: An Introduction to Dogon Religious Ideas*. Translated from the French by Ralph Butler,

Audrey I. Richards, and Beatrice Hooks. London: Oxford University Press, 1965.

Grossinger, Richard. *Planet Medicine: From Stone-Age Shamanism to Post-Industrial Healing,* revised edition. Boulder, Colorado: Shambhala Publications, 1982.

_____. *The Night Sky: The Science and Anthropology of the Stars and Planets.* San Francisco: Sierra Club Books, 1981.

_____, ed. *The Alchemical Tradition in the Late Twentieth Century.* Berkeley: North Atlantic Books, 1983.

_____, ed. *Dreams (Io/8).* Berkeley: North Atlantic Books, 1971.

_____, ed. *Ecology and Consciousness.* Berkeley: North Atlantic Books, 1978.

_____, ed. *Imago Mundi (Io/14).* Berkeley: North Atlantic Books, 1973.

Gurdjieff, G. I. *Views from the Real World: Early Talks of Gurdjieff.* New York: E. P. Dutton, 1975.

Haldane, J. B. S. "The Origin of Life." In *The Origin of Life,* edited by J. D. Bernal. Cleveland: World Publishing Company, 1967.

Hall, Manly P. *Man, Grand Symbol of the Mysteries: Thoughts in Occult Anatomy.* Los Angeles: The Philosophical Research Society, 1972.

Hanawalt, Philip C., and Robert H. Haynes, ed. *The Chemical Basis of Life.* San Francisco: W. H. Freeman and Co./*Scientific American,* 1973.

Haraway, Donna Jeanne. *Crystals, Fabrics, and Fields: Metaphors of Organicism in Twentieth-Century Developmental Biology.* New Haven: Yale University Press, 1976.

Herbert, Frank. *Dune.* New York: Berkley, 1965.

_____. *Dune Messiah.* New York: Berkley, 1969.

————.*God Emperor of Dune.* New York: Berkley, 1981.

Hillman, James. *Re-Visioning Psychology.* New York: Harper and Row, 1975.

————."Senex and Puer: An Aspect of the Historical and Psychological Present." In *Puer Papers.* Irving, Texas: Spring Publications, 1979.

Huxley, Aldous. *After Many a Summer Dies the Swan.* New York: Harper and Row, 1965.

Jantsch, Erich. *The Self-Organizing Universe.* New York: Pergamon Press, 1980.

Joyce, James. *Ulysses.* New York: Random House, 1934.

Jung, Carl Gustav. *The Archetypes and the Collective Unconscious.* Translated from the German by R. F. C. Hull. New York: Pantheon Books, 1959.

Keats, John. "Ode on a Grecian Urn" and "Ode to a Nightingale." In *Selected Poems.* Edited by Douglas Bush. Boston: Houghton Mifflin Co., 1959.

Keleman, Stanley. "Professional Colloquium: 29 October 1977." In *Ecology and Consciousness.* Edited by Richard Grossinger. Berkeley: North Atlantic Books, 1978.

Kelly, Robert. "Staccato for Tarots." In *The Alchemist to Mercury: An Alternate Opus, Uncollected Poems (1960–1980).* Edited by Jed Rasula. Berkeley: North Atlantic Books, 1981.

Kelso, A. J. *Physical Anthropology.* Philadelphia: J. B. Lippincott Co., 1970.

König, Karl. *Embryology and World Evolution.* Translated from the German by R. E. K. Meuss. In *British Homoeopathic Journal* 57 (1968).

Kushi, Michio. *The Book of Do-IN: Exercises for Physical and Spiritual Development.* Tokyo: Japan Publications, 1979.

Lacan, Jacques. *Écrits.* Translated from the French by Alan Sheridan. New York: W. W. Norton and Co., 1977.

Lederer, Laura. *Take Back the Night: Women on Pornography.* New York: Bantam Books, 1982.

Le Gros Clark, W. E. *The Antecedents of Man: An Introduction to the Evolution of the Primates.* New York: Harper and Row, 1963.

Lévi-Strauss, Claude. *The Elementary Structures of Kinship.* Translated from the French by James Harle Bell, John Richard von Sturmer, and Rodney Needham. Boston: Beacon Press, 1969.

_____. *The Raw and the Cooked: Introduction to a Science of Mythology,* Vol. 1. Translated from the French by John and Doreen Weightman. New York: Harper and Row, 1969.

Lo, Benjamin, Martin Inn, Robert Amacker, and Susan Foe, translators. *The Essence of T'ai Chi Ch'uan: The Literary Tradition.* Berkeley: North Atlantic Books, 1979.

Maeterlinck, Maurice. *The Life of the Bee.* New York: Dodd, Mead, and Co., 1936.

Mailer, Norman. *Ancient Evenings.* Boston: Little, Brown and Co., 1983.

_____. *The Executioner's Song.* Boston: Little, Brown, and Co., 1979.

Marais, Eugène. *The Soul of the White Ant.* London: Methuen and Co., 1937.

Marshack, Alexander. *The Roots of Civilization: The Cognitive Beginnings of Man's First Art, Symbol and Notation.* New York: McGraw-Hill Book Co. 1972.

Marvell, Andrew. "To His Coy Mistress." In *Seventeenth Century Poetry: The Schools of Donne and Jonson.* Edited by Hugh Kenner. New York: Holt, Rinehart and Winston, 1964.

Matthews, L. Harrison. *The Life of Mammals,* Vol. 1. New York: Universe Books, 1969.

Mauss, Marcel. *The Gift.* Translated by I. Cunnison. New York: Free Press, 1954.

McClure, Michael. "Wolf Net." In *Ecology and Consciousness.* Edited by Richard Grossinger. Berkeley: North Atlantic Books, 1978.

Montagu, M. F. Ashley, ed. *Culture and the Evolution of Man.* New York: Oxford University Press, 1962.

Moore, Keith L. *The Developing Human: Clinically Oriented Embryology,* 2d ed. Philadelphia: Saunders College Publishing, 1977.

Murchie, Guy. *The Seven Mysteries of Life: An Exploration in Science and Philosophy.* Boston: Houghton Mifflin Company, 1978.

Neumann, Erich. *The Great Mother: An Analysis of the Archetype.* Translated from the German by Ralph Manheim. Princeton: Princeton University Press, 1963.

Olson, Charles. *The Maximus Poems.* Berkeley: University of California Press, 1983.

Oparin, A. I. "The Origin of Life." In *The Origin of Life.* Edited by J. D. Bernal. Cleveland: World Publishing Company, 1967.

Ouspensky, P. D. *In Search of the Miraculous: Fragments of an Unknown Teaching.* New York: Harcourt, Brace and World, 1949.

Parvati, Jeannine. "Prenatal Care Guidelines." Unpublished paper, 1982.

———. "Psyche's Midwife." Unpublished paper, 1984.

Plotinus. *The Six Enneads.* Translated from the Latin by Stephen MacKenna. Chicago: Encyclopedia Britannica, 1952.

Poncé, Charles. *Papers Toward a Radical Metaphysics: Alchemy.* Berkeley: North Atlantic Books, 1984.

Poppelbaum, Hermann. *Man and Animal: Their Essential Difference.* Translated from the German. London: Anthroposophical Publishing Company, 1960.

Portugal, Franklin H., and Jack S. Cohen. *A Century of DNA: A History of the Discovery of the Structure and Function of the Genetic Substance.* Cambridge: MIT Press, 1977.

Reich, Wilhelm. *Ether, God and Devil—Cosmic Superimposition.* Translated from the German by Therese Pol. New York: Farrar, Straus and Giroux, 1972.

_____. *The Function of the Orgasm: Sex-Economic Problems of Biological Energy.* Translated from the German by Vincent R. Carfagno. New York: Farrar, Straus & Giroux, 1973.

_____. *The Murder of Christ: The Emotional Plague of Mankind.* New York: Farrar, Straus and Giroux, 1970.

Restak, Richard M. *The Brain: The Last Frontier.* New York: Warner Books, 1979.

Róheim, Géza. *The Riddle of the Sphinx, or Human Origins.* Translated from the German by R. Money-Kyrle. New York: Harper and Row, 1974.

Romer, A. S. *Man and the Vertebrates.* Baltimore: Penguin Books, 1954.

Rose, Steven. *The Conscious Brain.* New York: Alfred A. Knopf, 1974.

Russell-Hunter, W. D. *A Life of Invertebrates.* New York: Macmillan Co., 1979.

Sagan, Carl. *Cosmos.* New York: Random House, 1980.

Sardello, Robert J. "The Suffering Body of the City." In *Spring,* 1983 Annual Issue. Dallas, Texas.

Saunders, John W., Jr. *Patterns and Principles of Animal Development.* New York: Macmillan Co., 1970.

Schwenk, Theodor. *Sensitive Chaos: The Creation of Flowing Forms in Water and Air.* Translated from the German by Olive Whicher and Johanna Wrigley. London: Rudolf Steiner Press, 1965.

Sheldrake, Rupert. *A New Science of Life.* Los Angeles: J.P. Tarcher, 1982.

Sherrington, Sir Charles. *Man on His Nature.* Cambridge: Cambridge University Press, 1963.

Shklovskii, I. S., and Carl Sagan. *Intelligent Life in the Universe.* Shklovskii translated from the Russian by Paula Fern. New York: Dell Publishing Co., 1966.

Smith, Eric, Garth Chapman, R. B. Clar, David Nichols, and J. D. McCarthy. *The Invertebrate Panorama.* New York: Universe Books, 1971.

Spuhler, J. N., ed. *The Evolution of Man's Capacity for Culture.* Detroit: Wayne State University Press, 1965.

Stanley, Wendell M., and Evans G. Valens. *Viruses and the Nature of Life.* New York: E. P. Dutton, 1961.

Stanner, W. E. H. "The Dreaming." Indianapolis: Bobbs-Merrill Reprint Series in the Social Sciences, A-214, 1956.

Steiner, Rudolf. *An Outline of Occult Science.* Translated from the German by Henry B. Maud, and Lisa D. Monges. Spring Valley, New York: Anthroposophic Press, 1972.

Tansley, Arthur A. "The Deadly Ninja: Agents of Death." In *Fighting Arts Magazine*, 5, No. 3. Liverpool, England (1983).

Teilhard de Chardin, Pierre. *The Phenomenon of Man.* Translated from the French by Bernard Wall. New York: Harper and Row, 1959.

Thomas, Lewis. *The Lives of a Cell: Notes of a Biology Watcher.* New York: Viking Press, 1974.

Thompson, D'Arcy. *On Growth and Form.* Cambridge, Cambridge University Press, 1966.

Tillich, Hannah. *From Time to Time.* New York: Stein and Day, 1973.

Tillich, Paul. *The Meaning of Health: The Relation Between Religion and Health.* Berkeley: North Atlantic Books, 1981.

Todd, John. Interview by Richard Grossinger and Lindy Hough, November, 1982. Unpublished section of transcript published in *Omni,* July 1984.

Weil, Andrew. *The Marriage of the Sun and Moon: A Quest for Unity in Consciousness.* Boston: Houghton Mifflin Co., 1980.

Wendt, Herbert. *The Sex Life of Animals.* Translated from the German by Richard and Clara Winston. New York: Simon and Schuster, 1965.

Whitmont, Edward C. *Return of the Goddess: Femininity, Aggression and the Modern Grail Quest.* London: Routledge and Kegan Paul, 1983.

Wilson, Robert Anton. *Cosmic Trigger.* Berkeley: And/Or Press, 1977.

Wintrobe, M. M. *Clinical Hematology.* Philadelphia: Lea and Febiger, 1981.

Woodburne, Lloyd S. *The Neural Basis of Behavior.* Columbus, Ohio: Charles E. Merrill Books, 1967.

The Yellow Emperor's Classic of Internal Medicine. Translated from the Chinese by Ilza Veith. Berkeley: University of California Press, 1972.

Zimmer, Heinrich. "The Indian World Mother." In *The Mystic Vision.* Papers from the Eranos Yearbooks, 6. Princeton: Princeton University Press, 1968.

Richard Grossinger was born in New York City in 1944 and raised there; he attended Amherst College and the University of Michigan from which he received a Ph.D. in anthropology. He is the author of a number of books, including *Planet Medicine: From Stone-Age Shamanism to Post-Industrial Healing* and *The Night Sky;* in addition, he has edited *Nuclear Strategy and the Code of the Warrior, The Alchemical Tradition in the Late Twentieth Century, The Temple of Baseball,* and *Planetary Mysteries.* Currently, he is the Director of the Society for the Study of Native Arts and Sciences in Berkeley, California.

1491

(His knowledge is incomplete, his description & conclusions often in err & he is often confused)

87 - premammalian placenta
eggs may be as old as 50 yrs
when fertilized

93, 368, 399 - bibl.
(discovery of fertilization)

93 - an activated egg may develop
into a haploid embryo w/o
any sperm absorption

96 - fish kiss

368 (bibl.) Reich
102 - failure to understand life